INTRODUCTION TO
Probability
with
Mathematica®

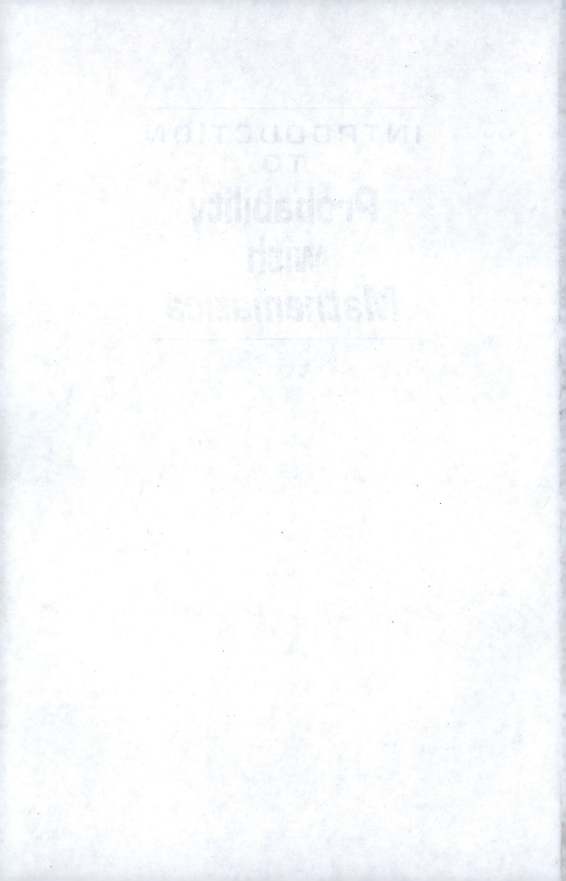

INTRODUCTION TO
Probability
with
Mathematica®

Kevin Hastings

Knox College
Galesburg, Illinois

CHAPMAN & HALL/CRC

Boca Raton   London   New York   Washington, D.C.

Library of Congress Cataloging-in-Publication Data

Hastings, Kevin J., 1955-
  Introduction to probability with Mathematica / by Kevin J. Hastings.

      p.   cm.
  Includes bibliographical references and index.
  ISBN 1-58488-109-7
  1. Title.
QA273.19.E4 H37 2000
519.2′0285′5369—dc21   00-046857

This book contains information obtained from authentic and highly regarded sources. Reprinted material is quoted with permission, and sources are indicated. A wide variety of references are listed. Reasonable efforts have been made to publish reliable data and information, but the author and the publisher cannot assume responsibility for the validity of all materials or for the consequences of their use.

Neither this book nor any part may be reproduced or transmitted in any form or by any means, electronic or mechanical, including photocopying, microfilming, and recording, or by any information storage or retrieval system, without prior permission in writing from the publisher.

The consent of CRC Press LLC does not extend to copying for general distribution, for promotion, for creating new works, or for resale. Specific permission must be obtained in writing from CRC Press LLC for such copying.

Direct all inquiries to CRC Press LLC, 2000 N.W. Corporate Blvd., Boca Raton, Florida 33431.

Trademark Notice: Product or corporate names may be trademarks or registered trademarks, and are used only for identification and explanation, without intent to infringe.

To my little Emily, Daddy will play with you again soon

Preface

The subject of probability has always been a springboard for many under-graduates to the study of mathematical statistics. For other students it gives background for subjects such as operations research and mathematical economics. The content of a probability (or first semester mathematical statistics) course has been relatively stable over the years, and pedagogy has been conservative as well. Yet, students often have difficulty with the main concepts of probability: event, sample space, random variable, distribution, and expectation. These ideas are crucial for statistics and other later courses.

The observations made above suggest that it is worthwhile to look at new pedagogical approaches whose goal is to increase understanding of concepts. This book is in that spirit.

In other courses, the most noteworthy example being calculus, technology has been used as a tool to let students visualize concepts better, to relieve computa-tional burden, and most importantly to extend the range of questions that can be asked beyond those whose answers can be found by closed form hand computation. Also, an active learning style in which students experiment, do, and interact not only makes a course more interesting, but leaves a more lasting impression. I believe that technology and an active learning style can work together to have a similar beneficial effect on the study of probability, for many reasons. Among them are:

(1) Simulation is a powerful tool to assist in understanding randomness in general, especially the conceptually difficult idea of a sampling distribution.

(2) Many probabilistic problems, especially those relating to stochastic processes, have naturally recursive solutions that are easy to obtain using a computer but not so easy to find in closed form by hand.

(3) Symbolic algebra-graphical packages are very effective at connecting graphical with symbolic relationships, particularly in the area of dependence of a solution on parameters.

(4) Integration, so frequently needed in probability but so infrequently mastered by calculus students, need not be a roadblock to understanding in the presence of machine computation of integrals.

(5) The use of the machine excites students, and rather surprisingly, can have the effect of producing better mathematical writing when high quality text processing is integrated with computations.

(6) Many students now gain a great deal of expertise using symbolic algebra-graphical packages in prior mathematics courses such as calculus, linear algebra, and differential equations, and consequently their adjustment to the system is not difficult.

(7) Frequent student use of the computer in class projects, exercises and/or laboratory settings is naturally active and interactive, as contrasted with the passive learning style of listening with one ear to a lecture.

I would like to elaborate on some of these points, and comment specifically on the choice of *Mathematica*® as supporting software. Nearly every college and university offers courses in probability and statistics at the junior-senior undergraduate level. Traditionally, the computational support in these courses has been supplied by statistics packages such as SAS, Minitab, SPSS and others. These programs have several advantages: they are specially geared to statistical analysis, they offer very broad capabilities, they are fairly easy to use, and for those students who pursue statistics later in their careers these packages or others like them will be encountered. But they are much better suited to the statistical analysis half of the sequence than to the probability half, where their usefulness is chiefly in the display of histograms and computation of table probabilities for the key distributions. Changes have been occurring in the last few years which make it reasonable to argue that the multipurpose symbolic-graphical program *Mathematica* is ready to take its place among the leading programs used for mathematical statistics courses.

First, as mentioned above, many students are already familiar with *Mathematica* from prior mathematics courses. Regardless, when version 3.0 appeared, *Mathematica* became easier to use for the student without much experience than it was in its earlier versions. Also, many institutions already have site licenses for *Mathematica*, hence availability is not as big an issue as it once was. And, the *Mathematica* developers have been slowly adding to the statistical capabilities of their product by updating the standard add-on packages, so that there are many ways in which students can use the program in the second term of the probability-statistics sequence as well as the first. Because of *Mathematica*'s extensive utility as a programming language, the trained *Mathematica* programmer can develop his or her own supplementary commands that may be important for a particular problem. *Mathematica*'s first rate symbolic processing, numerical algorithms, and function graphs fit in much better with the probability part of the course than do the current stat packages. One can see easily, for instance, how the gamma distribution changes shape as its defining parameters change. *Mathematica* is also very good as a general purpose simulation engine; for example, demonstrations of the weak and strong laws of large numbers are easy to put together and are powerful tools in understanding these theorems. And it is very important that *Mathematica* is now a complete environment not only for doing computations but for high quality presentation of technical writing. Students seem to take the job of writing a solution to a problem more seriously when they are doing it using the symbol processing capabilities provided in the front end of *Mathematica*.

The format of *Introduction to Probability with Mathematica* is unique and exciting. The print copy is accompanied by a downloadable electronic version which has the text and the live commands that are used in the text. The student may read the print copy in his/her room, then go to the computer and load up the section of interest as a *Mathematica* notebook in order to check something that was done in the text, reexecute a command using different arguments, produce an animation, solve a problem and submit it on-line, etc. The instructor has the printed resource for convenience (and because we all still just like to sit back in a chair and read a book), and also has the same electronic version to greatly simplify the task of creating classroom demonstrations and student laboratory experiences when appropriate. The text itself is written so as to suggest frequent explorations for the student to do as new material develops, and to rely less on hand computations in exercises and more on the technology. At the same time though, it should be added that there are many occasions when hand computation is very beneficial, and the technology must enhance and extend understanding of concepts, not replace it.

There are six chapters in this book, at least five of which should be comfortably done in a semester. Chapter 1 sets up the foundations of discrete probability, taking simulation and sampling as its recurring themes. In fact simulation and sampling continue to be highlights throughout the book, not only because they are important in themselves, but also because they are relatively concrete means of understanding the important concepts, such as events, randomness, and distributions of random variables. The reader encounters the axioms and elementary theorems of probability in this chapter, some combinatorial probability, and the related ideas of conditional probability and independence.

In Chapter 2 we stay in the discrete world, focusing on distributions of random variables. The main discrete distributions are studied, including the hypergeometric distribution, the binomial distribution, and the Poisson distribution. Here we also see mathematical expectation defined, and we expose the student to multivariate and conditional distributions.

Chapter 3 is a transitional chapter to the world of continuous probability. Here the notions of continuous distributions, spaces of events, and expectations are introduced, taking advantage of analogies with the discrete case. It is in the fourth chapter where the student meets the major continuous distributions, featuring the normal, bivariate normal, and the gamma family. A section on transformations focuses on the continuous simulation theorem, and the use of c.d.f.'s and moment-generating functions to find the distributions of certain transformed normal random variables. This sets up the last, important section of Chapter 4 on the chi-square, t, and F-distributions. There is more material here than usual on the use of these distributions for statistical inference. Interestingly, I do not provide tables of distributions in the printed text, preferring to rely on *Mathematica* to compute probabilities when necessary. It follows also that there is a shorter treatment than most texts have of using standardization to compute normal probabilities.

Chapter 5 was a lot of fun to write. It has two brief sections, one on the laws of large numbers and the other on the Central Limit Theorem. I decided that since these are the lynchpins of probability theory, it was a good opportunity to set the subject in historical context and describe the main events that led to them. So this chapter is much more discursive and less problem oriented than the previous chapters. Instructors might decide to assign this as reading for in-class discussion.

It was also interesting to write Chapter 6, which looks at some of the modern applications of probability: Markov chains, queueing theory, and mathematical finance. Many instructors will have to consider carefully whether they have time for this, but I think that it is a great opportunity to give students a glimpse of other subjects that one may spring off to from probability. Each section in this chapter is independent of the other sections, so the instructor can choose any of the possible subsets of the three sections to cover. I would recommend going through the rest of the book in order and without omission, with the possible exception of the section on the bivariate normal distribution.

A complete, detailed solutions manual is available upon request from the publisher. I have worked very hard to eliminate errors from that manual, but things do creep in unawares so I ask your forgiveness in advance if you find a mistake.

Finally I would like to emphasize that this is a living textbook. It has all the features of a stand-alone book; it can be read and understood without being at a machine and without being a *Mathematica* expert, but its intention is for the student to interact with it. Each section has several activities in it for the student to monitor progress, which can also be used to stimulate class discussion and problem solving. Opportunities abound for the student to rerun commands under different scenarios, check the variability of simulated experiments, and write programs to solve problems. At one time I was among the vast majority of mathematicians who thought it would be too ambitious to ask students to program anything non-trivial in a mathematics course. But the world has changed, mathematics has changed, and the students have also changed. Those who are in mathematics at this level are likely to continue on to a world in which closed form answers are rare, and in which they are forced to use, in fact take pleasure in using, technological tools to answer questions. But I have also found that the process of designing an algorithm to solve a problem develops important logical reasoning skills, and often clarifies the problem. So in the text itself, in the self-check activities, and in the exercises I have taken a lot of opportunities to have students see and write *Mathematica* programs. I hope that the instructor will include some of them, and that the students enjoy them as much as I enjoyed developing them.

Thanks go out to Bob Stern and the helpful folks at Chapman & Hall/CRC Press and at Wolfram Research for *Mathematica* advice. The reviewers have also been very helpful. I also need to thank my students past and present, and my colleagues here at Knox College for stimulating me, and always keeping me honest. And lastly I want to thank my wife Gay Lynn and daughter Emily for being so patient while Daddy spent all that time at his computer.

Kevin J. Hastings
July 28, 2000

Note on the Electronic Version and the KnoxProb`Utilities` Package

This book is accompanied by an electronic version, downloadable from the CRC Press website, which contains the complete text in the form of *Mathematica* notebooks, broken into sections. Copyright law applies here; you are no more allowed to distribute copies of the e-files freely to your friends than you are allowed to Xerox the printed book and give it out for free. The electronic text is identical to the printed text, except for page formatting and the fact that all output cells have been deleted to save space. Input cells that produced the output that you see in the printed text have all been tagged as initialization cells, so that you can just go up to the Kernel/Evaluation menu and have *Mathematica* execute all initialization cells at once to reproduce the output.

Also available at the website is the KnoxProb`Utilities` package that I have written to support the book. You must create a folder called KnoxProb in the AddOns\ExtraPackages subdirectory of your main *Mathematica* directory, and copy into it the file Utilities.m from the publisher's website in order to have access to this package and thereby to make the electronic book run properly.

A couple of naming problems are worth mentioning. KnoxProb`Utilities` preloads several standard *Mathematica* packages, and there are a few name conflicts among all these packages that I have worked around. I have included a Histogram command in KnoxProb`Utilities`, and I have arranged to have it overwrite the Histogram command in Graphics`Graphics`, which does not have the options I needed. Some pictures are produced using the Graphics`Arrow` package, which has names Absolute and Relative in it, as does KnoxProb`Utilities`, and so I have removed definitions in one before using those names in the other on certain occasions. This should be transparent to the user. And *Mathematica* itself has at least one name conflict in two of the packages that KnoxProb`Utilities` loads, but none of these conflicts has any bearing on the commands needed in this text. So you may very occasionally see a warning message about a name conflict, but you should experience no problems.

Table of Contents

Chapter 5 - Asymptotic Theory

Chapter 6 - Applications of Probability

Appendix

References

CHAPTER 1
DISCRETE PROBABILITY

1.1 The Cast of Characters

Why study probability? There are many answers, but perhaps none is more important than this. All around us is uncertainty and we must understand its nature in order to conduct ourselves in the world. We join the shortest waiting line in hopes that it has the best chance of allowing us to be served quickly. We invest our savings in assets that we believe are most likely to yield high returns. We buy the battery that we think is most likely to last long. Certainly gambling is very much a part of our society now, like it or, not, and the study of probability lets you understand the mechanics of games of chance; what is predictable and what is not, what to bet on (usually nothing) and what not to bet on (usually everything). In all of these cases we do not know for certain what will happen before the phenomenon is observed (we don't know how much time we will wait in line, what our opponent's blackjack hand will be, etc). But some common-sense assumptions, background knowledge, or a long history of observed data from similar experiments can enable us to quantify likelihoods. Several of these examples have actually given birth to significant subfields of probability or related areas of study, for example queueing theory (i.e., the study of waiting lines) and mathematical finance, which are introduced in Chapter 6. Probability also provides the theoretical basis for statistics, which is the art of using data obtained randomly to draw conclusions about social and scientific phenomena.

In addition to all this, probability exhibits many of the aesthetically appealing properties that attract people to mathematics in general. It proceeds logically from a few simple axioms to build an abstract structure which is straightforward, yet powerful. The cast of characters of probability is both foreign and tangible, which implies that you must work a little to get to know them, but there is hope in succeeding in the acquaintance.

This text will be somewhat different from others in the field. While the main ideas of probability have not changed much in at least fifty years, the world has, and learning in mathematics has changed with it. Many things are technologically possible today of which mathematics education should take advantage. This book is a living, interactive text which asks you to do activities as you read carefully to help you understand and discover the ideas of probability in new ways, on your own terms. You can obtain from the publisher's website the complete text in the form of *Mathematica* notebooks, one for each section. The notebooks are executable, so that you can reenter existing commands, modify them, or write your own supplementary commands to take the study in the direction you wish.

You will be building probability yourself in a kind of workshop, with *Mathematica* and your own insight and love of experimentation as your most powerful tools. I caution you against depending too much on the machine, however. Much of the time, traditional pencil, paper, and brainpower are what you need most to learn. What the computer does, however, is open up new problems, or let you get new insights into old problems. So we will be judiciously blending new technology with traditionally successful pedagogy to enable you to learn effectively. Consistent themes will be the idea of taking a sample randomly from a larger universe of objects in order to get information about that universe, and the computer simulation of random phenomena to observe patterns in many replications that speak to the properties of the phenomena.

The cast of characters in probability whose personalities we will explore in depth in the coming chapters include the following six principal players: (1) Event; (2) Sample Space; (3) Probability Measure; (4) Random Variable; (5) Distribution; and (6) Expectation. The rest of this section will be a brief intuitive introduction to these six.

Most of the time we are interested in assessing likelihoods of *events*, that is, things that we might observe as we watch a phenomenon happen. The event that AT&T's bid to acquire a local cable company is successful, the event that at least 10 patients arrive to an emergency room between 6:00 and 7:00 AM, the event that a hand of five poker cards forms a straight, the event that we hold the winning ticket in the lottery, and the event that two particular candidates among several are selected for jury duty are just a few examples. They share the characteristic that there is some uncertain experiment, sampling process, or phenomenon whose result we do not know "now". But there is a theoretical "later" at which we can observe the phenomenon taking place, and decide whether the event did or didn't occur.

A *simple event* (often called an *outcome*) is an event that cannot be broken down into some combination of other events. A *composite event* is just an event that isn't simple. For instance the event that we are dealt a poker hand that makes up a straight is composite because it consists of many simple events which completely specify the hand, such as 2 through 6 of hearts, 10 through ace of clubs, etc. This is where the *sample space* makes its entry: the sample space is the collection of all possible outcomes of the experiment or phenomenon, that is, all possible indecomposable results that could happen. For the poker hand example, the sample space would consist of all possible five-card hands. In specifying sample spaces, we must be clear about assumptions; for instance we might assume that the order in which the cards were dealt does not matter, and we cannot receive the same card twice, so that the cards are dealt without replacement.

Activity 1 How can you characterize the sample space for the AT&T example? Make up your own random phenomenon, informally describe the sample space, and give examples of events.

> **Activity 2**   In the emergency room example above, is the event that was described simple or composite?   If that event is composite, write down an example of a simple event relating to that phenomenon.

In random phenomena, we need a way of measuring how likely the events are to occur. This may be done by some theoretical considerations, or by experience with past experiments of the same kind. A *probability measure* assigns a likelihood, that is a number between 0 and 1, to all events. Though we will have to be more subtle later, for now we can intuitively consider the extremes of 0 and 1 as representing impossibility and complete certainty, respectively.

Much of the time in probability problems, probability measures are constructed so that the probability of an event is the "size" of the event as a proportion of the "size" of the whole sample space. "Size" may mean different things depending on the context, but the two most frequent usages are cardinality, that is number of elements of a set, and length (or area in two dimensions). For example, if you roll a fair die once, since there are six possible faces that could land on top the sample space is {1, 2, 3, 4, 5, 6}. Since three of the faces are odd, it seems reasonable that the probability that an odd number is rolled is 3/6. Size means length in the following scenario. Suppose that a real number is to be randomly selected in the interval [0, 4]. This interval is the sample space for the random phenomenon. Then the probability that the random real number will be in the interval [1, 3] is $2/4 = 1/2$, which is the length of the interval [1, 3] divided by the length of the whole sample space interval.

The first two chapters will concentrate on discrete probability, in which counting the number of elements in sets is very important. We will leave much of the detail for later sections, except to remind you of the intuitively obvious *multiplication principle* for counting. If outcomes of a random experiment consist of two stages, then the total number of outcomes of the experiment is the number of ways the first stage can occur times the number of ways that the second can occur. This generalizes to multiple stage experiments. For instance, if a man has 3 suits, 6 shirts, and 4 ties, and he doesn't care about color coordination and the like, then he has $3 \times 6 \times 4 = 72$ possible outfits.

It is possible to experimentally estimate a probability by repeating the experiment over and over again. The probability of the event can be estimated by the number of times that it occurs among the replications divided by the number of replications. Try this exercise. In the KnoxProb`Utilities` package that you should have installed is a *Mathematica* command called

DrawIntegerSample[populationsize, n]

which outputs a sequence of *n* numbers drawn at random, without reusing numbers previously in the sequence, from the positive integers in the range 1, ... , populationsize. (You should be sure that the package has been loaded into the ExtraPackages subdirectory within the *Mathematica* directory you are using on your computer.) First execute the command to load the package, then try repeating the experiment of drawing samples of two numbers from {1, ... , 5} a large number of times. I have shown one such sample below. Keep track of how frequently the number 1 appears in your samples. Empirically, what is a good estimate of the probability that 1 is in the random sample of size two? Talk with a group of your classmates, and see if you can set up a good theoretical foundation that supports your conclusion. (Try to specify the sample space of the experiment.) We will study this experiment more carefully later in this chapter. Incidentally, recall that *Mathematica* uses braces { } to delineate sequences as well as unordered sets. We hope that the context will make it clear which is meant; here the integer random sample is assumed to be in sequence, so that for example {1, 3} is different from {3, 1}.

```
Needs["KnoxProb`Utilities`"]
```

```
DrawIntegerSample[5, 2]
```

{3, 5}

We are halfway through our introduction of the cast of characters. Often the outcomes of a random phenomenon are not just numbers. They may be sequences of numbers as in our sampling experiment, or even non-numerical data such as the color of an M&M that we randomly select from a bag. Yet we may be interested in encoding the simple events by single numbers, or extracting some single numerical characteristic from more complete information. For example, if we are dealt a five card hand in poker, an outcome is a full specification of all five cards, whereas we could be interested only in the number of aces in the hand, which is a numerical valued function of the outcome. Such a function for transforming outcomes to numerical values is called a *random variable*, the fourth member of our cast of characters. Since the random variable depends on the outcome, and we do not know in advance what will happen, the value of the random variable is itself uncertain until after the experiment has been observed.

Activity 3 Use DrawIntegerSample again as below to draw samples of size five from a population of size 50. Suppose we are interested in the random variable which gives the minimum among the members of the sample. What values of this random variable do you observe in several replications?

```
outcome = DrawIntegerSample[50, 5]
X = Min[outcome]
```

The cell below contains a command I have called SimulateMinima for observing a number *m* of values of the minimum random variable from Activity 3. Read the code to understand how it works. To get an idea of the pattern of those sample minima, we plot a graph called a *histogram* which tells the proportion of times among the *m* replications of the experiment that the sample minimum fell into each of a given number of categories (we use 6 categories here). The Histogram command is also in the KnoxProb`Utilities` package. Its exact syntax is

Histogram[listofdata, numberof rectangles]

and it has several options described in the appendix on *Mathematica* commands. One output cell is shown below using 100 replications, six histogram rectangles, and once again using samples of size 5 from {1, ... , 50}. You should reexecute several times to see whether your replications give similar histograms. Should they be identical to this one? Why or why not? (If you want to suppress the list of minima and get only the histogram, put a semicolon between the SimulateMinima and Histogram commands.)

```
SimulateMinima[m_ , popsize_ , sampsize_ ] :=
    Table[
   Min[DrawIntegerSample[popsize, sampsize]],
   {i, 1, m}]
```

```
mins = SimulateMinima[100, 50, 5]
Histogram[mins, 6];
```

```
{2, 10, 5, 2, 28, 1, 7, 3, 27, 4, 3, 2, 9, 19, 3, 8, 3, 7,
14, 11, 2, 6, 10, 7, 9, 3, 4, 9, 9, 9, 1, 4, 8, 1, 7,
12, 2, 5, 9, 3, 3, 17, 24, 5, 12, 7, 8, 15, 2, 15, 3,
2, 14, 30, 10, 1, 6, 3, 1, 5, 5, 10, 10, 11, 4, 3, 9,
1, 13, 2, 1, 11, 3, 4, 8, 7, 9, 8, 2, 26, 7, 2, 7, 3,
11, 4, 4, 10, 5, 7, 14, 16, 20, 15, 9, 5, 15, 10, 6, 6}
```

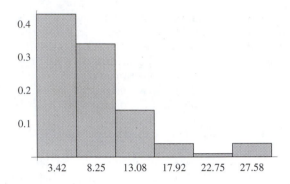

Figure 1 - Sample histogram of minimum of five values from {1, 2, ... , 50}

The histogram for the sample minimum that we just saw, together with your experiments, hints at a very deep idea, which is the fifth member of our cast. A random variable observed many times will give a list of values that follows a characteristic pattern, with some random variation. The relative frequencies of occurrence (that is the number of occurrences divided by the number of replications), which are the histogram rectangle heights, of each possible value of the random variable will stabilize around some theoretical probability. Much as a probability measure can be put on a sample space of simple events, a *probability distribution* can be put on the collection of possible values of a random variable. To cite a very simple example, if 1/3 of the applicants for a job are college graduates and 2/3 are not, if one applicant is selected randomly, and if we define a random variable

$$X = \begin{cases} 1 & \text{if applicant is a college grad} \\ 0 & \text{if applicant is not} \end{cases}$$

then the two possible values of X are 1 and 0, and the probability distribution for this X gives probability 1/3 to value 1 and 2/3 to value 0.

The last member of our cast is *expectation* or *expected value* of a random variable. Its purpose is to give a single number which is an average value that the random variable can take on. But if these possible values of the random variable are not equally likely, it doesn't make sense to compute the average as a simple arithmetical average. For example, suppose the number of 911 emergency calls during an hour is a random variable X whose values and probability distribution are as in the table below.

value of X	2	4	6	8
probability	$1/10$	$2/10$	$3/10$	$4/10$

What could we mean by the average number of calls per hour? The simple arithmetical average of the possible values of X is (2+4+6+8)/4 = 5, but 6 and 8 are more likely than 2 and 4, so they deserve more influence. In fact, the logical thing to do is to define the expected value of *X,* denoted by E[*X*], to be the weighted average of the possible values, using the probabilities as weights:

$$E[X] = \frac{1}{10} \times 2 + \frac{2}{10} \times 4 + \frac{3}{10} \times 6 + \frac{4}{10} \times 8 = \frac{60}{10} = 6$$

We will be looking at all of our characters again in a more careful way later, but I hope that the seeds of intuitive understanding have been planted. You should reread this introductory section from time to time as you progress to see how the intuition fits with the formality.

Mathematica for Section 1.1

Command	Location
DrawIntegerSample[popsize, n]	KnoxProb` Utilities`
Histogram[datalist, numberofrectangles]	KnoxProb` Utilities`
SimulateMysteryX[m]	KnoxProb` Utilities` (see Exercise 11)
SimulateMinima[m, popsize, sampsize]	Section 1.1

Exercises 1.1

1. Most state lotteries have a version of a game called Pick 3 in which you win an amount, say $500, if you correctly choose three digits in order. Suppose that the price of a Pick 3 ticket is $1. There are two possible sample spaces that might be used to model the random experiment of observing the outcome of the Pick 3 experiment, according to whether the information of the winning number is recorded, or just the amount you won. Describe these two sample spaces clearly, give examples of simple events for both, and describe a reasonable probability measure for both.

2. A grand jury is being selected, and there are two positions left to fill, with candidates A, B, C, and D remaining who are eligible to fill them. The prosecutor decides to select a pair of people by making slips of paper with each possible pair written on a slip, and then drawing one slip randomly from a hat. List out all simple events in the sample space of this random phenomenon, describe a probability measure consistent with the stated assumptions, and find the probability that A will serve on the grand jury.

3. In an effort to predict performance of entering college students in a beginning mathematics course on the basis of their mathematics ACT score, the following data were obtained. The entries in the table give the numbers of students among a group of 179 who had the indicated ACT score and the indicated grade.

Grade

		A	B	C	D	F
	23	2	5	16	31	3
ACT	24	6	8	12	15	2
	25	7	10	15	6	1
	26	8	15	13	4	0

(a) A student is selected randomly from the whole group. What is the probability that this student has a 25 ACT and received a B? What is the probability that this student received a C? What is the probability that this student has an ACT score of 23?

(b) A student is selected randomly from the group of students who scored 24 on the ACT. What is the probability that this student received an A?

(c) A student is selected randomly from the group of students who received B's. What is the probability that the student scored 23 on the ACT? Is it the same as your answer to the last question in part (a)?

4. Find the sample space for the following random experiment. Four people labeled A, B, C, and D are to be separated randomly into a group of 2 people and two groups of one person each. Assume that the order in which the groups are listed does not matter, only the contents of the groups.

5. Consider the experiment of dealing two cards from a standard 52 card deck (as in blackjack), one after another without replacing the first card. Describe the form of all outcomes, and give two examples of outcomes. What is the probability of the composite event "King on 1st card"? What is the probability of "Queen on 2nd card"?

6. In the Pick 3 lottery example of Exercise 1, consider the random variable X that gives the net amount you win if you buy one ticket (subtracting the ticket price). What is the probability distribution of X? What is the expected value of X?

7. An abstract sample space has outcomes a, b, c, d, e, and f as shown on the left in the figure. Each outcome has probability 1/6. A random variable X maps the sample space to the set {0,1} as in the diagram. What is the probability distribution of X? What is the expected value of X?

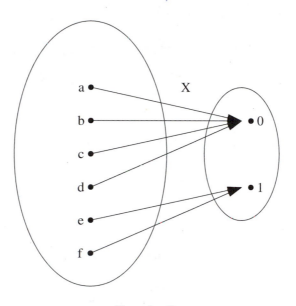

Exercise 7

8. Two fair coins are flipped at random and in succession. Write the sample space explicitly, and describe a reasonable probability measure on it. If X is the random variable that counts the number of heads, find the probability distribution of X and compute the expected value of X.

9. Using the data in Exercise 3, give an empirical estimate of the probability distribution of the random variable X that gives the ACT score of a randomly selected student. What is the expected value of X, and what is its real meaning in this context?

10. (*Mathematica*) Use the DrawIntegerSample command introduced in the section to build a command that draws a desired number of samples, each of a desired size, from the set {1, 2, ..., popsize}. For each sample, the command appends the maximum number in the sample to a list. This list can be used to create a histogram of the distribution of maxima. Repeat the experiment of taking 50 random samples of size 4 from {1, 2, ..., 7} several times. Describe in words the most salient features of the histograms of the maximum. Use one of your replications to give an empirical estimate of the probability distribution of the maximum.

11. (*Mathematica*) The command below will simulate a desired number m of values of a mystery random variable X. First load the KnoxProb`Utilities` package, then run the command several times using specific numerical values for m. Try to estimate the probability distribution of X and its expected value.

```
SimulateMysteryX[10]
```

12. (*Mathematica*) Write a command in *Mathematica* which can take as input a list of pairs $\{p_i, \ x_i\}$ in which a random variable X takes on the value x_i with probability p_i and return the expectation of X. Test your command on the random variable whose probability distribution is in the table below.

value of X	0	1	2	3
probability	1/4	1/2	1/8	1/8

1.2 Properties of Probability

We will begin this section with examples of the concrete construction of probability measures for sample spaces with finitely many outcomes. With our feet on good intuitive ground, we will then be ready to set up a formal axiom system which permits rigorous proof of some important properties of probability. This is the way that most mathematical research actually proceeds - from concrete to abstract rather than the other way around.

Example 1 If, as we said in Section 1.1, a probability measure gives likelihood to events, then at its most basic level it gives likelihood to simple events. As an example, I recently saw a breakdown of advanced placement scores for Illinois 12th graders on the subject test in English Language and Composition. Scores range from 1 to 5, and the numbers of test takers in these categories for the particular year I looked at were: 1: 74, 2: 293, 3: 480, 4: 254, 5: 130 for a total of 1231 students. Consider first the random experiment of drawing a student at random from this group and observing the test score.

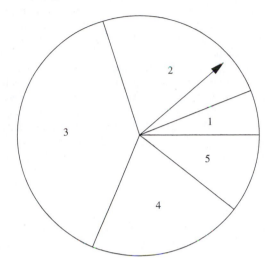

Figure 2 - Sampling from a discrete population

There are 5 possible simple events, namely the scores 1-5. We can think of the experiment of randomly sampling a student score as spinning the arrow on the pie chart of Figure 2, which has been constructed so that the angles of the circular wedges are proportional to the frequencies in the categories that were listed in the last paragraph.

This example illustrates several important features of the subject of probability. First there is an underlying physics that drives the motion and final resting position of the arrow. But differences in the starting point, the angular momentum imparted to the arrow, and even slight wind currents and unevenness of the surface of the spinner give rise to a non-deterministic or random phenomenon. We cannot predict with certainty where the arrow will land. It could be philosophically argued that if we knew all of these randomizing factors perfectly, then the laws of physics would perfectly predict the landing point of the arrow, so that the phenomenon is not really random. Perhaps this is true in most, if not all, phenomena that we call random. Nevertheless, it is difficult to know these factors, and the ability to measure likelihoods of events that we model as random is useful to have.

Second, this example shows the two main ways of arriving at an assignment of probabilities. We might make simplifying assumptions which lead to a logical assignment of probability. Or we might repeat the experiment many times and observe how frequently events occur. We were fortunate enough here to have the background data from which our spinner was constructed, which would indicate by simple counting that if the spinner was truly spun "randomly" we should assign probabilities

$$P[1] = \tfrac{74}{1231}, \ P[2] = \tfrac{293}{1231}, \ P[3] = \tfrac{480}{1231}, \ P[4] = \tfrac{254}{1231}, \ P[5] = \tfrac{130}{1231} \qquad (1)$$

to the five outcomes. But step back for a moment and consider what we would have to do with our spinner if we weren't blessed with the numbers. We would have no recourse but to perform the experiment repeatedly. If, say, 1000 spins produced 70 1's, then we would estimate P[1] empirically by $70/1000 = .07$, and similarly for the other outcomes.

Third, we can attempt to generalize conclusions. This data set itself can be thought of as a sort of repeated spinning experiment, in which each student represents a selection from a larger universe of students who could take the test. The probabilities in formula (1), which would be exact probabilities for the experiment of selecting one student from this particular batch, become estimates of probabilities for a different experiment; that of selecting 1231 students from the universe of all students, past and present, who could take the test. Whether our particular data set is a "random" sample representing that universe without bias is a very serious statistical question, best left for a statistics course with a good unit on samples and surveys.

Using the probabilities in (1), how should probability be assigned to composite events like "1 or 3"? The logical thing to do is to pool together the probabilities of all simple events that are contained in the composite event. Then we would have

$$P[1 \text{ or } 3] = P[1] + P[3] = \tfrac{74}{1231} + \tfrac{480}{1231} = \tfrac{554}{1231} \tag{2}$$

So the probability of an event is the total of all probabilities of the outcomes that make up the event.

Two implications follow easily from the last observation: the empty event $\emptyset$ which contains no outcomes has probability zero. Also, the event that the spinner lands somewhere is certain, that is, it has probability 1 (100%). Another way to look at it is that the sample space, denoted by Ω say, is the composite event consisting of all simple events. In this case we have:

$$\begin{aligned} P[\Omega] &= P[1 \text{ or } 2 \text{ or } 3 \text{ or } 4 \text{ or } 5] \\ &= P[1] + P[2] + \cdots + P[5] \\ &= \tfrac{74}{1231} + \tfrac{293}{1231} + \cdots + \tfrac{130}{1231} \\ &= 1 \end{aligned} \tag{3}$$

In summary, so far we have learned that outcomes, or simple events, can be given probabilities between 0 and 1 using theoretical assumptions or empirical methods, and probabilities for composite events are the totals of the outcome probabilities for outcomes in the event. Also $P[\emptyset] = 0$ and $P[\Omega] = 1$. Try the following activity, which suggests another property. You will learn more about that property shortly.

Activity 1 For the spinner experiment we looked at P[1 or 3]. The events {1} and {3} are distinct simple events, in particular they have no outcomes in common. Consider now the two events "at least 3" and "between 1 and 3 inclusive". What is the probability of the event {"at least 3" or "between 1 and 3 inclusive"}? How does this probability relate to the individual event probabilities? Can you deduce a general principle regarding the probability of one event or another occurring?

Sometimes we are interested in the set theoretic complement of an event, i.e., the event that the original event does not occur. Does the probability of the complementary event relate to the probability of the event? For the spinner, look at the events "at least 3" and "less than 3", which are complements. We have, using formula (1):

$$P[\text{at least 3}] = P[3 \text{ or } 4 \text{ or } 5] = \tfrac{480}{1231} + \tfrac{254}{1231} + \tfrac{130}{1231} = \tfrac{864}{1231}$$

$$(4)$$

$$P[\text{less than 3}] = P[1 \text{ or } 2] = \tfrac{74}{1231} + \tfrac{293}{1231} = \tfrac{367}{1231}$$

Notice that the two probabilities add to one, that is for these two events E and E^c:

$$P[E] + P[E^c] = 1 \tag{5}$$

This is not a coincidence. Together E and E^c contain all simple events, and they have no simple events in common. The total of their probabilities must therefore equal 1, which is the probability of the sample space. ∎

Example 2. Let's look at another random phenomenon. Recall from Section 1.1 the experiment of selecting a sequence of two numbers at random from the set {1, 2, 3, 4, 5} with no repetition of numbers allowed. Such an ordered selection without replacement is known as a *permutation*, in this case a permutation of 2 objects from 5. *Mathematica* can display all possible permutations as follows. The command

KPermutations[list, k]

in the KnoxProb`Utilities` package returns a list of all ordered sequences of length k from the list given in its first argument, where a list member, once selected, cannot be selected again.

```
Needs["KnoxProb`Utilities`"]
```

```
KPermutations[{1, 2, 3, 4, 5}, 2]
```

```
{{1, 2}, {2, 1}, {1, 3}, {3, 1}, {1, 4}, {4, 1},
 {1, 5}, {5, 1}, {2, 3}, {3, 2}, {2, 4}, {4, 2}, {2, 5},
 {5, 2}, {3, 4}, {4, 3}, {3, 5}, {5, 3}, {4, 5}, {5, 4}}
```

You can count that there are 20 such permutations. These 20 are the outcomes that form the sample space of the experiment. The randomness of the drawing makes it quite reasonable to take the theoretical approach to assigning probabilities. Each simple event should have the same likelihood. Since the total sample space has probability 1, the common likelihood of the simple events, say x, satisfies

$$x + x + \cdots + x \,(20 \text{ times}) = 1 \Longrightarrow 20\,x = 1 \Longrightarrow x = \tfrac{1}{20} \qquad (6)$$

In other words, our probability measure on the sample space is defined on outcomes as:

$$P[\{1, 2\}] = P[\{2, 1\}] = P[\{1, 3\}] == \cdots = P[\{5, 4\}] = \tfrac{1}{20} \qquad (7)$$

and for composite events, we can again total the probabilities of outcomes contained in them.

Activity 2 In the example of selecting two positive integers from the first five, find the probability that 1 is in the sample. Explain why your result is intuitively reasonable.

For instance,

P[either 1 or 2 is in the sample]
= P[{1, 2}, {2, 1}, {1, 3}, {3, 1}, {1, 4}, {4, 1}, {1, 5}, {5, 1},
      {2, 3}, {3, 2}, {2, 4}, {4, 2}, {2, 5}, {5, 2}]

= $\tfrac{14}{20}$

and

P[2 is not in sample]

= P[{1, 3}, {3, 1}, {1, 4}, {4, 1}, {1, 5}, {5, 1}, {3, 4},
{4, 3}, {3, 5}, {5, 3}, {4, 5}, {5, 4}]

= $\frac{12}{20}$

By counting outcomes you can check that P[2 in sample] = 8/20, which is complementary to P[2 is not in sample] = 12/20 from above. Also notice that

$$P[1 \text{ in sample}] = \frac{8}{20} = P[2 \text{ in sample}]$$

however,

$$P[1 \text{ in sample or 2 in sample}] = \frac{14}{20} \neq \frac{8}{20} + \frac{8}{20}$$

We see that the probability of the event "either 1 is in the sample or 2 is in the sample" is not simply the sum of the individual probability that 1 is in the sample plus the probability that 2 is in it. The reason is that in adding $8/20 + 8/20$ we let the outcomes {1, 2} and {2, 1} contribute twice to the sum. This overstates the true probability, but we now realize that it is easy to correct for the double count by subtracting 2/20 which is the amount of probability contributed by the doubly counted outcomes. (See Theorem 2 below.) ■

Activity 3 Use *Mathematica* to display the ordered samples of size 3 without replacement from {1, 2, 3, 4, 5, 6}. What is the probability that 1 is in the sample? What is the probability that neither 5 nor 6 is in the sample?

Armed with the experience of our examples, let us now develop a mathematical model for probability, events, and sample spaces that proceeds from a minimal set of axioms to derive several of the most important basic properties of probability. A few other properties are given in the exercises.

Definition 1. A *sample space* Ω of a random experiment is a set whose elements ω are called *outcomes*.

Definition 2 below is provisional - we will have to be more careful about what an event is when we come to continuous probability.

Definition 2. An *event A* is a subset of the sample space Ω; hence an event is a set of outcomes. The collection of all events is denoted by $\mathcal{H}$.

Suppose for example we consider another sampling experiment in which three letters are to be selected at random, without replacement and without regard to order from the set of letters {*a*, *b*, *c*, *d*}. The DiscreteMath`Combinatorica` package (which is loaded automatically when you load KnoxProb`Utilities`) has the commands

KSubsets[list, k]   and   Subsets[universe]

which, respectively, return all subsets of size k taken from the given list of objects, and all subsets of the given universal set. We can use these to find the sample space of the experiment, and the collection of events $\mathcal{H}$.

```
Ω = KSubsets[{a, b, c, d}, 3]
```

{{a, b, c}, {a, b, d}, {a, c, d}, {b, c, d}}

```
𝓗 = Subsets[Ω]
```

{{}, {{a, b, c}}, {{a, b, c}, {a, b, d}}, {{a, b, d}},
  {{a, b, d}, {a, c, d}}, {{a, b, c}, {a, b, d}, {a, c, d}},
  {{a, b, c}, {a, c, d}}, {{a, c, d}},
  {{a, c, d}, {b, c, d}}, {{a, b, c}, {a, c, d}, {b, c, d}},
  {{a, b, c}, {a, b, d}, {a, c, d}, {b, c, d}},
  {{a, b, d}, {a, c, d}, {b, c, d}},
  {{a, b, d}, {b, c, d}}, {{a, b, c}, {a, b, d}, {b, c, d}},
  {{a, b, c}, {b, c, d}}, {{b, c, d}}}

Notice that the empty set {} is the first subset of Ω that is displayed, the next is the simple event {{*a*, *b*, *c*}}, the next is the composite event {{*a*, *b*, *c*}, {*a*, *b*, *d*}}, etc.

In all subsequent work we will continue a practice that we have done so far, which is to blur the distinction between *outcomes* ω, which are members of the sample space, and *simple events* {ω} which are sets consisting of a single outcome. This will enable us to speak of the "probability of an outcome" when in fact probability is a function on events, as seen in the next definition.

Definition 3. A *probability measure* P is a function taking the family of events $\mathcal{H}$ to the real numbers such that

    (i) $P[\Omega] = 1$;

    (ii) For all $A \in \mathcal{H}$, $P[A] \geq 0$;

    (iii) If A_1, A_2, $\cdots$ is a sequence of pairwise disjoint events then

$$P[A_1 \cup A_2 \cup \cdots] = \sum_i P[A_i]$$

Properties (i), (ii), and (iii) are called the *axioms of probability*. Our mathematical model so far corresponds with our intuition: a random experiment has a set of possible outcomes called the sample space, an event is a set of these outcomes, and probability attaches a real number to each event. Axiom (i) requires the entire sample space to have probability 1, axiom (ii) says that all probabilities are non-negative numbers, and axiom (iii) says that when combining *disjoint* events, the probability of the union is the sum of the individual event probabilities. For finite sample spaces this axiom permits us to just define probabilities on outcomes, and then give each event a probability equal to the sum of its (disjoint) outcome probabilities. We will encounter various ways of constructing probability measures along the way, which we must verify to satisfy the axioms. Once having done so, other properties of probability listed in the theorems below are ours for free. They follow from the axioms alone, and not from any particular construction, which of course is the huge advantage of basing a mathematical model on the smallest possible collection of axioms.

Theorem 1. If A is an event with complement A^c, then

$$P[A] + P[A^c] = 1 \tag{8}$$

Therefore $P[A] = 1 - P[A^c]$ and $P[A^c] = 1 - P[A]$.

Proof. The events A and A^c are pairwise disjoint, by definition of complementation. By the third axiom,

$$P[A \cup A^c] = P[A] + P[A^c]$$

Also, $A \cup A^c = \Omega$. By the first axiom, $P[\Omega] = 1$, therefore

$$1 = P[\Omega] = P[A \cup A^c] = P[A] + P[A^c]$$

Theorem 1 was anticipated earlier. It is generally used in cases where the complement of an event of interest is easier to handle than the original event. If we have the probability of one of the two, then we have the other by subtracting from 1.

The next result is based on the idea that the total probability of a union can be found by adding the individual probabilities and correcting if necessary by subtracting the probability in the double-counted intersection region.

Theorem 2. If A and B are arbitrary events, then

$$P[A \cup B] = P[A] + P[B] - P[A \cap B] \qquad (9)$$

In particular if A and B are disjoint, then

$$P[A \cup B] = P[A] + P[B]$$

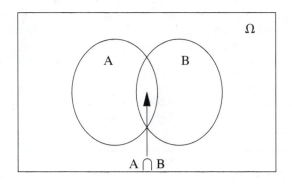

Figure 3 - The probability of a union is the total of the event probabilities minus the intersection probability

Proof. (see Figure 3) Since A can be expressed as the disjoint union $A = (A \cap B) \cup (A \cap B^c)$, and B can be expressed as the disjoint union $B = (A \cap B) \cup (B \cap A^c)$ we have, by axiom (iii),

$$
\begin{aligned}
P[A] + P[B] &= P[A \cap B] + P[A \cap B^c] + P[A \cap B] + P[B \cap A^c] \\
&= P[A \cap B^c] + P[B \cap A^c] + 2\,P[A \cap B]
\end{aligned}
$$

Therefore,

$$P[A] + P[B] - P[A \cap B] = P[A \cap B^c] + P[B \cap A^c] + P[A \cap B] \qquad (10)$$

But it is easy to see from Figure 3 that the events $(A \cap B^c)$, $(B \cap A^c)$, and $(A \cap B)$ are pairwise disjoint and their union is $A \cup B$. Thus, by axiom (iii) of probability again,

$$P[A \cap B^c] + P[B \cap A^c] + P[A \cap B] = P[A \cup B] \qquad (11)$$

Combining (10) and (11) proves the first assertion. The second assertion follows directly as a specialization of axiom (iii) to two events. Alternatively, if A and B are disjoint, then $A \cap B = \emptyset$ and by Exercise 11, $P[\emptyset] = 0$.

Activity 4 Try to devise an analogous thoerem to Theorem 2 for a three set union. (Hint: if you subtract all paired intersection probabilities, are you doing too much?) See Exercise 14 for a statement of the general result.

You may be curious that the axioms only require events to have non-negative probability. It should also be clear that events must have probability less than or equal to 1, because probability 1 represents certainty. In the exercises you are asked to give a rigorous proof of this fact, stated below as Theorem 3.

Theorem 3. For any event A, $P[A] \leq 1$.

Another intuitively obvious property is that if an event B has all of the outcomes in it that another event A has, and perhaps more, then B should have higher probability than A. This result is given as Theorem 4, and the proof is Exercise 12.

Theorem 4. If A and B are events such that $A \subset B$, then $P[A] \leq P[B]$.

Theorems 3 and 4 help to reassure us that our theoretical model of probability corresponds to our intuition, and they will be useful in certain proofs. But Theorems 1 and 2 on complements and unions are more useful in computations. We would like to finish this section by adding a third very useful computational result, usually called the *Law of Total Probability*. Basically it states that we can "divide and conquer", that is break up the computation of a complicated probability $P[A]$ into a sum of easier probabilities of the form $P[A \cap B]$. The imposition of condition B in addition to A presumably simplifies the computation by limiting or structuring the outcomes in the intersection. (See Figure 4.) All of our computational results will be used regularly in the rest of the text, and the exercise set for this section lets you begin to practice with them.

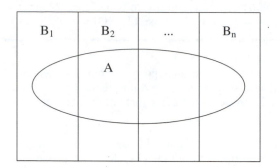

Figure 4 - The Law of Total Probability

Theorem 5. Let A be an arbitrary event, and assume that B_1, B_2, B_3, ..., B_n is a partition of the sample space Ω, that is, the sets B_i are pairwise disjoint and $\bigcup_{i=1}^{n} B_i = \Omega$. Then

$$P[A] = \sum_{i=1}^{n} P[A \cap B_i] \tag{12}$$

Proof. We will use induction on the number of events B_i beginning with the base case $n = 2$. In the base case, by the partition properties it is clear that $B_2 = B_1^c$. We have that $A = (A \cap B_1) \cup (A \cap B_1^c)$, and the two sets in this union are disjoint. By Theorem 2,

$$P[A] = P[A \cap B_1] + P[A \cap B_1^c] = P[A \cap B_1] + P[A \cap B_2]$$

The anchor step of the induction is complete.

Now assume that for all partitions consisting of n events formula (12) is true. Let B_1, B_2, B_3, ..., B_{n+1} be a partition of $n + 1$ events; we must show (12) with the upper index of summation replaced by $n + 1$. Define the events C_i by:

$$C_1 = B_1, \ C_2 = B_2, \ \cdots, \ C_{n-1} = B_{n-1}, \ C_n = B_n \cup B_{n+1}$$

Then the C_i's are also a partition of Ω (see Activity 5 below). By the inductive hypothesis

$$
\begin{aligned}
P[A] &= \sum_{i=1}^{n} P[A \cap C_i] \\
&= \sum_{i=1}^{n-1} P[A \cap B_i] + P[A \cap (B_n \cup B_{n+1})] \\
&= \sum_{i=1}^{n-1} P[A \cap B_i] + P[(A \cap B_n) \cup (A \cap B_{n+1})] \\
&= \sum_{i=1}^{n-1} P[A \cap B_i] + P[A \cap B_n] + P[A \cap B_{n+1}] \\
&= \sum_{i=1}^{n+1} P[A \cap B_i]
\end{aligned}
\tag{13}
$$

This finishes the inductive step of the proof.

Activity 5 Why are the C_i's in the proof of the Law of Total Probability a partition of Ω? Why is the fourth line of (13) true?

Mathematica for Section 1.2

Command	Location
KSubsets[fromlist, k]	DiscreteMath` Combinatorica`
Subsets[universe]	DiscreteMath` Combinatorica`
KPermutations[fromlist, k]	KnoxProb` Utilities`

Exercises 1.2

1. Consider the experiment of rolling one red die and two white dice randomly. Observe the number of the face that lands up on the red die and the sum of the two face up numbers on the white dice.
(a) Describe the sample space, giving the format of a typical outcome and two specific examples of outcomes.
(b) Define a probability measure on this sample space. Argue that it satisfies the three axioms of probability.

2. This is a continuation of Exercise 1. A simulated baseball game is played in such a way that each batter has a card like the one shown in the figure. One red die and two white dice are rolled. The outcome of the batter's plate appearance is decided by finding the column corresponding to the red die and the row corresponding to the sum of the white dice. The entry in that row and that column could be a base hit (1b), a double (2b), a triple (3b), a home run (HR) or a blank, which would mean that the batter is out. For the batter shown, what is the probability that he will get a hit of any kind? If you were designing such a card to simulate a hitter who historically hits about a home run every 20 plate appearances, where would you place HR symbols on the card?

red die

white sum	1	2	3	4	5	6
2						3 b
3	2 b		1 b			
4				1 b		
5			1 b		2 b	
6				1 b		
7		1 b			HR	
8	1 b			1 b		
9		2 b				1 b
10				1 b		
11			1 b		1 b	
12	2 b					

Exercise 2

3. On a driver's education test there are three questions that are considered most important and most difficult. The questions are multiple choice, with four alternative answers labeled a, b, c, and d on each question. Exactly one answer is correct for each question. A test taker guesses randomly on these three questions.
(a) Describe the sample space, giving the format of a typical outcome and two specific examples of outcomes.
(b) Define a probability measure on this sample space. Argue that it satisfies the three axioms of probability.
(c) What is the probability that the random guesser gets at least two questions right?

4. (*Mathematica*) Use *Mathematica* to display the sample space of all possible random samples without order and without replacement of three names from the list of names {Al, Bubba, Celine, Delbert, Erma}. How many events are in the collection of all events $\mathcal{H}$? Describe a probability measure for this experiment, and find the probability that either Bubba or Erma is in the sample.

5. Explain why each of the following attempts P_1 and P_2 is not a good definition of a probability measure on the space of outcomes $\Omega = \{a, b, c, d, e, f, g\}$.

	P_1	P_2
a	.12	.16
b	.05	.20
c	.21	.20
d	.01	.12
e	.04	.08
f	.28	.15
g	.13	.12

6. (*Mathematica*) Use *Mathematica* to display the sample space of all possible random samples in succession and without replacement of two numbers from the set $\{1, 2, ..., 10\}$. Describe a probability measure, and find the probability that 10 is not in the sample.

7. A 1999 Internal Revenue Service study showed that 44.9 million U.S. taxpayers reported earnings of less than $20,000, 20.4 million earned between $20,000 and $30,000, 16.2 million earned between $30,000 and $40,000, 41.9 million earned between $40,000 and $100,000, and 10.5 million earned more than $100,000.
(a) Write the outcomes for the experiment of sampling one taxpayer at random, and define a probability measure. In such a random sample of one taxpayer, what is the probability that the reported earnings is at least $30,000?
(b) Write the outcomes for the experiment of sampling two taxpayers at random and with replacement, that is, once a taxpayer has been selected that person goes back into the pool and could be selected again. Define a probability measure. (Hint: for a succession of two events which do not depend on one another, it makes sense to compute the probability that they both occur as the product of their probabilities. You can convince yourself of this by considering a very simple experiment like the flip of two coins.) Find the probability that at least one of the two earned more than $100,000.

8. Suppose that a point with random x and y coordinates in $[-2, 2]$ is selected. By this we mean that the probability that the random point lies in a region R inside the square is the proportion of the square's area that R has.
(a) Argue that this means of defining a probability measure on the square $[-2, 2] \times [-2, 2]$ satisfies the axioms of probability.
(b) Find the probability that the random point will be in the third quadrant.
(c) Find the probability that the random point is not in the circle of radius 2 shown in the picture.
(d) Find the probability that the random point is either in the third quadrant or not in the circle of radius 2.

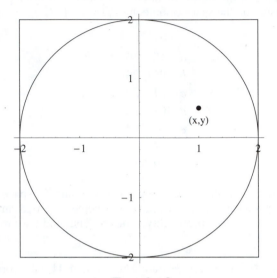

Exercise 8

9. If A and B are events such that $P[A \cap B] = .24$, $P[A] = .61$, and $P[B] = .37$, find $P[(A \cup B)^c]$.

10. Prove Theorem 3.

11. Prove that $P[\emptyset] = 0$.

12. Prove Theorem 4.

13. List out all events for the sample space $\Omega = \{\omega_1, \omega_2, \omega_3\}$. Find the probability of each simple event if $P[\{\omega_1, \omega_2\}] = .6$, $P[\{\omega_2, \omega_3\}] = .8$, and $P[\{\omega_1, \omega_3\}] = .6$.

14. Prove that for arbitrary events A, B, and C:

$$P[A \cup B \cup C] = P[A] + P[B] + P[C] - (P[A \cap B] + P[A \cap C] + P[B \cap C]) + P[A \cap B \cap C]$$

15. The general result for the probability of a union, which has Theorem 2 and Exercise 14 as special cases, is called the *law of inclusion and exclusion*. Let A_1, A_2, A_3, ..., A_n be a collection of n events. To simplify notation, let A_{ij} stand for a typical intersection of two of them $A_i \cap A_j$, let A_{ijk} stand for a three-fold intersection $A_i \cap A_j \cap A_k$, etc. The law states that

$$P[A_1 \cup A_2 \cup A_3 \cup ... \cup A_n] = \sum_i P[A_i] - \sum_{i<j} \sum P[A_{ij}] + \sum_{i<j<k} \sum \sum P[A_{ijk}]$$

$$- ...$$

$$+ (-1)^{n+1} P[A_1 \cap A_2 \cap A_3 \cap ... \cap A_n]$$

In other words, one alternately adds and subtracts all k-fold intersection probabilities as k goes from 1 to n. Use this result in the following classical problem. Four men leave their hats at a hat check station, then on departing take a hat randomly. Find the probability that at least one man takes his own hat.

16. Recall Exercise 5 of Section 1.1 (the blackjack deal). Use the Law of Total Probability, Theorem 5, to find P[queen on 2nd] in a well-justified way.

1.3 Simulation

To simulate means to reproduce the results of a real phenomenon by artificial means. We are interested here in phenomena that have a random element. What we really simulate are the outcomes of a sample space and/or the states of a random variable in accordance with the probability measure or distribution that we are assuming. Computers give a fast and convenient way of carrying out these sorts of simulations.

There are several reasons to use simulation in problems. Sometimes we do not yet have a good understanding of a phenomenon, and observing it repeatedly may give us intuition about a question related to it. In such cases, closed form analytical results might be derivable, and the simulation mostly serves the purpose of suggesting what to try to prove. But many times the system under study is complex enough that analysis is either very difficult or impossible. (Queueing problems - see Section 6.2 - are good illustrations of both situations.) Then simulation is the only means of getting approximate information, and the tools of probability and statistics also allow us to measure our confidence in our conclusions.

Another reason to simulate is somewhat surprising. In some inherently non-probabilistic problems we can use a simulated random target shooting approach to gain information about physical measurements of size and shape. A simple example of such a problem solving method (traditionally called *Monte Carlo simulation* in reference to the European gaming center) is the example on area of a region in Exercise 8.

I have chosen to give the subject emphasis and early treatment in this book for another reason. There is more at stake for you intellectually than gaining intuition about systems by observing simulations. Building an algorithm to solve a problem by simulation helps to teach you about the subtle concepts of the subject of probability, such as probability measures and random variables. Thus, here and elsewhere in this book I will be asking you to carefully study the details of programs I have written, and to write some simulation programs in *Mathematica* yourself. You will invest some time and mental effort, but the payoff will be great.

We have already experienced simulation in the form of selecting ordered samples without replacement of two numbers from the set {1, 2, 3, 4, 5}. We even looked at the empirical histogram of the minimum of the two numbers. We will treat sampling simulations again in more detail in the next section. For the time being let us see how to use *Mathematica*'s tools for generating streams of random numbers to solve problems.

Most of the kind of simulations that we will do are adaptations of the simple process of choosing a random real number in the interval [0, 1]. The word "random" here could be replaced by the more descriptive word "uniform", because the assumption will be that probability is smeared uniformly through [0, 1], with sets getting large probability only if their size is large, not because they happen to be in some favored location in [0, 1]. More precisely, we suppose that the sample space of the random experiment is $\Omega = [0, 1]$, the collection of events $\mathcal{H}$ consists of all subsets of [0, 1] for which it is possible to define their length, and the probability measure P gives probability to a set equal to its length. Then for instance:

$$P[\,[0,\,1]\,] = 1, \quad P\left[\left(\frac{1}{2},\,\frac{3}{4}\right]\right] = \frac{1}{4},$$

$$P\left[\left[0,\,\frac{1}{4}\right] \cup \left[\frac{2}{3},\,1\right]\right] = \frac{1}{4} + \frac{1}{3} = \frac{7}{12},$$

$$P\left[\left\{\frac{5}{6}\right\}\right] = 0$$

Activity 1 Try to describe a uniform distribution of probability on the interval [2,5]. What would be the probability associated to the event [2,3]? [9/4, 13/4] $\cup$ (4,5]? {2.5}?

```
Needs["Statistics`ContinuousDistributions`"]
```

In *Mathematica*'s standard Statistics`ContinuousDistributions` package there are several probability distribution objects, including one for the uniform distribution, and also a means of sampling one or more random numbers from a distribution. Fittingly,

UniformDistribution[a,b]

is the syntax for the uniform distribution on the interval [a,b], and

Random[distribution]   and   RandomArray[distribution, n]

return, respectively, one or a list of *n* random numbers from the given distribution. Here are some examples of their use.

```
dist = UniformDistribution[2, 5];
Random[dist]
Random[dist]
RandomArray[dist, 5]
```

4.22601

4.76283

{4.82717, 2.61044, 2.34392, 3.17102, 2.95292}

Notice that new random numbers in [2, 5] are returned each time that Random and RandomArray are called. Incidentally, Random may be used with an empty argument when the distribution being sampled from is the uniform distribution on [0, 1], which is what we will usually need. The Random[ ] command is in the *Mathematica* kernel; therefore if the uniform distribution on [0, 1] is all you will need in a given situation, then it is not necessary to load the ContinuousDistributions package.

How do *Mathematica* and other programs obtain such random numbers? There are several means that are possible, but the most common type of random number generator is the *linear congruential generator* which we describe now. The ironic fact is that our supposedly random stream of numbers is only simulated in a deterministic fashion in such a way as to be indistinguishable statistically from truly random numbers. Because of this the stream of numbers is called *pseudo-random*. The algorithm is as follows. Begin with a single non-negative integer called a *seed*. (Computers can use internal contents such as collections of digits in their clocks as seeds, which change frequently.) To get the first pseudo-random number, compute a new value of the seed by:

new value of *seed* = (*multiplier* × *seed* + *increment*) mod *modulus*   (1) where *multiplier*, *increment*, and *modulus* are positive integer constants. Then using the newly computed *seed*, return the number

$$rand = seed/modulus \qquad\qquad (2)$$

Subsequent random numbers are found in the same way, updating *seed* by formula (1) and dividing by *modulus*. Notice that because the initial *seed* is non-negative and all of the constants are positive, *seed* will stay non-negative, which makes *rand* ≥ 0 as well. And since the mod operation returns a value less than *modulus*, *seed* will be less than *modulus*, hence *rand* will be less than 1. So each pseudo-random number that we produce will lie in [0, 1).

Now the larger the value of *modulus*, the more possible values of *seed* there can be, hence the more real numbers that can be returned when formula (2) is applied. In fact, there would be exactly *modulus* number of possible *seed* values. But *modulus*, *multiplier*, and *increment* must be carefully chosen to take advantage of this, otherwise the sequence of pseudo-random numbers could cycle with a very small period. As a simple example, if the constants were as follows:

$$\text{new } seed = (2 \times seed + 2) \bmod 8$$

then possible *seed* values are 0, 1, 2, 3, 4, 5, 6, and 7, but it is easy to check that for these seeds the new seeds are, respectively, 2, 4, 6, 0, 2, 4, 6, 0, so that after the first value of *seed*, subsequent seeds can only take on one of four values. Even worse, should *seed* ever equal 6, then new seeds will continue to be 6 forever.

Therefore the choice of the constants is critical to making a stream of numbers that has the appearance of randomness. There is a lot of number theory involved here which we cannot hope to cover. The book by Rubinstein (see References) gives an introduction. It is reported there that the values

$$multiplier = 2^7 + 1, \quad increment = 1, \quad modulus = 2^{35}$$

have been successful, but these are not the only choices. We can build our own function similar to the RandomArray command in *Mathematica* to generate a list of uniform[0,1] numbers using these values as follows.

```
MyRandomArray[initialseed_, n_] :=
   Module[
   {seed, rand, thearray, mult, inc, modul, i},
        mult = 2^7 + 1;
        inc = 1;
        modul = 2^35;
        thearray = {};
        seed = initialseed;
        Do[
   seed = Mod[mult * seed + inc, modul];
             rand = N[ seed / modul];
             AppendTo[thearray, rand],
   {i, 1, n}];
        thearray]
```

The module begins with the list of local variables used in the subsequent lines, and sets the values above for the multiplier, here called *mult*, the increment *inc*, and the modulus *modul*. After initializing an empty list of random numbers and setting the initial value of *seed* according to the input *initialseed*, the module enters a loop. In each of the n passes through the loop, three things are done. The *seed* is updated by the linear congruence relation (1), the new random number *rand* is computed by formula (2), and then *rand* is appended to the array of random numbers. Here is a sample run. You should try running the command again with the same initial seed to see that you get the same sequence of pseudo-random numbers, then try a different seed.

```
MyRandomArray[2^24 + 19, 10]
```

```
{0.0629884, 0.125497, 0.189176, 0.403682, 0.0749859,
 0.673181, 0.840295, 0.398108, 0.355958, 0.918557}
```

Activity 2 Revise the random number generating command above using the following constants, taken from a popular computer science text.

   multiplier = 25173   increment = 13849   modulus = 65536

Load the KnoxProb`Utilities` package and use the command Histogram[datalist,numrectangles] to plot a histogram of 50 random numbers using 4 rectangles. Do you see what you expected?

One advantage to the pseudo-random approach is that simulation experiments are reproducible. If one starts with the same seed, then the same stream of random numbers will ensue, which will produce the same experimental results. This is advantageous because you may want to compare the performance of similar systems using the same random inputs to ensure fairness. *Mathematica* permits this by using the command

SeedRandom[n]

which sets the initial seed value as the integer *n*. If you use an empty argument to SeedRandom, it resets the seed using the computer's clock time. In the sequence of commands below, the seed is set to 79, two random numbers are generated, then the seed is reset to 79, and the same random number .735532 is returned as was returned the first time. We then reset the seed "randomly" using the computer clock, and the first random number that is returned is something other than .735532.

```
SeedRandom[79];
Random[]
Random[]
```

0.735532

0.47383

```
SeedRandom[79];
Random[]
```

0.735532

```
SeedRandom[];
Random[]
```

0.245812

> **Activity 3** Perform a similar test with SeedRandom using a seed of 591 and requesting a RandomArray[UniformDistribution[0,1], 5]. Reset the seed by computer clock time and call for another random array to see that different random numbers are returned. (By the way, RandomArray is not analogous to Random[]; without a distribution argument in the first position it does not work.)

Example 1 It is time to build a simulation model for an actual system. A *random walk* on the line is a random process in which at each time instant 0, 1, 2, ... an object occupies a point with integer coordinates shown as circled numbers in Figure 5.

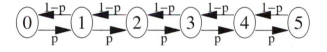

Figure 5 - Schematic diagram of a random walk

The position on the line cannot be foretold with certainty, but it satisfies the condition that if the position is n at one time instant, then it will either be $n + 1$ with probability p or $n - 1$ with probability $1 - p$ at the next time instant, as indicated by the annotated arrows in the figure. Variations are possible, including allowing the object to wait at its current position with some probability, enforcing boundary behavior such as absorption or reflection to confine the random walk to a bounded interval, and increasing the dimension of the set on which the random walk moves to 2, 3, or higher. Random walks have been used as models for diffusion of heat, the motion of economic quantities such as stock prices, the behavior of populations, and the status of competitive games among many other applications, and their fascinating properties have kept probabilists busy for many decades.

For our purposes suppose we are interested in estimating the probability that a random walk which starts at a point $n \geq 0$ reaches a point $M \geq n$ before it reaches 0. This is a famous problem called the *gambler's ruin problem*, which really asks how likely it is that a gambler's wealth that follows a random walk reaches a target level of M before bankruptcy.

We have an initial position n followed by a sequence of random positions $X_1, X_2, X_3, \dots$. The underlying randomness is such that at each time a simple random experiment determines whether the move is to the right (probability p) or the left (probability $1 - p$). So to simulate the random walk we only need to be able to iteratively simulate the simple experiment of adding 1 to the current position with probability p and subtracting 1 with probability $1 - p$. We will stop when position 0 or position M is reached, and make note of which of the two was the final position. By running such a simulation many times and observing the proportion of times that the final state was M, we can estimate the probability that the final state is *M*.

To implement the plan in the preceding paragraph, we need a way of randomly returning +1 with probability *p* and −1 with probability 1 − *p*. But since we know how to simulate uniform real random numbers on [0,1], and the probability of an interval is equal to its length, let's say that if a simulated uniform random number is less than or equal to *p* then the result is +1, otherwise −1. Then we have the desired properties of motion, because

$$P[\text{move right}] = P[\,[0, p]\,] = \text{length}\,[\,[0, p]\,] = p$$
$$P[\text{move left}] = P[\,(p, 1]\,] = \text{length}[\,(p, 1]\,] = 1 - p$$

Here is a *Mathematica* function that does the job. A uniform random number is selected and compared to *p*. If the ≤ comparison is true, then +1 is returned, else −1 is returned. So that we can simulate random walks with different *p* values, we make *p* an argument. One function call with *p* = .5 is displayed; you should try a few others.

```
StepSize[p_] := If[Random[] ≤ p, 1, -1]
```

```
StepSize[.5]
```

−1

To simulate the process once, we let the initial position *n* be given, as well as the right step probability *p* and the boundary point *M*. We repeatedly move by the random StepSize amount until the random walk reaches 0 or *M*. The function below returns the complete list of states that the random walk occupied. Look carefully at the code to see how it works. The variable *statelist* records the current list of positions occupied by the random walk, the variable *done* becomes true when either the boundary 0 or *M* is reached, and variables *currstate* and *nextstate* keep track of the current and next position of the random walk. (The || syntax means logical "or".)

```
SimulateRandomWalk[p_, n_, M_] :=
    Module[
    {statelist, done, currstate, nextstate},
        statelist = {n};
        currstate = n;
        done =
    (currstate ≤ 0) || (currstate ≥ M);
        While[Not[done],
            nextstate =
    currstate + StepSize[p];
            AppendTo[statelist, nextstate];
            currstate = nextstate;
            done =
    (currstate ≤ 0) || (currstate ≥ M)];
    statelist]
```

Below is an example run of the command. We include a list plot of the sequence of positions to help your understanding of the motion of the random walk. You should rerun this command, changing parameters as desired to get more intuition about the behavior of the random walk. A more systematic investigation is suggested in Exercise 6. We set a random seed first for reproducibility, but if you are in the electronic version of the book and want to see different paths of the random walk, just delete the SeedRandom command.

```
SeedRandom[68354];
thewalk = SimulateRandomWalk[.5, 4, 8]
ListPlot[thewalk, PlotStyle -> PointSize[.02],
    DefaultFont → {"Times-Roman", 8}];
```

{4, 3, 2, 3, 4, 3, 4, 3, 2, 1, 2, 1, 0}

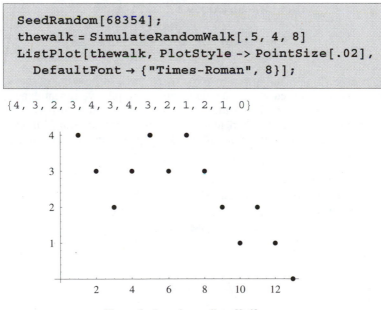

Figure 6 – A random walk on [0, 8]

Activity 4 Roughly how would you describe the dependence of the probability that the random walk reaches *M* before 0 on the value of *p*? How does it depend on the value of *n*? How could you use SimulateRandomWalk to check your intuition about these questions?

Continuing the random walk example, to estimate the probability of hitting *M* before 0 we must repeat the simulation many times, with new streams of random numbers. We merely take note of the last element of the simulated list of positions, which will be either 0 or *M*, count the number of times it is *M*, and divide by the number of repetitions to estimate the probability. Here is a command that does this. Any input values we like for *p*, *n*, and *M* can be used. We also let the number of replications of the simulation be a fourth argument.

```
AbsorptionPctAtM[p_, n_, M_, numreps_] :=
    Module[
    {absatM, empprob, poslist, lastpos, i},
        absatM = 0;
        Do[
    poslist = SimulateRandomWalk[p, n, M];
            lastpos = Last[poslist];
            If[lastpos == M,
    absatM = absatM + 1, Null],
            {i, 1, numreps}];
        empprob = N[absatM / numreps]]
```

The variable *absatM*, initialized to zero, keeps track of how many times so far the walk has been absorbed at *M*. The algorithm is simple: we repeat exactly *numreps* times the actions of simulating the list of positions (called *poslist*) of the random walk, then pick out the last element *lastpos*, and increment *absatM* if necessary. For *p* = .4, *n* = 3, and *M* = 5 here are the results of two runs of 100 simulated random walks each.

```
SeedRandom[342197];
AbsorptionPctAtM[.4, 3, 5, 100]
AbsorptionPctAtM[.4, 3, 5, 100]
```

0.37

0.36

There is good consistency in the results of our runs. About 1/3 of the time, for these parameters, the random walk reaches $M = 5$ before it hits 0. There will be a way of finding the probability exactly by analytical methods later.

Mathematica for Section 1.3

Command	Location
UniformDistribution[*a*, *b*]	Statistics` ContinuousDistributions`
Random[distribution]	Statistics` ContinuousDistributions`
also in :	Statistics` DiscreteDistributions`
RandomArray[distribution, length]	Statistics` ContinuousDistributions`
also in :	Statistics` DiscreteDistributions`
Random[]	kernel
SeedRandom[seedvalue]	kernel
SeedRandom[ ]	kernel
MyRandomArray[initseed, n]	Section 1.3
StepSize[*p*]	Section 1.3
SimulateRandomWalk[p, n, M]	Section 1.3
AbsPctAtM[p, n, M, numreps]	Section 1.3

Exercises 1.3

1. For a uniform distribution on the interval $[a, b]$ of real numbers, find the open interval probability P[(c, d)], the closed interval probability P[$[c, d]$], and the probability of a single point P[$\{d\}$], where $a < c < d < b$.

2. (*Mathematica*) Use the MyRandomArray command from the section with initial seeds of 2, 5, 11, and 25 to generate lists of 10 pseudo–random numbers. Look carefully at your lists. (It may help to ListPlot them.) Do the sequences appear random? Where is the source of the problem, and what might be done to correct it?

3. (*Mathematica*) Write a command to simulate an outcome from a finite sample space with outcomes a, b, c, and d with probabilities p_1, p_2, p_3, p_4, respectively.

4. (*Mathematica*) Write a command to simulate a two–dimensional random walk starting at (0, 0). Such a random process moves on the grid of points in the plane with integer coordinates such that wherever it is now, at the next instant of time it moves to one of the immediately adjacent grid points: right with probability r, left with probability l, up with probability u, and down with probability d, where $r + l + u + d = 1$.

5.   Consider a linear congruential pseudo–random number generator whose sequence of seed values satisfies the recursive relation

$$X_{n+1} = (aX_n + c) \bmod m$$

Assume that $m = 2^p$ for some positive integer p. The generator is called *full period* if given any initial seed all of the values $0, \ldots, m-1$ will appear as seed values before repetition occurs. Show that if the generator is full period, then c must be an odd number.

6. (*Mathematica*) For the random walk with $p = .3$ and $M = 6$, use the Absorption-PctAtM command in the section to try to estimate the functional dependence of the probability of absorption at M on the starting state n.

7.   (*Mathematica*) Use *Mathematica* to simulate ten samples of size 8 from the uniform distribution on $[0, 10]$. Study the samples, come up with a concrete measure of the variability of a sample, and compute and compare the variabilities of your ten samples.

8. (*Mathematica*) While simulation is most productively used with random systems such as the random walk, it is interesting to note that certain numerical approximations of physical quantities like area and volume can be done by Monte Carlo techniques. Develop a *Mathematica* command to approximate the area between the function $f(x) = e^{-x^2}$ and the x-axis on the interval $[0,1]$, which is graphed below. Compare the answers you get from repeated simulations to a numerical integration using *Mathematica*'s NIntegrate command. (Hint: The square $[0,1] \times [0,1]$ contains the desired area A, and the area of the square is 1. If you figuratively shoot a random arrow at the square, it should fall into the shaded region a proportion A of the time. Simulate a large number of such random arrows and count the proportion that fall into A.)

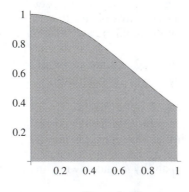

Exercise 8

9. Referring to Exercise 8, why would you not be able to use Monte Carlo simulation to find the area bounded by the x and y-axes, the line $x = 1$, and the curve $y = 1/x$? Could you do it for the curve $y = 1/\sqrt{x}$?

10. (*Mathematica*) A *Galton-Watson board* is a triangular grid of pegs such as the one shown, encased in plastic or glass and stood on end. A large number of small marbles escape from a hole in the top exactly in the middle of the board as shown, bounce from side to side with equal probability as they fall down the board, and eventually land in the bins in the bottom. Build a Galton-Watson simulator for the small five level board shown using arbitrarily many marbles, run it several times, and comment on the shape that is typically formed by the stacks of marbles that gather in the bottom bins.

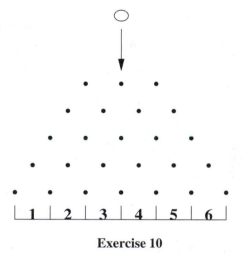

Exercise 10

11. (*Mathematica*) A dam on a 50-foot deep lake is operated to let its vents pass water through at a constant rate so that 1 vertical inch of lake depth is vented per week. The water begins at a level of 5 feet below the top of the dam. A rainy season begins in which each week a random and uniformly distributed depth of water between 2 and 5 inches is added to the lake. Simulate this process, and study how long it takes for the water to overflow the top of the dam. Do the simulations match your intuition and expectations about the situation?

12. (*Mathematica*) A device uses two components in such a way that the device will no longer function if either of the components wears out. Suppose that each component lifetime is random and uniformly distributed on the interval [10, 20]. Write a command to simulate the lifetime of the device many times over and investigate the probability law of the device lifetime empirically. About what is the average lifetime?

1.4 Random Sampling

In this section we focus on a particular kind of random experiment that we have glimpsed before, namely sampling randomly from a universal set. In so doing, we are led to the area of mathematics known as *combinatorics*, the study of counting. The reason why combinatorics and random sampling are so important is that one of a statistician's best tools for obtaining information about a population inexpensively and reliably is by taking a random sample from it.

There are several main types of sampling situations, depending on whether the objects being sampled are taken in sequence or just in a batch, and whether the objects are replaced as sampling proceeds, and so become eligible to be sampled again, or the sampling is done without replacement.

Sampling in Sequence without Replacement

If the order in which sampled items are taken matters and sampling is done without replacing previously sampled items then the sample is a *permutation*. In the following discussion let there be *n* objects in the universe and suppose the sample is to be of size *k*. We can utilize the command KPermutations[list, k] (see Section 1.2) in the KnoxProb`Utilities` package to display all of the permutations. Here for example are the permutations of 2 objects chosen from the set {*a*, *b*, *c*, *d*, *e*}.

```
Needs["KnoxProb`Utilities`"];
KPermutations[{a, b, c, d, e}, 2]
```

```
{{a, b}, {b, a}, {a, c}, {c, a}, {a, d}, {d, a},
 {a, e}, {e, a}, {b, c}, {c, b}, {b, d}, {d, b}, {b, e},
 {e, b}, {c, d}, {d, c}, {c, e}, {e, c}, {d, e}, {e, d}}
```

The same package also contains the command

RandomKPermutation[list, k]

which randomly selects one of the permutations.

```
RandomKPermutation[{a, b, c, d, e}, 3]
```

{a, d, c}

Do the following activity to begin the problem of finding the number of permutations of *k* objects from *n*.

Activity 1 Without utilizing *Mathematica*, how many permutations of 1 object from *n* are there? Now use KPermutations to count the permutations of 2 objects from 2, 2 objects from 3, 2 from 4, 2 from 5, and 2 from 6. Can you detect a pattern and speculate about the number of permutations of 2 objects from *n*?

It is not hard to find the number of permutations of 2 objects from *n* mathematically, and the line of reasoning carries over to permutations of *k* objects from *n*. Let $P_{n,k}$ denote this number of *k*-permutations. To determine a unique 2-permutation, you must pick the first sampled item, which can be done in *n* ways, and then the second, which can be done in *n* − 1 ways. The result is therefore *n*(*n* − 1) by the multiplication principle. Here is the general theorem, which can be proved similarly (see Exercise 3).

Theorem 1. The number of ways of selecting *k* objects in sequence and without replacement from a population of *n* objects is

$$P_{n,k} = n\,(n-1)\,(n-2)\cdots(n-k+1) \;=\; \frac{n!}{(n-k)!} \tag{1}$$

where $n! = n(n-1)(n-2)\cdots 2\cdot 1$.

Example 1 Recall the command called DrawIntegerSample[n, k] in the KnoxProb`Utilities` package which presumes that the universe being sampled from is {1, 2, ..., *n*}. By default it will select a random permutation from this space, but you can also include the option Ordered→False to sample without regard to order (see the next subsection) and/or the option Replacement→True to draw a sample with replacement. We can use this command to study aspects of drawing random samples more deeply, in the way that we looked at the minimum of the sampled values in Section 1.1. Exercise 6 asks you to explore the average of the sample members. Here is a similar experiment on the *range* of the data, that is, the largest value minus the smallest.

For instance if 10 numbers are randomly selected from the set {1, 2, ..., 50} and the sample elements turn out to be 42, 30, 2, 45, 27, 6, 15, 31, 29, and 40 then the range is 45 − 2 = 43. The command below initializes an empty list of ranges, then for a desired number of replications, it draws a new sample, computes the range, and appends this new range to the list of ranges. The final list of ranges is returned.

```
SimulateRange[numreps_, popsize_, sampsize_] :=
    Module[{thesample, rangelist, i, newrange},
        rangelist = {};
        Do[
        thesample =
        DrawIntegerSample[popsize, sampsize];
            newrange = Max[thesample] -
        Min[thesample];
            AppendTo[rangelist, newrange],
    {i, 1, numreps}];
            rangelist]
```

Below is one run of the command for 100 replications of a sample of 10 integers from the first 50. You should try sampling again a few times. You will see that the histogram is rather consistent in its shape: it is asymmetrical with a longer left tail than right tail, it has a tall peak in the low 40's, which is probably about where the average range is located, and a very high percentage of the time the range is at least 30. ∎

```
SeedRandom[124];
ranges = SimulateRange[100, 50, 10]
```

```
{28, 48, 44, 47, 45, 48, 42, 43, 49, 45, 47, 44, 48, 40, 44,
 37, 45, 49, 36, 41, 41, 42, 42, 41, 48, 37, 47, 44, 43, 43,
 46, 39, 42, 41, 42, 46, 43, 34, 23, 45, 40, 40, 47, 46,
 49, 40, 36, 40, 35, 33, 46, 33, 47, 40, 32, 47, 41, 41,
 39, 48, 49, 38, 40, 45, 42, 47, 45, 39, 46, 42, 47, 45,
 48, 48, 42, 40, 39, 43, 46, 46, 48, 44, 44, 35, 48, 40,
 36, 46, 48, 41, 33, 32, 42, 44, 44, 46, 22, 39, 46, 39}
```

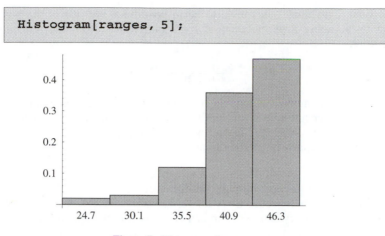

Figure 7 - Histogram of ranges

Example 2 In a random sample of given size, with what likelihood does each individual appear in the sample? Recall we asked this question in Section 1.1 about a sample of two numbers from the set {1, 2, 3, 4, 5}. If the universe being sampled from is $U = \{1, 2, 3, ..., n\}$, and a sample of k individuals is drawn in sequence and without replacement, then the sample space is

$$\Omega = \{(x_1, x_2, \cdots, x_k): x_i \in U \text{ for all } i, \text{ and } x_i \neq x_j \text{ for all } i \neq j\}$$

and by Theorem 1, the cardinality of Ω is

$$n(\Omega) = n(n-1)(n-2)\cdots(n-k+1)$$

We suppose that the randomness assumption means that all outcomes are equally likely and hence the probability of an event E is its cardinality $n(E)$ divided by this $n(\Omega)$.

Consider the event A that individual 1 is in the sample. That individual could have been sampled first, or second, or third, etc., so that A can be broken apart into k disjoint subsets

$$A = B_1 \bigcup B_2 \bigcup \cdots \bigcup B_k$$

where B_i = "1 occurs on ith draw". Therefore by the third axiom of probability, $P[A] = \sum_{i=1}^{k} P[B_i]$. But also, for each i,

$$P[B_i] = (n-1)(n-2)\cdots((n-1)-(k-1)+1)/n(\Omega)$$

since once individual 1 is known to be in position i, the rest of the positions are filled by selecting a permutation of $k - 1$ individuals from the remaining $n - 1$. Therefore,

$$P[A] = \frac{k(n-1)(n-2)\cdots(n-k+1)}{n(n-1)(n-2)\cdots(n-k+1)} = \frac{k}{n}$$

The same reasoning holds for every other individual in the population. So we have shown the rather intuitive result that if a sample in sequence and without replacement of k objects from n is taken, every individual in the universe has an equal chance of k/n of appearing in the sample. ∎

Activity 2 Try to carry through similar reasoning to that of Example 2 if the sample is drawn in sequence with replacement. Does the argument break down? If it does, try another approach (such as complementation).

Sampling in a Batch without Replacement

Now let us look at sampling without replacement when the order in which sample values are drawn is of no consequence, i.e., the sample may as well have been drawn all at once in a batch. Most card games follow this prescription, but there are some other very important situations of this type too. A sample without order or replacement of k objects from n is called a *combination* of k chosen from n.

Recall the *Mathematica* command KSubsets[list, k] in the package DiscreteMath`Combinatorica` which returns all subsets of size k from the given list. (This package is loaded automatically when you load KnoxProb`Utilities`.) It therefore returns the sample space of all combinations of k chosen from n. Here are the three-element combinations taken from $\{a, b, c, d, e\}$.

```
KSubsets[{a, b, c, d, e}, 3]
```

```
{{a, b, c}, {a, b, d}, {a, b, e}, {a, c, d}, {a, c, e},
 {a, d, e}, {b, c, d}, {b, c, e}, {b, d, e}, {c, d, e}}
```

The same package has a command

        RandomKSubset[list, k]

which returns one such combination selected at random. Below are a couple of instances, and as we would anticipate, we get a different randomly selected subset the second time than the first time.

```
SeedRandom[67653];
RandomKSubset[{a, b, c, d, e}, 3]
RandomKSubset[{a, b, c, d, e}, 3]
```

{b, c, d}

{a, b, c}

Following along the lines of Activity 1, let's use KSubsets to try to find a pattern for the numbers

$$C_{n,k} = \text{number of combinations of } k \text{ chosen from } n$$

The Length function in the *Mathematica* kernel can be used to return the length of a list, that is, the cardinality of the set represented by the list. First, for $k = 1$ and $n = 3, 4$, and 5,

```
{Length[KSubsets[{1, 2, 3}, 1]],
Length[KSubsets[{1, 2, 3, 4}, 1]],
Length[KSubsets[{1, 2, 3, 4, 5}, 1]]}
```

{3, 4, 5}

As we should have known immediately, $C_{n,1}$ = number of subsets of size 1 from a universe of n elements = n. What about $C_{n,2}$? For $n = 2, 3, 4, 5, 6$, and 7 the results are below.

```
{Length[KSubsets[{1, 2}, 2]],
Length[KSubsets[{1, 2, 3}, 2]],
Length[KSubsets[{1, 2, 3, 4}, 2]],
Length[KSubsets[{1, 2, 3, 4, 5}, 2]],
Length[KSubsets[{1, 2, 3, 4, 5, 6}, 2]],
Length[KSubsets[{1, 2, 3, 4, 5, 6, 7}, 2]]}
```

{1, 3, 6, 10, 15, 21}

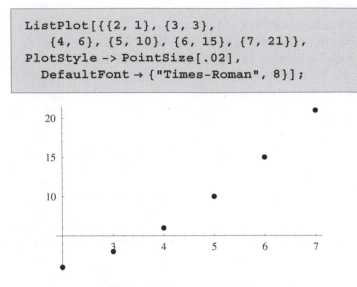

```
ListPlot[{{2, 1}, {3, 3},
   {4, 6}, {5, 10}, {6, 15}, {7, 21}},
PlotStyle -> PointSize[.02],
   DefaultFont → {"Times-Roman", 8}];
```

Figure 8 - $C_{n,2}$ as a function of n.

The list plot of $C_{n,2}$ against n above suggests what could be a quadratic shape. If $C_{n,2}$ is a quadratic function of n, then since both $C_{0,2} = 0$ and $C_{1,2} = 0$ (there are no subsets of size 2 from a set of 0 or 1 elements) we suspect a function of the form $C_{n,2} = a\, n(n-1)$. Using the fact that $C_{2,2} = 1$, it is easy to compute $a = 1/2$, which also fits all the data points computed above. So we are guessing that for all n,

$$C_{n,2} = \frac{n\,(n-1)}{2}$$

It would be instructive for you to do Exercise 8 now, which continues this line of investigation to find a formula for $C_{n,3}$. But for efficiency of exposition let us now try to abstract to general n and k to prove a simple result that covers all cases.

If at first k objects are selected in a batch from n, then it is possible to think of subsequently putting them into a sequence. By doing so we would create a unique permutation of k objects from n. The number of possible permutations is known from Theorem 1. And it is fairly easy to see that there are $k! = k(k-1)\,(k-2)\cdots 2\cdot 1$ ways of sequencing the k selected objects, by the multiplication principle. Therefore, by the multiplication principle again,

#permutations = #combinations · # ways of sequencing the combination

$$\implies P_{n,k} = C_{n,k} \cdot k!$$

If we now bring in the formula for $P_{n,k}$ from Theorem 1, we have the following result.

Theorem 2. The number of combinations of k objects selected in a batch without replacement from n is

$$C_{n,k} = \frac{n!}{k! \cdot (n-k)!} \tag{2}$$

You may recognize $C_{n,k}$ as the same quantity as the binomial coefficient $\binom{n}{k}$ from the binomial theorem and other contexts. This coefficient is usually read as "n choose k", and we now know why. Incidentally, *Mathematica* can compute this quantity using the Binomial[n, k] function, which is in the kernel.

```
C6choose3 = Binomial[6, 3]
```

20

Activity 3 Check to see whether empirically the probability that a particular element of the universe is sampled is dependent on the element. For a sample of k from n in a batch without replacement, what does this probability seem to be? (See also Exercise 9.) You may use the command below, which inputs the population size n (assuming the population is coded as {1, 2, ..., n}), the sample size k, the population member m between 1 and n to be checked, and the number of replications *numreps*. It outputs the proportion of the replications for which m was in the sample.

```
ProportionMInSample[n_, k_, m_, numreps_] :=
    Module[{mcount, simsample, i},
        mcount = 0;
        Do[simsample =
    DrawIntegerSample[n, k, Ordered -> False];
            If[MemberQ[simsample, m],
    mcount = mcount + 1, Null],
            {i, 1, numreps}];
        mcount / numreps]
```

Example 3 Theorem 2 is extremely useful in answering probability questions. For example let us investigate whether it is reasonable to suppose that a population of 1000 tax returns has no more than 20% with errors on them, if a random sample of 100 returns from this population yielded 40 with errors. The proportion in the sample is twice that which is assumed for the population. Is it likely that such an anomaly could happen by chance?

The 20% population figure, and not 19%, 18%, or anything less than 20%, is the most likely to have caused a large number of erroneous returns in the sample (see Exercise 11), so we will recast the question in the following way. If there are exactly 200 among the 1000 returns that have errors, how likely is it that a random sample of 100 has 40 erroneous returns or more? We say 40 *or more* because any large number of errors, not just identically 40, gives evidence against the 20% hypothesis. If the probability of 40 or more errors is quite small, then there is strong evidence in the observed data against the 20% hypothesis.

Theorem 2 allows us to write the probability of exactly k erroneous returns in the sample, hence $100 - k$ good returns, as

$$\frac{\binom{200}{k}\binom{800}{100-k}}{\binom{1000}{100}}, \ k = 0, 1, 2, \ ... \ , \ 100$$

This is because there are $\binom{1000}{100}$ equally likely samples of size 100 in the sample space, and there are $\binom{200}{k}$ ways of selecting k erroneous returns from the group of 200 of them, and $\binom{800}{100-k}$ ways of selecting the remaining correct returns from the 800 correct returns in the population.

We can calculate the probability of 40 or more erroneous returns by summing these numbers from 40 to 100, or equivalently by complementation, summing from 0 to 39 and subtracting from 1. We use the latter approach to reduce computation.

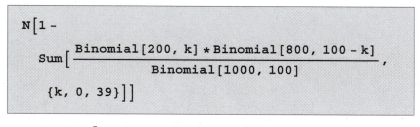

```
N[1 -
     Sum[ Binomial[200, k] * Binomial[800, 100 - k]
          ──────────────────────────────────────────── ,
                    Binomial[1000, 100]
     {k, 0, 39}]]
```

8.85353×10^{-7}

Since the probability is only on the order of 10^{-6}, the hypothesized 20% error rate in the population of tax returns is in very serious doubt. ∎

Example 4 As indicated earlier, Theorem 2 has implications of importance for card players. In poker, a hand of 2-pair means that there are two cards of one rank (such as 4's or 7's), two cards of a different rank, and a single card that doesn't match either of the first two ranks. A hand of 3-of-a-kind consists of three cards of one rank (such as 6's), and two unmatching singletons (such as one king and one ace). Which hand is rarer, and hence more valuable?

For the experiment of drawing a batch of 5 cards without replacement from a standard deck of 52 cards we want to compare the probabilities of the 2-pair event and the 3-of-a-kind event. The sample space has $C_{52,5}$ elements. For the 2-pair event, to determine a unique hand we must first select in no particular order two ranks for the pairs. This can be done in $C_{13,2}$ ways, since there are 13 possible ranks 2, 3, ... , Ace. For each rank we must select two of the suits (hearts, diamonds, clubs, or spades) from among the four possible suits to determine the exact content of the pairs. Then we must select a final card to fill the hand from among the $52 - 4 - 4 = 44$ cards left in the deck which do not match our selected ranks. Thus, the following is the number of 2-pair hands:

```
num2pairhands = Binomial[13, 2] *
   Binomial[4, 2] * Binomial[4, 2] * 44
```

```
123552
```

For 3-of-a-kind, we must pick a rank from among the 13 for our 3-of-a-kind card, then we select 3 from among the 4 suits for that rank, and then two other different ranks from the 12 remaining ranks, and a suit among the four possible for each of those unmatched cards. Therefore this expression gives the number of 3-of-a-kind hands:

```
num3ofakindhands = Binomial[13, 1] *
   Binomial[4, 3] * Binomial[12, 2] * 4 * 4
```

```
54912
```

Below are the two probabilities. Since the 2-pair hand is over twice as likely to be dealt as the 3-of-a-kind hand, the 3-of-a-kind hand is more valuable. ∎

```
{Ptwopair, P3ofkind} =
  {num2pairhands / Binomial[52, 5],
  num3ofakindhands / Binomial[52, 5]}
```

$$\left\{ \frac{198}{4165}, \frac{88}{4165} \right\}$$

Other Sampling Situations

There are four possible sampling scenarios that allow for the cases of sequential or batch sampling and replacement or no replacement. The two scenarios in which there is no replacement have been covered. To sample with replacement but in a batch is an unusual scenario that we will not consider here. But the final possibility of sampling in sequence and with replacement is an important one that deserves a few words.

Once again let there be n objects in the population and let the sample be of size k. Since we can now reuse previously sampled objects, there are n possibilities for each of the k positions in the sample. Directly from the multiplication principle there are

$$n \times n \times \cdots \times n \, (k \text{ times}) = n^k \tag{3}$$

possible such random samples. For example, if $\{1, 2, \ldots, n\}$ is the population, the probability that the first sample object is a 3 is

$$\frac{1 \cdot n^{k-1}}{n^k} = \frac{1}{n}$$

since the restriction that 3 is in position 1 means that there is just one way of filling that position, and still there are n ways of filling each of the other $k - 1$ positions. There is nothing special about object 3, nor is there anything special about position 1. Each population member has a probability of $1/n$ of appearing in each position.

The special case $n = 2$ comes up frequently, and will in fact be the subject of a section in Chapter 2. If there are two possibilities, call them 1 and 2, for each of the k sample positions, then there are 2^k possible sequences of length k of these 1's and 2's. Try this activity, which will be very useful later.

Activity 4 How many sequences of 1's and 2's are there such that there are exactly m 2's in the sample, for m between 0 and k? First try writing down all sequences with $k = 3$ and counting how many of them have m 2's for $m = 0, 1, 2, 3$. Next move to $k = 4$, and then try to generalize. (Hint: Think what must be done to uniquely determine a sequence of k terms with m 2's in it.)

Another sampling situation arises in statistical experimentation in which a researcher wants to split a population of subjects at random into several groups, each of which will receive a different treatment. This is called *partitioning* the test group. The random assignment of subjects to groups is done so as to reduce the possibilty of systematic *confounding*, in which members within one or more of the groups share some common feature apart from the fact that they receive the same treatment. For example, if all of the members of one group were over 60 years of age, and all members of another were under 30 years of age, then the researcher would have a hard time distinguishing whether a difference in response to treatment between groups was due to the different treatments they received or due to the confounding factor of age.

Example 5 A medical researcher wants to divide 15 healthy adult male cigarette smokers numbered 1, 2, 3, ... , 15 randomly into three equally sized groups. Group 1 will be given a placebo and told that it is a drug to reduce the impulse to smoke, group 2 will be given a nicotine patch, and group 3 will be given nicotine gum to chew. The intent is to let some time pass and then measure the effectiveness of the smoke reduction methods, as well as other physical indicators of nicotine related effects. In how many ways can the smokers be partitioned? If after partitioning, the placebo group contains subjects 1, 2, 3, 4, and 5, is there cause to doubt the efficacy of the randomization procedure?

The sample space of all partitions of 15 subjects into three groups has the form below, which is a triple of sets of size 5 drawn from the set $\{1, 2, ..., 15\}$.

$$\Omega = \{(\{x_1, x_2, x_3, x_4, x_5\}, \{x_6, x_7, x_8, x_9, x_{10}\}, \{x_{11}, x_{12}, x_{13}, x_{14}, x_{15}\}):$$
$$x_i \in \{1, 2, ..., 15\} \text{ and } x_i \neq x_j \text{ for all } i \neq j\}$$

This suggests a way of counting the elements in Ω. First we must pick a set of 5 subjects from 15 in a batch and without replacement to form the placebo group. Then from the remaining 10 subjects we sample 5 more for the nicotine patch group, and finally we are forced to use the remaining 5 subjects for the nicotine gum group. By the multiplication principle there are

$$\binom{15}{5}\binom{10}{5}\binom{5}{5} = \frac{15!}{5!\,10!}\,\frac{10!}{5!\,5!}\,\frac{5!}{0!\,5!} = \frac{15!}{5!\,5!\,5!}$$

elements of Ω. (Notice that the rightmost factor in the product of binomial coefficients is just equal to 1.) *Mathematica* tells us how many this is:

```
Binomial[15, 5] * Binomial[10, 5]
```

756756

If a partition is selected at random, then each outcome receives probability 1/756,756 and events have probability equal to their cardinality divided by 756,756. Now the number of partitions in the set of all partitions for which the placebo group is fixed at {1, 2, 3, 4, 5} can be found similarly: we must sample 5 from 10 for the patch group, and we are left with a unique set of 5 for the gum group. Thus, the probability that the placebo group is {1, 2, 3, 4, 5} is

$$\frac{\binom{10}{5}}{\binom{15}{5}\binom{10}{5}} = \frac{1}{\binom{15}{5}}$$

which is the following number

```
1 / Binomial[15, 5]
```

$$\frac{1}{3003}$$

There is less than a one in 3000 chance that this placebo group assignment could have arisen randomly, but we should not be too hasty. Some placebo group subset must be observed, and we could say the same thing about the apparently random subset {3, 5, 8, 12, 13} as we did about {1, 2, 3, 4, 5}: it is only one in 3000 likely to have come up. Should we then doubt any partition we see? The answer is not a mathematical one. Since {1, 2, 3, 4, 5} did come up, we should look at the sampling process itself. For instance if slips of paper were put into a hat in order and the hat wasn't sufficiently shaken, the low numbers could have been on top and therefore the first group would tend to contain more low numbers. If numbers are sampled using a reputable random number generator on a computer, one would tend to be less suspicious. ■

For a population of size n and k groups of sizes n_1, n_2, ..., n_k you should check that the reasoning of Example 5 extends easily to yield the number of partitions as

$$\frac{n!}{n_1! \, n_2! \cdots n_k!} \tag{4}$$

In Exercise 15 you are asked to try to code the algorithm for a function RandPartition[poplist, numgroups, sizelist], which returns a random partition of the given population, with the given number of groups, whose group sizes are as in the list *sizelist*. Again you will find the reasoning of Example 5 to be helpful.

Mathematica for Section 1.4

Command	Location
KPermutations[list, size]	KnoxProb` Utilities`
RandomKPermutation[list, size]	KnoxProb` Utilities`
DrawIntegerSample[n, k]	KnoxProb` Utilities`
Histogram[list, numrecs]	KnoxProb` Utilities`
KSubsets[list, k]	DiscreteMath` Combinatorica`
RandomKSubset[list, k]	DiscreteMath` Combinatorica`
SimulateRange[reps, popsize, sampsize]	Section 1.4
ProportionMInSample[n, k, m, reps]	Section 1.4
Binomial[n, k]	kernel

Exercises 1.4

1. A police lineup consists of five individuals who stand in a row in random order. Assume that all lineup members have different heights.
(a) How many possible lineups are there?
(b) Find and justify carefully the probability that the tallest lineup member is in the middle position.
(c) Find the probability that at least one lineup member is taller than the person in the first position.

2. A bank machine PIN number consists of four digits. What is the probability that a randomly chosen PIN will have at least one digit that is 6 or more? (Assume that digits can be used more than once.)

3. Use mathematical induction to prove Theorem 1.

4. (*Mathematica*) (This is a famous problem called the *birthday problem*.) There are *n* people at a party, and to liven up the evening they decide to find out if anyone there has the same birthday as anyone else. Find the probability that this happens, assuming that there are 365 equally likely days of the year on which each person at the party could be born. Report your answer in the form of a table of such probabilities for values of *n* from 15 to 35.

5. (*Mathematica*) Use DrawIntegerSample to study the distribution of the maximum sample member in a batch random sample without replacement of five integers from {1,2,...,40}. Discuss the most important features of the distribution.

6. (*Mathematica*) The *mean* of a random sample X_1, X_2, ..., X_n is the arithmetical average $(X_1 + X_2 + \cdots + X_n)/n$ of the sample values. *Mathematica* can compute the sample mean with the command Mean[list]. Use DrawIntegerSample to compare the spread of the distribution of sample means of a sequenced random sample with replacement to that of sample means of a sequenced random sample without replacement. Is there a difference, and if so, is there a reasonable explanation for it?

7. How many total subsets of a set of size n are there? Devise a counting approach that shows that this number is a power of 2.

8. (*Mathematica*) As demonstrated in the subsection on batch sampling without replacement, use the KSubsets command to find an expression for the number $C_{n,3}$ of samples of 3 items from n as a function of n.

9. For a random sample of k subjects chosen from n in a batch without replacement, show that the probability that a particular subject is in the sample is k/n.

10. Which of these two poker hands is more valuable: a full house (3 cards of one rank and 2 cards of another) or a flush (5 cards of the same suit)?

11. (*Mathematica*) In Example 3, argue that the sample of 100 returns with 40 errors is more likely to happen if the true proportion of erroneous returns in the population is 20% than if it is some $p < 20\%$. You may need *Mathematica* to exhaustively check all of the possible values of p.

12. A major cookie manufacturer has an inspection plan that determines whether each carton it prepares will be shipped out for sale under its brand name or will be rejected and repackaged and sold under a generic name. A carton contains 50 packages of cookies, and a random sample of 10 packages will be removed from the carton and carefully inspected. If one or more sample packages are found that have significant cookie breakage, the whole carton will be sold as generic.
(a) If a particular carton has 3 bad packages, what is the probability that it will be sold as generic?
(b) (*Mathematica*) Under the current inspection plan, how many bad packages would be necessary in the carton in order that the probability of rejecting the carton should be at least 90%?
(c) (*Mathematica*) For a carton with 3 bad packages, what should the sample size be so that the probability of rejecting the carton is at least 75%?

13. For a partition of n population subjects into k groups of sizes n_1, n_2, ... , n_k, find and carefully justify the probability that subject number 1 is assigned to the first group.

14. In how many distinguishable ways can the letters of the word Illinois be arranged?

15. (*Mathematica*) Write the *Mathematica* command RandPartition described at the end of the section.

16. We looked at pseudo-random number generation in Section 1.3, in which a sequence of numbers x_1, x_2, x_3, x_4, ... is generated which simulates the properties of a totally random sequence. One test for randomness that is sensitive to upward and downward trends is to consider each successive pair, (x_1, x_2), (x_2, x_3), (x_3, x_4), ... and generate a sequence of symbols U (for up) or D (for down) according to whether the second member of the pair is larger or smaller than the first member of the pair. A sequence such as UUUUUUUUDDDDDDD with a long upward trend followed by a long downward trend would be cause for suspicion of the number generation algorithm. This example sequence has one so-called *run* of consecutive U's followed by one *run* of D's for a total of only 2 runs. We could diagnose potential problems with our random number generator by saying that we will look at a sequence such as this and reject the hypothesis of randomness if there are too few runs. For sequences that are given to have 8 U symbols and 7 D symbols, if our random number generator is working well, what is the probability that there will be 3 or fewer runs?

17. In large sampling problems, such as polling registered voters in the United States, it is not only impractical to take an exhaustive census of all population members, it is even impractical to itemize all of them in an ordered list $\{1, 2, ... , n\}$ so as to take a simple random sample from the list. In such cases a popular approach is to use *stratified sampling*, in which population members are classified into several main strata (such as their state of residence), which may not be equally sized, and then further into one or more substrata (such as their county, township, district, etc.) that are nested inside the main strata. A stratified sample is taken by sampling at random a few strata, then a few substrata within it, etc. until the smallest strata level is reached at which point a random sample of individuals is taken from that.

   Here suppose there are three main strata, each with three substrata. Each of the nine substrata will also have the same number of individuals, namely 20. Two main strata will be sampled at random in a batch without replacement, then in each of the sampled strata two substrata will be sampled in a batch without replacement, and then four individuals from each substratum will be sampled in a batch without replacement to form the stratified sample. Find the probability that the first individual in the first substrata of the first stratum will be in the stratified sample. Is the probability of being in the sample the same for all individuals in this case?

1.5 Conditional Probability

In many problems, especially in random phenomena that take place over time, we are interested in computing probabilities of events using partial knowledge that some other event has already occurred. These kinds of probabilities are called *conditional probabilities*, and we will define them and study their properties and applications in this section.

A simple and intuitive way to begin is with the following example. For quite a few years the state of Illinois has been engaged in a sometimes bitter battle over a proposal for a third airport in the Chicago metropolitan area. The goal is to help relieve the burden on O'Hare and Midway airports. South suburban Peotone is often mentioned as a possible site for the new airport. Suppose that a survey drew responses to the proposal from four regions: north and northwest suburban Chicago, the city itself, south and southwest suburban Chicago, and "downstate" Illinois (a strange usage of the term intended to mean anything outside of the Chicago area). The numbers of people jointly in each region and opinion category are in the table below, and totals are computed for each row and column.

	in favor	opposed	don't care	total
north suburbs	85	34	20	139
Chicago	126	28	64	218
south suburbs	12	49	5	66
downstate	43	72	25	140
total	266	183	114	563

An individual is sampled at random from the group of 563 survey respondents. Because there are 183 who oppose the Peotone airport proposal, the probability that the sampled individual opposes the new airport is $183/563 \approx .33$. But suppose that it is known that the sampled individual is from the south suburbs. Then only the 66 people in line 3 of the table are eligible, among whom 49 oppose the proposal. Therefore the conditional probability that the sampled individual opposes the Peotone airport given that he or she lives in the south suburbs is $49/66 \approx .74$, which is much higher than the unconditional probability. In Exercise 1 you are asked to compute several other conditional probabilities in a similar way.

The main idea is that if an event B is known to have happened, the sample space reduces to those outcomes in B, and every other event A reduces to its part on B, that is $A \cap B$. This motivates the following definition.

Definition 1. If B is an event with $P[B] > 0$, and A is an event, then the *conditional probability* of A given B is

$$P[A \mid B] = \frac{P[A \cap B]}{P[B]} \tag{1}$$

In the Peotone airport example, A is the event that the sampled individual opposes the Peotone airport, B is the event that he or she is from the south suburbs, and $A \cap B$ is the event that the individual both opposes the proposal and is from the south suburbs. Thus, $P[A \cap B]$ = 49/563, $P[B]$ = 66/563, and the conditional probability is

$$\frac{49/563}{66/563} = \frac{49}{66}$$

as if the group of survey respondents in the south suburbs were the only ones in a reduced sample space for the experiment. Be aware though that there is nothing about the definition expressed by formula (1) that requires outcomes to be equally likely.

Example 1 Among a flat of 50 eight-foot 2×4 pine boards there are ten that are in a condition unsuitable for building. A small pile of 10 boards from the flat is assembled, and it is obvious that the top 2 are bad ones. What is the probability that the pile will have at least 1 other bad board?

The universe being sampled from is the entire collection of 50 boards, 10 of which are bad and 40 good. We will suppose that our sample of 10 boards is obtained without order or replacement. The probability of the event that there are k bad boards in the sample and $10 - k$ good ones is therefore

$$\frac{\binom{10}{k}\binom{40}{10-k}}{\binom{50}{10}}$$

by the methods of Section 1.4. We want to compute the probability of at least 3 bad boards in the sample given that at least 2 bad boards are there. By the definition of conditional probability, this is

$$P[\text{at least 3 bad} \mid \text{at least 2 bad}] \;=\; \frac{P[\text{at least 3 bad} \cap \text{at least 2 bad}]}{P[\text{at least 2 bad}]}$$

$$=\; \frac{P[\text{at least 3 bad}]}{P[\text{at least 2 bad}]}$$

$$=\; \frac{1-P[0,1,\text{ or 2 bad}]}{1-P[0 \text{ or } 1 \text{ bad}]}$$

$$=\; \frac{1-\left(\binom{10}{0}\binom{40}{10}+\binom{10}{1}\binom{40}{9}+\binom{10}{2}\binom{40}{8}\right)/\binom{50}{10}}{1-\left(\binom{10}{0}\binom{40}{10}+\binom{10}{1}\binom{40}{9}\right)/\binom{50}{10}}.$$

Using *Mathematica*, this comes out to be the following

```
N[(1 - (Binomial[10, 0] * Binomial[40, 10] +
      Binomial[10, 1] * Binomial[40, 9] +
      Binomial[10, 2] * Binomial[40, 8]) /
   Binomial[50, 10]) /
  (1 - (Binomial[10, 0] * Binomial[40, 10] +
      Binomial[10, 1] * Binomial[40, 9]) /
   Binomial[50, 10])]
```

0.482722

■

> **Activity 1**   Referring to Example 1, if the sample of 10 boards is drawn in sequence instead of in a batch, find the general expression for the probability of *k* bad boards in the sample. Try to show that it is the same as the case of sampling in a batch.

Multiplication Rules

The definition of conditional probability can be rewritten as

$$P[A \cap B] = P[B]\cdot P[A \mid B] \tag{2}$$

which can be used to compute intersection probabilities. One thinks of a chain of events in which *B* happens first; the chance that *A* also happens is the probability of *B* times the probability of *A* given *B*. For instance, if we draw two cards in sequence and without replacement from a standard deck, the probability that both the first and second are aces is $\frac{4}{52}\cdot\frac{3}{51}$, that is, the probability that the first is an ace times the conditional probability that the second is an ace given that the first is an ace. For three aces in a row, the probability would become $\frac{4}{52}\cdot\frac{3}{51}\cdot\frac{2}{50}$, in which the third factor is the conditional probability that the third card is an ace given that both of the first two cards are aces.

The example of the three aces suggests that formula (2) can be generalized as in the following theorem. You are asked to prove it in Exercise 6. Notice that from stage to stage one must condition not only on the immediately preceding event but all past events.

Theorem 1. Let B_1, B_2, ..., B_n be events such that each of the following conditional probabilities is defined. Then

$$P[B_1 \cap B_2 \cap \cdots \cap B_n] =$$
$$P[B_1] \cdot P[B_2 \mid B_1] \cdot P[B_3 \mid B_1 \cap B_2] \cdots P[B_n \mid B_1 \cap \cdots \cap B_{n-1}] \tag{3}$$

Example 2 Exercise 4 in Section 1.4 described the following famous problem. Suppose you ask a room of n people for their birthdays, one by one. What is the probability that there will be at least one set of matching birthdays?

In the exercise you were expected to use combinatorics to solve the problem. Here we take a conditional probability approach using formula (3). It is convenient to study the complementary event of no matching birthdays. Let B_1 be the event that there is no match after the first person is asked, which has probability 1. Let B_2 be the event that there is no match after the first two people, which has conditional probability 364/365 given B_1 since one birthday is disallowed. Similarly, if B_3 is the event that there is no match after the first three people, then B_3 has conditional probability 363/365 given the event $B_1 \cap B_2$ of no previous matches. For n people we have

$$
\begin{aligned}
\text{P[at least 1 match]} \ &= \ 1 - \text{P[no matches after } n \text{ people]} \\
&= \ 1 - P[B_1 \cap B_2 \cap \cdots \cap B_n] \\
&= \ 1 - \tfrac{364}{365} \cdot \tfrac{363}{365} \cdots \cdot \tfrac{365-n+1}{365}
\end{aligned}
$$

Here is a *Mathematica* function to compute that probability.

```
f[n_] := 1 - Product[ (365 - i)/365 , {i, 0, n - 1}]
```

For example for, respectively, 20 and 50 people,

```
{N[f[20]], N[f[50]]}
```

{0.411438, 0.970374}

If you have not done the exercise, you might be surprised at how large these probabilities are. Here is a plot that shows how the probability depends on the number of people *n*. The halfway mark is reached at about 23 people. You should try finding the smallest *n* for which there is a 90% probability of a match. ∎

```
birthdays = Table[N[{n, f[n]}], {n, 5, 50}];
ListPlot[birthdays,
    PlotStyle -> PointSize[.02],
AxesLabel -> {"n", "prob. at least 1 match"},
    DefaultFont → {"Times-Roman", 8}];
```

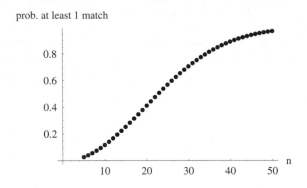

Figure 9 - The probability of matching birthdays as a function of *n*

Activity 2 Try this similar problem. Suppose the *n* people are each asked for the hour and minute of the day when they were born. How many people would have to be in the room to have at least a 50% chance of matching birth times?

The next proposition, though quite easy to prove, is very powerful. It goes by the name of the *Law of Total Probability*. We saw an earlier version of it in Section 2 which only talked about intersection probabilities. The proof simply takes that result and adds on the multiplication rule (2).

Theorem 2. If B_1, B_2, ..., B_n is a partition of the sample space, that is, the B's are pairwise disjoint and their union is Ω, then

$$P[A] = \sum_{i=1}^{n} P[A \mid B_i] \, P[B_i] \qquad (4)$$

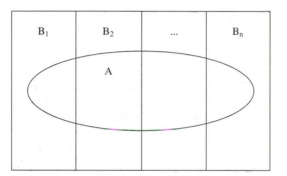

Figure 10 - The Law of Total Probability

Proof. The events $A \cap B_i$ are pairwise disjoint, and their union is A (see Figure 10). Thus, by the Law of Total Probability for intersections and the multiplication rule (2),

$$P[A] = \sum_{i=1}^{n} P[A \cap B_i] = \sum_{i=1}^{n} P[A \mid B_i] \, P[B_i] \qquad (5)$$

(Note: It is easy to argue that the conclusion still holds when the B's are a partition of the event A but not a partition of all of Ω. Try it.)

The Law of Total Probability is very useful in staged experiments where A is a second stage event and it is difficult to directly find its probability; however, conditional probabilities of A given first stage events are easier. Formula (4) lets us condition A on a collection of possible first stage events B_i, and then "un-condition" by multiplying by $P[B_i]$ and adding over all of these possible first stage cases. The next example illustrates the idea.

Example 3 Each day a Xerox machine can be in one of four states of deterioration labeled 1, 2, 3, and 4 from the best condition to the worst. The conditional probabilities that the machine will change from each possible condition on one day to each possible condition on the next day are shown in the table below.

		tomorrow			
		1	2	3	4
	1	3/4	1/8	1/8	0
today	2	0	3/4	1/8	1/8
	3	0	0	3/4	1/4
	4	0	0	0	1

For instance, P[machine in state 2 tomorrow | machine in state 1 today] = 1/8. Given that the machine is in state 1 on Monday, let us find the probability distribution of its state on Thursday.

    A helpful device to clarify the use of the law of total probability is to display the possible outcomes of a stage of a random phenomenon as branches on a tree. The next stage emanates from the tips of the branches of the current stage. The branches can be annotated with the conditional probabilities of making those transitions. For the Monday to Tuesday transition, the leftmost level of the tree in Figure 11 illustrates the possibilities. The leftmost 1 indicates the known state on Monday. From state 1 we can go to state 1 on Tuesday with probability 3/4, and the other two states each with probability 1/8. From each possible machine state 1, 2, and 3 on Tuesday we can draw the possible states on Wednesday at the next level. Then from Wednesday, the Thursday states can be drawn to complete the tree. (We omit labeling the transition probabilities on the Wednesday to Thursday branches to reduce clutter in the picture; they can be found from the table.)

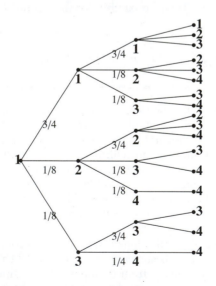

Figure 11 - Monday to Thursday transition probabilities

To see how the law of total probability works in this example, consider first the simpler problem of finding the probability that the state on Wednesday is 2. We condition and uncondition on the Tuesday state, which can be 1, 2, or 3 (however, if it is 3 then Wednesday's state cannot be 2).

$$P[\text{Wed} = 2] = \quad P[\text{Wed} = 2 \mid \text{Tues} = 1]\,P[\text{Tues} = 1]$$
$$+ P[\text{Wed} = 2 \mid \text{Tues} = 2]\,P[\text{Tues} = 2]$$
$$= \tfrac{1}{8} \cdot \tfrac{3}{4} + \tfrac{3}{4} \cdot \tfrac{1}{8} = \tfrac{3}{16}$$

Notice that this is the sum of path probabilities in Figure 11 for paths ending in 2 on Wednesday. A path probability is the product of the branch probabilities on that path. Applying the same reasoning for Thursday,

```
ThursdayProb1 =  3   3   3
                 — * — * —
                 4   4   4

ThursdayProb2 =

  3   3   1   3   1   3   1   3   3
  — * — * — + — * — * — + — * — * —
  4   4   8   4   8   4   8   4   4
```

$$\frac{27}{64}$$

$$\frac{27}{128}$$

Similarly you can check that the probability that the state on Thursday is 3 equals 63/256, and the probability that it is 4 equals 31/256. Notice that these four numbers sum to 1 as they ought to do. ■

> **Activity 3** For the previous example, finish the computation of the Wednesday probabilities. Look closely at the sum of products in each case, and at the table of transition probabilities given at the start of the problem. Try to discover a way of using the table directly to efficiently calculate the Wednesday probabilities, and check to see if the idea extends to the Thursday probabilities. For more information, see Section 6.1 on Markov chains, of which the machine state process is an example.

Bayes' Formula

The law of total probability can also be used easily to derive an important formula called *Bayes' Formula*. The setting is as above, in which a collection of events B_1, B_2, ..., B_n partitions an event A (or all of Ω). Then $P[A]$ can be expressed in terms of the conditional probabilities $P[A \mid B_i]$, which we suppose are known. The question we want to consider now is: is it possible to reverse these conditional probabilities to find $P[B_i \mid A]$ for any $i = 1, 2, ..., n$? The importance of being able to do so is illustrated by the next example. Roughly the idea is that diagnostic tools such as lie detectors, medical procedures, etc. can be pre-tested on subjects who are known to have a condition, so that the probability that they give a correct diagnosis can be estimated. But we really want to use the diagnostic test later on unknown individuals and conclude something about whether the individual has the condition given a positive test result. In the pretesting phase the event being conditioned on is that the individual has the condition, but in the usage phase we know whether or not the test is positive, so the order of conditioning is being reversed.

All we need to derive *Bayes' formula*, which is (6) below, is the definition of the conditional probability $P[B_i \mid A]$ and the law of total probability:

$$P[B_i \mid A] \ = \ \frac{P[A \cap B_i]}{P[A]} \ = \ \frac{P[A \mid B_i] \, P[B_i]}{\sum\limits_{j=1}^{n} P[A \mid B_j] \, P[B_j]} \tag{6}$$

We close the section with a nice application.

Example 4 Early estimates of the number of people in the U.S. who were infected with the HIV virus by the year 1997 were around 640,000 among an overall population of 270,000,000, which is a proportion of about $p = .0024$ of the population. In considering policy issues like the institution of universal HIV screening, one must be aware of the possibility of needlessly frightening uninfected people with a positive screening result. Also, a person with a positive screening result could then be subjected unnecessarily to further costly, time-consuming, but more accurate diagnostic procedures to confirm the condition. Suppose for our example that a screening procedure is accurate at a level q (near 1) for people who do have HIV, and it gives an incorrect positive test with probability r (near 0), for people who do not have HIV. If a randomly selected individual has a positive screening result, what is the probability that this person actually has HIV? If the screening could be redesigned to improve performance, which would be the most important quantity to improve: q or r?

We introduce the following notation, suggested by the problem statement:

$$p = P[\text{randomly selected person has HIV}]$$
$$q = P[\text{test positive} \mid \text{person has HIV}]$$
$$r = P[\text{test positive} \mid \text{person does not have HIV}]$$

By Bayes' formula,

$$P[\text{person has HIV} \mid \text{test positive}] = \frac{q \cdot p}{q \cdot p + r(1-p)}$$

We would like to see how best to increase this probability: by improving q (making it closer to 1) or r (making it closer to 0). We should also investigate whether the decision is different for high p vs. low p. Let us study some graphs of the function above for values of p near 0 (due to the relative rarity of the disease), q near 1, and r near 0.

```
f[p_, q_, r_] :=  p * q
                 ─────────────────
                 p * q + r (1 - p)
```

```
Show[GraphicsArray[{{Plot[f[.001, q, .01],
    {q, .8, 1}, AxesLabel -> {"q", "f"},
    DefaultFont → {"Times-Roman", 8},
    DisplayFunction → Identity],
   Plot[f[.001, .99, r], {r, 0, .20},
    AxesLabel -> {"r", "f"},
    DefaultFont → {"Times-Roman", 8},
    DisplayFunction → Identity]}}],
  DisplayFunction → $DisplayFunction];
```

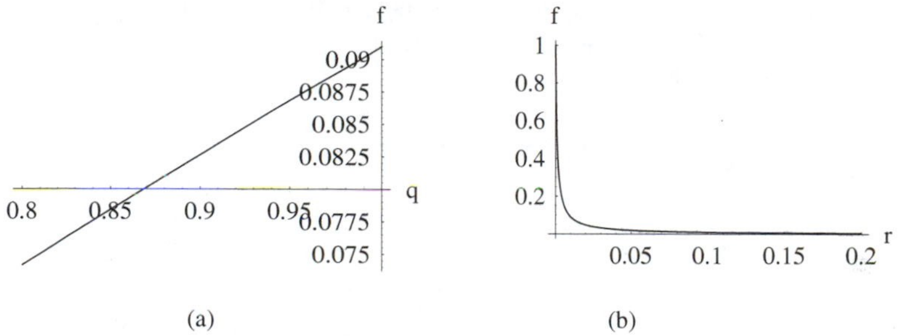

Figure 12 - (a) Probability of HIV given positive test, as function of q; (b) Probability of HIV given positive test, as function of r ($p = .001$ in both cases)

The results are very striking for a small p value of .001, and would be more so for even smaller p's: increasing q toward 1 produces only roughly linear increases in our objective f, to a level even less than .1. In other words, even if the test was made extremely accurate on persons with HIV, it is still unlikely that someone with a positive HIV screening actually is infected. This may be due to the very small p in the numerator, which is overcome by the term $r(1 - p)$ in the denominator. By contrast, the rate of increase of f as r decreases toward 0 is much more rapid, at least once the screening reaches an error rate of less than .05. When the probability of an incorrect positive test on a person without HIV is sufficiently reduced, the probability that truly HIV is present when the screening test reads positive dramatically increases toward 1. A 3D plot shows this effectively. (The code to produce it is in the closed cell above the figure.) Change in f due to increasing q is almost imperceptible in comparison to reducing r, at least when r is small enough.

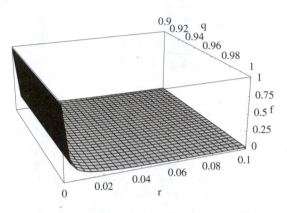

Figure 13 - Probability of HIV given positive test, as function of q and r ($p = .001$)

When p is not quite so small, as for instance in the case where the disease under study is a cold or flu passing quickly through a small population instead of HIV, the preference for decreasing r to increasing q is not quite as dramatic. Try repeating these graphs for some other parameter values, such as $p = .2$ to continue the investigation. (See also Exercise 13.)

Mathematica **for Section 1.5**

No new *Mathematica* commands.

Exercises 1.5

1. In the example on the Peotone airport proposal at the start of the section, find
    (a) P[in favor | Chicago]
    (b) P[downstate | don't care]
    (c) P[north suburbs or Chicago | in favor or don't care]

2. Under what condition on $A \cap B$ will P[$A \mid B$] = P[A]? Try to interpret the meaning of this equation.

3. In the Peotone airport example, a sample of size 4 in sequence and without replacement is taken from the group of 563 survey respondents. Find the conditional probability that all four sample members oppose the proposal given that at least three of them are from the south suburbs.

4. Find $P[A^c \cap B^c]$ if $P[A] = .5$, $P[B] = .4$, and $P[A \mid B] = .6$.

5. In a small group of voters participating in a panel discussion, four are Republicans and three are Democrats. Three different people will be chosen in sequence to speak. Find the probability that
      (a) all are Republicans
      (b) at least two are Republicans given that at least one is a Republican

6. Prove Theorem 1.

7. Show that if B is a fixed event of positive probability, then the function $Q[A] = P[A \mid B]$ taking events A into $\mathbb{R}$ satisfies the three defining axioms of probability.

8. Blackjack is a card game in which each of several players is dealt one card face down which only that player can look at, and one card face up which all of the players can see. Players can draw as many cards as they wish in an effort to get the highest point total until their total exceeds 21 at which point they lose automatically. (An ace may count 1 or 11 points at the discretion of the player, all face cards count 10 points, and other cards of ranks 2-10 have point values equal to their rank.)
      Suppose that you are in a game of blackjack with one other person. Your opponent's face up card is a 7, and you are holding an 8 and a 5. If you decide to take exactly one more card and your opponent takes no cards, what is the probability that you will beat your opponent's total?

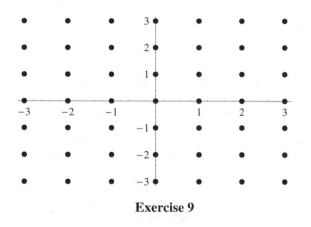

Exercise 9

9. A two-dimensional random walk moves on the integer grid shown in the figure. It begins at (0,0) and on each move it has equal likelihood of going up, down, right, or left to points adjacent to its current position. What is the probability that after three moves it is at the point (0,1)?

10. (*Mathematica*) Write a conditional probability simulator for the following situation. The sample space Ω has six outcomes a, b, c, d, e, and f, with probabilities 1/6, 1/12, 1/12, 1/6, 1/3, and 1/6, respectively. Define event A as $\{a, b, c\}$ and event B as $\{b, c, d, e\}$. For a given number of trials n, your simulator should sample an outcome from Ω according to the given probabilities, and it should return the proportion of times, among those for which the outcome was in B, that the outcome was also in A. What should this proportion approach as the number of trials becomes very large?

11. (*Mathematica*) Recall the random walk example from Section 1.3 in which we used the command AbsorptionPctAtM to empirically estimate the probabilty that the random walk, starting at state 3, would be absorbed at state 5 for a right step probability of .4. Let $f(n)$ be the probability of absorption at 5 given that the random walk starts at n, for $n = 0, 1, 2, 3, 4, 5$. Use the law of total probability to get a system of equations for $f(n)$, and solve that system in *Mathematica*. Check that the solution is consistent with the simulation results.

12. A coin drops down a pegboard as shown in the figure, bouncing right or left with equal probability. If it starts in postion 3 in row 1, find the probability that it lands in each of the slots 1, 2, 3, and 4 in the fourth row.

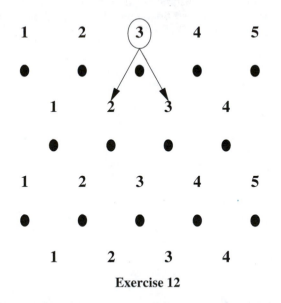

Exercise 12

13. (*Mathematica*) In the Bayes' Theorem investigation of HIV testing at the end of this section, suppose that instead our goal is to minimize the probability that a person who tests negative actually has the disease. Now what is the most important quantity to improve: q or r?

14. I am responsible for administering our college mathematics placement exam for first-year students. Suppose that students from some past years broke down into precalculus grade and placement exam score categories as follows:

		grade				
		A	B	C	D	F
placement	below 10	2	3	14	8	5
	10 -20	7	20	41	5	1
	above 20	9	15	18	4	2

Assuming that the next incoming class behaves similarly, use the table directly, then use Bayes' formula to estimate the probability that someone who scores between 10 and 20 gets at least a C. Do you get the same result using both methods?

15. A credit card company studies past information on its cardholders. It determines that an index of riskiness on a scale from 1 to 5, based on factors such as number of charge accounts, total amount owed, family income, and others, might help identify new card applicants who will later default on their debts. They find that among their past defaulters, the distribution of index values was: 1: 2%, 2: 15%, 3: 22%, 4: 28%, 5: 33%. Among non-defaulters the distribution of index values was: 1: 28%, 2: 25%, 3: 18%, 4: 16%, 5: 13%. It is known that about 5% of all card holders default. What is the probability that a new card applicant with an index of 5 will default? Repeat the computation for the other index values 4 through 1.

1.6 Independence

The approach to probability that we have been taking so far draws heavily on random sampling and simulation. These concepts are also very helpful in understanding the subject of this section: *independence* of events in random experiments. For example, what is the crucial difference between sampling five integers in sequence from {1, 2, ..., 10} without replacement as compared to sampling them with replacement? In the no replacement scenario, once a number, say 4, is sampled, it cannot appear again, and other numbers are a little more likely to appear later in the sequence because 4 is no longer eligible. When sampling with replacement, 4 is just as likely to appear again later as it was the first time, and other numbers also have the same likelihood of appearing, regardless of whether 4 was sampled. In this case, the occurrence of an event like: "4 on 1st" has no effect on the probability of an event like: "7 on 2nd". This is the nature of independence, and sampling with replacement permits it to happen while sampling without replacement does not.

Another way to look at independence involves the computation of intersection probabilities. Take the example above and use combinatorics to compute the probability that the first two sampled numbers are 4 and 7. Since we still must account for the other three members of the sample, the probability of this event, assuming no replacement, is

$$\text{P[4 on 1st, 7 on 2nd]} = \frac{1\cdot1\cdot8\cdot7\cdot6}{10\cdot9\cdot8\cdot7\cdot6} = \frac{1}{10}\cdot\frac{1}{9} = \frac{1}{90}$$

The probability of the same event, assuming replacement, is

$$\text{P[4 on 1st, 7 on 2nd]} = \frac{1\cdot1\cdot10\cdot10\cdot10}{10\cdot10\cdot10\cdot10\cdot10} = \frac{1}{10}\cdot\frac{1}{10} = \frac{1}{100}$$

which differs from the answer obtained under the assumption of no replacement. Moreover, assuming again that the sample is drawn with replacement,

$$\text{P[4 on 1st]} = \frac{1\cdot10\cdot10\cdot10\cdot10}{10\cdot10\cdot10\cdot10\cdot10} = \frac{1}{10} \;,\; \text{P[7 on 2nd]} = \frac{10\cdot1\cdot10\cdot10\cdot10}{10\cdot10\cdot10\cdot10\cdot10} = \frac{1}{10}$$

hence

$$\text{P[4 on 1st, 7 on 2nd]} = \frac{1}{10}\cdot\frac{1}{10} = \text{P[4 on 1st]}\cdot\text{P[7 on 2nd]} \;.$$

You should check that for sampling without replacement, the joint probability does not factor into the product of individual event probabilities.

Factorization can occur when three or more events are intersected as well. For instance when the sample is taken with replacement,

$$\text{P[4 on 1st, 7 on 2nd, 4 on 3rd]} = \frac{1\cdot1\cdot1\cdot10\cdot10}{10\cdot10\cdot10\cdot10\cdot10} = \frac{1}{10}\cdot\frac{1}{10}\cdot\frac{1}{10}$$

$$\text{P[4 on 1st]} = \frac{1\cdot10\cdot10\cdot10\cdot10}{10\cdot10\cdot10\cdot10\cdot10} = \frac{1}{10} = \text{P[7 on 2nd]} = \text{P[4 on 3rd]}$$

Therefore,

$$\text{P[4 on 1st, 7 on 2nd, 4 on 3rd]} = \text{P[4 on 1st]}\cdot\text{P[7 on 2nd]}\cdot\text{P[4 on 3rd]}$$

These three events are independent, and in fact it is easy to show that any subcollection of two of the events at a time satisfies this factorization condition also.

Having laid the groundwork for the concept of independence in the realm of sampling, let us go back to the general situation for our definition.

Definition 1. Events B_1, B_2, ... B_n are called *mutually independent* if for any subcollection of them B_{i_1}, B_{i_2}, ... B_{i_k}, $2 \le k \le n$,

$$P[B_{i_1} \cap B_{i_2} \cap \dots \cap B_{i_k}] = P[B_{i_1}]\, P[B_{i_2}] \cdots P[B_{i_k}] \tag{1}$$

In particular for two independent events A and B, $P[A \cap B] = P[A] \cdot P[B]$, which means that

$$P[B \mid A] = \frac{P[A \cap B]}{P[A]} = \frac{P[A] \cdot P[B]}{P[A]} = P[B] \tag{2}$$

In words, if A and B are independent, then the probability of B does not depend on whether A is known to have occurred.

> **Activity 1** For a roll of two fair dice in sequence, check that the events $B_1 =$ "1 on 1st" and $B_2 =$ "6 on 2nd" are independent by writing out the sample space and finding $P[B_1 \cap B_2]$, $P[B_1]$, and $P[B_2]$. Compare $P[B_2 \mid B_1]$ to $P[B_2]$.

Example 1 Four customers arrive in sequence to a small antique store. If they make their decisions whether or not to buy something independently of one another, and if they have individual likelihoods of 1/4, 1/3, 1/2, and 1/2 of buying something, what is the probability that at least three of them will buy?

The problem solving strategy here is of the *divide and conquer* type; <u>divide</u> the large problem into subproblems, and <u>conquer</u> the subproblems using independence. First, the event that at least three customers buy can be expressed as the disjoint union of five subevents, from which we can write:

$$
\begin{aligned}
P[\text{at least 3 buy}] \quad = \quad & P[\text{customers 1, 2, 3 buy and not 4}] \\
& + P[1, 2, 4 \text{ buy and not 3}] \\
& + P[1, 3, 4 \text{ buy and not 2}] \\
& + P[2, 3, 4 \text{ buy and not 1}] \\
& + P[1, 2, 3, 4 \text{ buy}]
\end{aligned}
$$

Each subevent is an intersection of independent events, so by formula (1) and the given individual purchase probabilities,

$$P[\text{at least 3 buy}] = \tfrac{1}{4} \cdot \tfrac{1}{3} \cdot \tfrac{1}{2} \cdot \tfrac{1}{2} + \tfrac{1}{4} \cdot \tfrac{1}{3} \cdot \tfrac{1}{2} \cdot \tfrac{1}{2} + \tfrac{1}{4} \cdot \tfrac{1}{3} \cdot \tfrac{1}{2} \cdot \tfrac{2}{3}$$
$$+ \tfrac{1}{3} \cdot \tfrac{1}{2} \cdot \tfrac{1}{2} \cdot \tfrac{3}{4} + \tfrac{1}{4} \cdot \tfrac{1}{3} \cdot \tfrac{1}{2} \cdot \tfrac{1}{2} = \tfrac{1}{6} \quad \blacksquare$$

Example 2 One frequently sees the notion of independence come up in the study of two-way tables called *contingency tables*, in which individuals in a population are classified according to each of two characteristics. Such a table appeared in Section 1.5 in reference to the Peotone airport proposal. The first characteristic was the region in which the individual lived (4 possible regions), and the second characteristic was the individual's opinion about the airport issue (3 possible opinions). If one person is sampled at random from the 563 in this group, you can ask whether such events as "the sampled person is from the north suburbs" and "the sampled person is opposed to the new airport" are independent. Using the counts in the table, we can check that $P[\text{north and opposed}] \approx .0604$, $P[\text{north}] \approx .2469$, $P[\text{opposed}] \approx .3250$, and $P[\text{north}] \cdot P[\text{opposed}] \approx .0803$. So the joint probability $P[\text{north and opposed}]$ is not equal to the product of the individual probabilities

P[north]·P[opposed]. Hence we conclude that these two events are not independent. There may be something about being from the north suburbs which changes
the likelihood of being opposed to the airport. (You can conjecture that north
suburban people might be more inclined to favor the airport, since it is not in their
own backyard.)

      However, this is only a small sample from a much larger population of
Illinois residents. If you want to extend the conclusion about independence to the
whole population of Illinois, you have the problem that probabilities like .0604,
.2469, etc. are only estimates of the true probabilities of "north and opposed",
"north", etc. for the experiment of sampling an individual from the broader universe. Sampling variability alone could account for some departures from equality
in the defining condition for independence $P[A \cap B] = P[A]\,P[B]$. We pursue this
idea further in the next example. ∎

Example 3 Suppose that each individual in a sample of size 80 taken in sequence
and with replacement from a population can be classified according to two characteristics A and B as being one of two types for each characteristic. Assume that there
are 30 people who are of type 1 for characteristic A, 50 of type 2 for A, and 40 each
of types 1 and 2 for characteristic B. Draw one individual at random from the
sample of 80. Find the unique frequencies x, y, w, and z in the table below which
make events of the form "individual is type i for A", $i = 1,2$, independent of events
of the form "individual is type j for B", $j = 1,2$.

		char B		
		1	2	total
char A	1	x	y	30
	2	w	z	50
	total	40	40	80

      The interesting thing in this example is that the answer is unique; the
sample of 80 can only come out in one way in order to make characteristic A events
independent of characteristic B events. Without the independence constraint, x
could take many values, from which y, w, and z are uniquely determined. If $x = 2$
for instance, y must be 28, w must be 38, and z must be 12 in order to produce the
marginal totals in the table. In general, letting x be the free variable, the table
entries must be:

		char B		
		1	2	total
char A	1	x	$30 - x$	30
	2	$40 - x$	$10 + x$	50
	total	40	40	80

For independence, we want in addition,

$$P[A = 1, B = 1] = P[A = 1]P[B = 1]$$
$$\implies \frac{x}{80} = \frac{30}{80} \cdot \frac{40}{80} \implies x = 15$$

The unique set of table entries is then $x = 15$, $y = 15$, $w = 25$, and $z = 25$. You can check (see also Exercise 3) that $\{A = 1\}$ is independent of $\{B = 2\}$ for these table entries, and also $\{A = 2\}$ is independent of $\{B = 1\}$ and of $\{B = 2\}$.

But it seems a lot to expect for the data in a sample to line up so perfectly if the real question of interest is whether in the whole population characteristic A is independent of characteristic B. There is a well-known test in statistics for independence of characteristics in contingency tables like ours, but here let us just use simulation to give us an idea of the degree of variability we can expect in the table.

Suppose our sample of 80 is taken from a universe of 200 individuals, and the universe is set up so that the characteristics are independent. One such configuration of the universe is given by the table on the left below. Let us shorten the notation a bit by using the code $A1$ to stand for the event $A = 1$, $A1B1$ for the event $A = 1$ and $B = 1$, etc. Each individual in the sample can be of type $A1B1$ with probability $60/200 = .3$, and similarly can be of type $A1B2$ with probability $.3$, type $A2B1$ with probability $.2$ and type $A2B2$ with probability $.2$. A sample of size 80 would be expected to have these proportions of the 80 sample members in the four categories. The expected counts in the sample are in the table on the right below; for example the expected number of $A1B1$ individuals in the sample is $80 \cdot (.3) = 24$.

		char B					char B		
		1	2	total			1	2	total
char A	1	60	60	120	char A	1	24	24	48
	2	40	40	80		2	16	16	32
	total	100	100	200		total	40	40	80
		universe					**sample**		

The next *Mathematica* function simulates a sample of a desired size (such as 80) from the population categorized as in the universe table, tallies up the frequencies in each category for the sample, and presents the results in tabular format. After initializing the counter variables to zero, it selects a random number between 0 and 1, and based on its value, increments one of the counters. The cutoffs used in the Which function are chosen so that the category probabilities of .3, .3, .2, and .2 are modeled. The marginal totals are then computed for the tabular display, and the output is done.

```
SimContingencyTable[sampsize_] :=
    Module[{A1, A2, B1, B2, A1B1, A1B2,
  A2B1, A2B2, nextsample, i, outtable},
        A1B1 = 0; A1B2 = 0; A2B1 = 0; A2B2 = 0;
        Do[nextsample = Random[];
            Which[nextsample < .3,
  A1B1 = A1B1 + 1, .3 ≤ nextsample < .6,
  A1B2 = A1B2 + 1, .6 ≤ nextsample < .8,
  A2B1 = A2B1 + 1, .8 ≤ nextsample ≤ 1,
  A2B2 = A2B2 + 1], {i, 1, sampsize}];
        A1 = A1B1 + A1B2;
        A2 = A2B1 + A2B2;
        B1 = A1B1 + A2B1;
        B2 = A1B2 + A2B2;
        outtable = {{"A/B", 1, 2, "Total"},
  {1, A1B1, A1B2, A1}, {2, A2B1, A2B2, A2},
  {"Total", B1, B2, sampsize}};
        TableForm[outtable]]
```

Here are three runs of the command. For replicability, we set the seed of the random number generator.

```
SeedRandom[34567];
SimContingencyTable[80]
SimContingencyTable[80]
SimContingencyTable[80]
```

```
A/B       1       2       Total
1         28      26      54
2         14      12      26
Total     42      38      80

A/B       1       2       Total
1         20      27      47
2         20      13      33
Total     40      40      80

A/B       1       2       Total
1         27      27      54
2         13      13      26
Total     40      40      80
```

The expected numbers under the independence assumption (24's in the first row and 16's in the second) are not always very close to what is observed. Try running the commands again, and you will regularly find discrepancies in the frequencies of at least 3 or 4 individuals. But the more important issue is this. The characteristics are independent in the population, but one must be rather lucky to have perfect independence in the sample. In the third table, by good fortune the independence condition in the sample does hold (for example, the $A1B1$ event has probability 27/80, while the probability of $A1$ times the probability of $B1$ is (54/80)(40/80) = 27/80), though the category frequencies do not match the expected ones. The statistical test for independence of categories that we referred to above must acknowledge that contingency tables from random samples have sampling variability, and should reject independence only when the sample discrepancies from independence are "large enough". ■

Activity 2 What configuration of the universe in the above example would lead to independence of characteristics A and B and totals of 100 in each of $A1$, $A2$, $B1$, and $B2$? Revise the SimContingencyTable command accordingly, and do a few runs using sample sizes of 100 to observe the degree of variability of the table entries.

Example 4 Electrical switches in a circuit act like gates which allow current to flow or not according to whether they are closed or open, respectively. Consider a circuit as displayed in Figure 14 with three switches arranged in parallel.

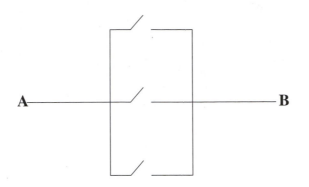

Figure 14 - A parallel circuit

Current coming from point A may reach point B if and only if at least one of the switches is closed. If the switches act independently of one another, and they have probabilities p_1, p_2, and p_3 of being closed, what is the probability that current will flow?

This is an easy one. The event that current will flow is the complement of the event that it will not flow. The latter happens if and only if all three switches are open. Switch i is open with probability $1 - p_i$. Thus, by independence,

P[current flows] $= 1 -$ P[current doesn't] $= 1 - (1 - p_1)(1 - p_2)(1 - p_3)$ ∎

Example 5 Our last example of this section again deals with simulation and contingency tables. A good pseudo-random number generator as described in Section 3 ought to have the property that each pseudo-random number that is generated appears to be independent of its predecessors (even though it is of course a deterministic function of its immediate predecessor). Checking for validity of the generator therefore requires a check for independence of each member from each other in a stream of numbers. There are several ways to do this, and here is one that is sensitive to unwanted upward or downward trends in the stream of numbers.

We can be alerted to a possible violation of independence if some property that independence implies is contradicted by the data. For example, in a stream of 40 numbers we can classify each number according to two characteristics: whether it is in the first group of 20 or the second, and whether it is above some fixed cutoff or below it. If the numbers are truly random and independent of one another, the group that a number is in should have no effect on whether that number is above or below the cutoff. If the two-by-two contingency table that we generate in this way shows serious departures from the tallies that we would expect under independence, then the assumption of independence is in doubt.

Here is a stream of 40 random integers between 1 and 20, simulated using DrawIntegerSample.

```
Needs["KnoxProb`Utilities`"];
SeedRandom[11753];
datalist =
  DrawIntegerSample[20, 40, Replacement -> True]
```

{15, 16, 6, 9, 7, 18, 7, 19, 10, 6, 4, 15, 20,
 12, 4, 12, 14, 14, 11, 18, 12, 8, 20, 6, 8, 12, 10,
 15, 8, 13, 14, 2, 18, 18, 16, 18, 17, 8, 14, 17}

Among the first 20 numbers, 8 are less than or equal to 10 and 12 are greater than 10. Among the second group of 20 numbers, 7 are less than or equal to 10 and 13 are greater. We therefore have the following observed contingency table.

		size		
		≤ 10	> 10	total
group	1	8	12	20
	2	7	13	20
	total	15	25	40

Equal numbers of 10 observations are to be expected in each category. Interestingly, this particular table is coming very close to independence, because the fact that the row totals are 20 means that the 15 and 25 observations in the two columns should split evenly between the two rows, which they do. There is very little evidence against independence in this run of the experiment, although a different problem, namely a preference for high numbers to low numbers, seems to be indicated. Do the following activity to check if that is a general problem. ∎

Activity 3 Repeat the drawing of a sample of 40 from {1, 2, ..., 20} a few more times. Do you observe many tables that are far from expectations? Try doubling the sample size to 80, and using four groups of 20 instead of two. What category counts are expected? Do you see striking departures from them in your simulations?

Mathematica for Section 1.6

Command	Location
Random[]	kernel
SeedRandom[]	kernel
DrawIntegerSample[sampsize, popsize]	KnoxProb` Utilities`
SimContingencyTable[sampsize]	Section 1.6

Exercises 1.6

1. A student guesses randomly on ten multiple choice quiz questions with five alternative answers per question. Assume that there is only one correct response to each question, and the student chooses answers independently of other questions. What is the probability that the student gets no more than one question right?

2. Consider the experiment of randomly sampling a single number from {1, 2, ..., 10}. Is the event that the number is odd independent of the event that it is greater than 4?

3. If A and B are independent events, prove that each of the following pairs of events are independent of one another: (a) A and B^c; (b) A^c and B; (c) A^c and B^c.

4. If events A and B have positive probability and are disjoint, is it possible for them to be independent?

5. Prove that if events A, B, C, and D are independent of one another, then so are the groups of events (a) A, B, and C; (b) A and B.

6. Prove that both Ω and $\emptyset$ are independent of every other event $A \subset \Omega$.

7. (*Mathematica*) Referring to Example 3, run the SimContingencyTable command a few times each for sample sizes of 40, 80, and 120. Compare the variabilities for the three sample sizes.

8. A sample space has six outcomes: a, b, c, d, e, and f. Define the event A as {a, b, c, d} and the event B as {c, d, e, f}. If each of a, b, e, and f has probability 1/8, find probabilities on c and d that make A and B independent events, or explain why it is impossible to do so.

9. Three coins are flipped in succession.
(a) If outcomes are equally likely, show that "H on 1st" is independent of "T on 2nd" and show that "H on 2nd" is independent of "T on 3rd".
(b) Put probabilities on the sample space consistent with the assumptions of

independence of different flips and an unfair weighting such that tails is 51% likely to occur on a single flip.

10. Below is a part of an electrical circuit as in Example 4, except that two parallel groups of switches are connected in series. Current must pass through both groups of switches to flow from A to B. If each switch in the first group has probability p of being closed, each switch in the second group has probability q of being closed, and the switches operate independently, find the probability that current will flow from A to B.

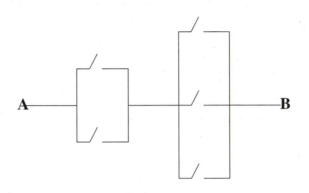

Exercise 10

11. (*Mathematica*) A system has three independent components which work for a random, uniformly distributed amount of time in $[10, 20]$ and then fail. Each component can take on the workload of the other, so the system fails when the final component dies. Simulate 1000 such systems, produce a histogram of the system failure times, and estimate the probability that the system fails by time 15. What is the exact value of that probability?

12. In Example 4 we were able to find the probability that current flows from A to B easily by complementation. Compute the same probability again without complementation.

13. (*Mathematica*) In Example 5 we set up the integer sampling process so that replacement occurs. If the pseudo-random number generator behaves properly we would not expect any evidence against randomness. But the question arises: is this test sensitive enough to detect real independence problems? Try drawing integer samples again, this time of 40 integers from $\{1, 2, ..., 80\}$ without replacement. Using characteristics of group (1st 20, 2nd 20) and size (40 or below, more than 40), simulate some contingency tables to see whether you spot any clear departures from independence. Would you expect any?

14. If events B_1, B_2, B_3, and B_4 are independent, show that
    (a) $P[B_1 \mid B_3 \cap B_4] = P[B_1]$
    (b) $P[B_1 \cap B_2 \mid B_3 \cap B_4] = P[B_1 \cap B_2]$
    (c) $P[B_1 \cup B_2 \mid B_3] = P[B_1 \cup B_2]$

15.   Two finite or countable families of events $\mathbf{A} = \{A_1, A_2, \ldots\}$ and $\mathbf{B} = \{B_1, B_2, \ldots\}$ are called *independent* if every event $A_i \in \mathbf{A}$ is independent of every event $B_j \in \mathbf{B}$. Suppose that two dice are rolled one after the other. Let $\mathbf{A}$ consist of basic events of the form "1st die = n" for $n = 1, 2, \ldots, 6$, together with all unions of such basic events. Similarly let $\mathbf{B}$ be the events that pertain to the second die. Show that each basic event in $\mathbf{A}$ is independent of each basic event in $\mathbf{B}$, and use this to show that the families $\mathbf{A}$ and $\mathbf{B}$ are independent.

CHAPTER 2
DISCRETE DISTRIBUTIONS

2.1 Discrete Random Variables, Distributions, and Expectations

We begin Chapter 2 by consolidating many of the ideas that you encountered in Chapter 1, and applying them in the context of discrete random variables. Here are the two main concepts.

Definition 1. A *discrete random variable X* is a function taking the sample space Ω of a random phenomenon to a discrete (i.e., finite or countable) set E called its *state space*. For each outcome $\omega \in \Omega$, the point $x = X(\omega)$ is called a *state* of the random variable.

Definition 2. A discrete random variable X is said to have *probability mass function* (abbr. p.m.f.) $f(x)$ if for each state x,

$$f(x) = P[X = x] = P[\{\omega \in \Omega : X(\omega) = x\}]$$

So a random variable X maps outcomes to states, which are usually numerical valued. The p.m.f. of the random variable gives the likelihoods that X takes on each of its possible values. Notice that by Axiom 2 of probability, $f(x) \geq 0$ for all states $x \in E$, and also by Axiom 1, $\sum_{x \in E} f(x) = 1$. These are the conditions for a function f to be a valid probability mass function.

Example 1 Exercise 7 in Section 1.1 illustrates the ideas well. We repeat the diagram of that exercise here for convenience as Figure 1. Assuming that all outcomes $\{a, b, c, d, e, f\}$ in Ω are equally likely, and X maps outcomes to states $\{0, 1\}$ as shown, then the probability mass function of X is

$$f(0) = P[X = 0] = P[\{a, b, c, d\}] = \tfrac{4}{6} = \tfrac{2}{3}$$
$$f(1) = P[X = 1] = P[\{e, f\}] = \tfrac{2}{6} = \tfrac{1}{3} \qquad \blacksquare$$

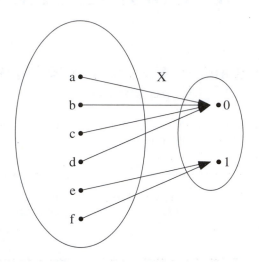

Figure 1 - A discrete random variable X

Example 2 Suppose that a list of grades for a class is in the table below, and the experiment is to randomly draw a single student from the class.

grade	A	B	C	D	F	total
number of students	4	8	6	2	1	21

The sample space Ω is the set of all students, and the random variable X operates on Ω by returning the grade of the student selected. Then the state space of X is $E = \{A, B, C, D, F\}$ and the probability mass function is

$$f(A) = \tfrac{4}{21}; \ f(B) = \tfrac{8}{21}; \ f(C) = \tfrac{6}{21}; \ f(D) = \tfrac{2}{21}; \ f(F) = \tfrac{1}{21} \tag{1}$$

The probability that the randomly selected student has earned at least a B for instance is

$$P[X = A \text{ or } X = B] = P[X = A] + P[X = B] = f(A) + f(B) = \tfrac{12}{21}$$

∎

The last computation in Example 2 illustrates a general rule. If X has p.m.f. f, and S is a set of states, then by the third axiom of probability,

$$P[X \in S] = \sum_{x \in S} P[X = x] = \sum_{x \in S} f(x) \tag{2}$$

The p.m.f. therefore completely characterizes how probability distributes among the states, so that the probability of any interesting event can be found. We often use the language that *f* characterizes (or is) the *probability distribution of X* for this reason.

Activity 1 Encode the letter grades in Example 2 by the usual integers 4, 3, 2, 1, and 0 for *A-F*, respectively. Write a description of, and sketch a graph of, the function $F(x) = P[X \le x]$ for $x \in \mathbb{R}$.

Example 3 If you did Exercise 3 of Section 1.3 you discovered how to simulate values of a random variable with a known p.m.f. on a finite state space. Let us apply the ideas here to write a command to repeatedly simulate from the grade distribution in Example 2.

This can be done by encoding simulated uniform[0,1] observations in accordance with the grade distribution in formula (1). If, for example, such a uniform number takes a value in [0, 4/21) we can call the result an *A*. Since the probability that a simulated uniform[0,1] observation falls into an interval is equal to the length of the interval, state *A* will have probability 4/21 as desired. State *B* must have probability 8/21, so we can call the result a *B* if the uniform number falls into [4/21, 12/21), which has length 8/21. Proceeding similarly for states *C*, *D*, and *F*, we have the following command. (Encode *A* as 4, *B* as 3, etc. as in Activity 1.)

```
SimulateGrades[numreps_] :=
    Module[{therand, i, gradelist, nextgrade},
        gradelist = {};
        Do[therand = Random[];
            nextgrade =
    Which[therand < 4 / 21, 4,
            4 / 21 ≤ therand < 12 / 21, 3,
            12 / 21 ≤ therand < 18 / 21, 2,
            18 / 21 ≤ therand < 20 / 21, 1,
            therand ≥ 20 / 21, 0];
    AppendTo[gradelist, nextgrade],
    {i, 1, numreps}];
    gradelist]
```

```
Needs["KnoxProb`Utilities`"];
SeedRandom[98921];
Histogram[SimulateGrades[100],
   5, Distribution → Discrete];
```

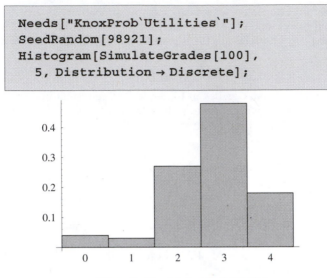

Figure 2 - Simulated grade distribution

Since the probabilities of states 0-4 are .05, .09, .29, .38, and .19, this particular empirical histogram fits the theoretical distribution fairly well. You might have noticed that although we are interested in simulating the selection of a random student, what we actually did in the command was to simulate a random variable Y taking [0, 1] to {4, 3, 2, 1, 0} as in Figure 3, with the same distribution as X. To produce the picture we have used *Mathematica*'s Which function to define the step function, and a utility contained in the KnoxProb`Utilities` package called

PlotStepFunction[function, domain, jumplist]

which plots a right continuous step function on an interval with the given domain, with jumps at the given list of x values. It has a DotSize option as shown to control the size of the dots, and also accepts the AxesOrigin and PlotRange options, as does the Plot command. ∎

```
F[x_] := Which[x < 4 / 21, 4,
                4 / 21 ≤ x < 12 / 21, 3,
                12 / 21 ≤ x < 18 / 21, 2,
                18 / 21 ≤ x < 20 / 21, 1,
                         x ≥ 20 / 21, 0];
PlotStepFunction[F[x], {x, 0, 1},
  {4 / 21, 12 / 21, 18 / 21, 20 / 21},
  AxesOrigin → {0, -.1},
  PlotRange → {-.1, 4.1}, DotSize → .02,
  DefaultFont → {"Times-Roman", 8}];
```

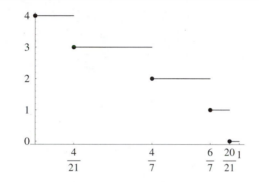

Figure 3 - Random variable Y taking $[0,1]$ to numerical grades

There is an alternative characterization of the distribution of a random variable as given in the definition below.

Definition 3. A discrete random variable X with state space $E \subset \mathbb{R}$ is said to have *cumulative distribution function* (abbr. c.d.f.) $F(x)$ if for all $x \in \mathbb{R}$,

$$F(x) = P[X \le x] = P[\{\omega \in \Omega : X(\omega) \le x\}] \qquad (3)$$

The c.d.f. accumulates total probability lying to the left of and including point x. It also characterizes the complete distribution of probability, because one can move back and forth from it to the probability mass function. For instance, suppose that a random variable has state space $E = \{.2, .6, 1.1, 2.4\}$ with probability masses $1/8$, $1/8$, $1/2$, and $1/4$, respectively. Then for the four states,

$$
\begin{aligned}
F(.2) &= P[X \le .2] = f(.2) = \tfrac{1}{8} \\
F(.6) &= P[X \le .6] = f(.2) + f(.6) = \tfrac{1}{8} + \tfrac{1}{8} = \tfrac{1}{4} \\
F(1.1) &= P[X \le 1.1] = f(.2) + f(.6) + f(1.1) = \tfrac{1}{8} + \tfrac{1}{8} + \tfrac{1}{2} = \tfrac{3}{4} \\
F(2.4) &= P[X \le 2.4] = f(.2) + f(.6) + f(1.1) + f(2.4) \\
&\qquad\qquad = \tfrac{1}{8} + \tfrac{1}{8} + \tfrac{1}{2} + \tfrac{1}{4} = 1
\end{aligned}
$$

Between the states there is no probability, so that a full definition of F is

$$
F(x) = \begin{cases}
0 & \text{if} \quad x < .2 \\
\tfrac{1}{8} & \text{if } .2 \le x < .6 \\
\tfrac{1}{4} & \text{if } .6 \le x < 1.1 \\
\tfrac{3}{4} & \text{if } 1.1 \le x < 2.4 \\
1 & \text{if} \quad x \ge 2.4
\end{cases}
\tag{4}
$$

It is clear that if we know f we can then construct F, but notice also that the probability masses on the states are the amounts by which F jumps; for example $f(.6) = 1/8 = 1/4 - 1/8 = F(.6) - F(.6^-)$, where $F(x^-)$ denotes the left-hand limit of F at the point x. Hence, if we know F we can construct f. In general, the relationships between the p.m.f. $f(x)$ and the c.d.f. $F(x)$ of a discrete random variable are

$$
F(x) = \sum_{t \le x} f(t) \quad \text{for all } x \in \mathbb{R}; \qquad f(x) = F(x) - F(x^-) \text{ for all states } x.
\tag{5}
$$

Figure 4 displays the c.d.f. F for this example. Notice that it begins at functional value 0, jumps by amount $f(x)$ at each state x, and reaches 1 at the largest state.

```
F[x_] := Which[x < .2, 0, .2 ≤ x < .6, 1/8,
    .6 ≤ x < 1.1, 1/4, 1.1 ≤ x < 2.4, 3/4, x ≥ 2.4, 1];
PlotStepFunction[F[x], {x, 0, 2.6},
    {.2, .6, 1.1, 2.4}, DotSize → .02,
    AxesOrigin → {0, -.01}, PlotRange → {-.01, 1.01},
    DefaultFont → {"Times-Roman", 8}];
```

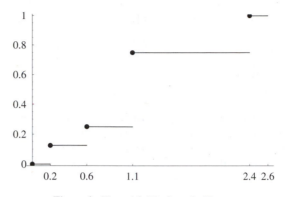

Figure 4 - The c.d.f. *F* in formula (5)

Activity 2   Consider a p.m.f. on a countably infinite state space $E = \{0, 1, 2, ...\}$ of the form $f(x) = \frac{1}{2^{x+1}}$. Argue that f is a valid mass function and find an expression for its associated c.d.f. F.   What is the limit of $F(x)$ as $x \longrightarrow \infty$?

Example 4   The c.d.f. is sometimes a better device for getting a handle on the distribution of a random variable than the p.m.f. For instance, remember that in Chapter 1 we simulated random samples, for each of which we calculated a maximum or minimum sample value. We then used a list of such extrema to give a histogram of their empirical distributions. Here we will find the exact theoretical distribution of a maximum sample value.

Suppose that the random experiment is to draw five numbers randomly in sequence and with replacement from $\{1, 2, ..., 20\}$. Let X be the maximum of the sample values, e.g., if $\omega = \{5, 2, 10, 4, 15\}$ then $X(\omega) = 15$. Though it is not clear immediately what is the p.m.f. of X, we can compute the c.d.f. easily. In order for the largest observation X to be less than or equal to a number x, all five sample values (call them X_1, X_2, X_3, X_4, X_5) must be less than or equal to x. Since sampling is done with replacement, we may assume that the five sample values are mutually independent; hence for $x \in \{1, 2, ..., 20\}$,

$$F(x) \;=\; P[X \le x] \;=\; P[X_1 \le x, X_2 \le x, X_3 \le x, X_4 \le x, X_5 \le x]$$
$$= \;\; \textstyle\prod_{i=1}^{5} P[X_i \le x]$$
$$= \;\; \left(\tfrac{x}{20}\right)^5$$

The last line follows because for each sampled item, the chance that it is less than or equal to x is the number of such numbers (x) divided by the total number of numbers (20) that could have been sampled. The p.m.f. of X is therefore

$$f(x) \;=\; P[X = x] \;=\; P[X \le x] - P[X \le x - 1]$$
$$= \; (\tfrac{x}{20})^5 - (\tfrac{x-1}{20})^5 \tag{6}$$

for $x \in \{1, 2, \ldots, 20\}$. ∎

Activity 3 Use formula (6) above to make a list plot in *Mathematica* of the values of f with the plot points joined. Superimpose it on a histogram of 500 simulated values of X using the DrawIntegerSample command with option Replacement→True in the KnoxProb`Utilities` package to observe the fit of the sample to the theoretical distribution.

Mean, Variance, and Other Moments

In Section 1.1 we briefly introduced the notion of expected value. Now we shall give a good definition in the context of discrete random variables, and explore certain special expectations of powers of the random variable called *moments* of the probability distribution. Remember that expected value is the average of the states, with states being weighted by their probabilities.

Definition 4. If X is a discrete random variable with p.m.f. $f(x)$, then the expected value of X is

$$E[X] = \sum_{x \in E} x \cdot f(x) \tag{7}$$

where the sum is taken over all states. Similarly, the expected value of a function g of X is

$$E[g(X)] = \sum_{x \in E} g(x) \cdot f(x) \tag{8}$$

In the case of the expectation of a function $g(X)$ note that the values of the function $g(x)$ are again weighted by the state probabilities $f(x)$ and summed, so that $E[g(X)]$ is a weighted average of the possible values $g(x)$. Often $E[X]$ is called the *mean of the distribution of X* (or just the *mean of X*), and is symbolized by μ.

Example 5 Suppose that on a particular day, a common stock price will change by one of the increments: $-1/4$, $-1/8$, 0, $1/8$, $1/4$, or $3/8$, with probabilities .06, .12, .29, .25, .18, and .10, respectively. Then the expected price change is

$$\mu = (-1/4)\ (.06) + (-1/8)\ (.12) + (0)\ (.29) +$$
$$(1/8)\ (.25) + (1/4)\ (.18) + (3/8)\ (.10)$$

0.08375

Let us also find the expected absolute deviation of the price change from the mean. Symbolically, this is $E[\,|X - \mu|\,]$, and it is of importance because it is one way of measuring the spread of the probability distribution of X. It is calculated as the weighted average $\sum_{x \in E} |x - \mu| \cdot f(x)$. In particular for the given numbers in this example,

$$\text{Abs}[(-1/4) - \mu] * (.06) + \text{Abs}[(-1/8) - \mu] * (.12) +$$
$$\text{Abs}[0 - \mu] * (.29) + \text{Abs}[1/8 - \mu] * (.25) +$$
$$\text{Abs}[1/4 - \mu] * (.18) + \text{Abs}[3/8 - \mu] * (.10)$$

0.138725

∎

The so-called *moments* of X play a key role in describing its distribution. These are expectations of powers of X. We have already met the mean $\mu = E[X]$, which is the *first moment about 0*. In general the *rth moment about a point a* is $E[(X - a)^r]$. Next to the mean, the most important moment of a distribution is the second moment about μ, i.e.

$$\sigma^2 \;=\; \text{Var}(X) \;=\; E[(X - \mu)^2] \tag{9}$$

which is called the *variance of the distribution of X* (or just the *variance of X*). The variance gives a weighted average of squared distance between states and the average state μ; therefore it is a commonly used measure of spread of the distribution. Its square root $\sigma = \sqrt{\sigma^2}$ is referred to as the *standard deviation* of the distribution.

For the probability distribution of Example 5, where we computed that $\mu = .08375$, the variance and standard deviation are

```
σsquared = (-1 / 4 - μ)² (.06) + (-1 / 8 - μ)² (.12) +
    (0 - μ)² (.29) + (1 / 8 - μ)² (.25) +
    (1 / 4 - μ)² (.18) + (3 / 8 - μ)² (.10)
σ = √σsquared
```

0.0278297

0.166822

Besides the mean and the variance, the moment of most importance is the third moment about the mean $E[(X - μ)^3]$. It is similar to the variance, but because of the cube it measures an average signed distance from the mean, in which states to the right of $μ$ contribute positively, and states to the left contribute negatively. So for example, if there is a collection of states of non-negligible probability which are a good deal greater than $μ$, which are not offset by states on the left side of $μ$, then we would expect a positive value for the third moment. On the left side of Figure 5 is such a probability distribution. Its states are {1, 2, 10, 20}, with probabilities {.25, .50, .125, .125}. Because of the long right tail we say that this distribution is *skewed to the right*. To produce the plot we have used the command

ProbabilityHistogram[statelist, problist]

in KnoxProb`Utilities`, which requires the list of states and the corresponding list of probabilities for the states. Certain stylistic options (exactly those of Generalized-BarChart in Graphics`Graphics`) are available to control the look of the diagram.

```
g1 = ProbabilityHistogram[
    {1, 2, 10, 20}, {1 / 4, 1 / 2, 1 / 8, 1 / 8},
    DisplayFunction → Identity,
    DefaultFont → {"Times-Roman", 8}];
g2 = ProbabilityHistogram[{1, 2, 15, 20},
    {1 / 8, 1 / 8, 1 / 2, 1 / 4},
    DisplayFunction → Identity,
    DefaultFont → {"Times-Roman", 8}];
Show[GraphicsArray[{{g1, g2}}],
    DisplayFunction → $DisplayFunction];
```

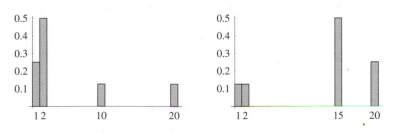

Figure 5 - Two skewed probability mass functions

We find the mean and third moment about the mean to be

$$\mu = (.25)\ (1)\ +\ (\ .50)\ (2)\ +$$
$$(.125)\ (10)\ +\ (.125)\ (20)$$
$$\text{thirdmoment} = (.25)\ (1 - \mu)^3\ +\ (\ .50)\ (2 - \mu)^3\ +$$
$$(.125)\ (10 - \mu)^3\ +\ (.125)\ (20 - \mu)^3$$

5.

408.

The third moment about the mean is positive, as expected. Do the next activity to convince yourself that when the distribution has a long left tail, the third moment about the mean will be negative, in which case we call the distribution *skewed to the left*.

> **Activity 4** Compute the mean and the third moment about the mean of the distribution shown on the right of Figure 5. (The states are {1, 2, 15, 20} and their probabilities are {1/8, 1/8, 1/2, 1/4}.)

Below are some important results about expectation and variance. The first two parts will require some attention to joint probability distributions of two random variables before we can prove them. We postpone these until later in this chapter, but a preview is given in Exercise 18. We will supply proofs of the other parts.

Theorem 1. Let X and Y be random variables (with finite means and variances wherever they are referred to). Then

(a) $E[X + Y] = E[X] + E[Y]$

(b) For any constants a and b, $E[a\,X + b\,Y] = a\,E[X] + b\,E[Y]$

(c) If a is a constant then $E[a] = a$

(d) $\mathrm{Var}(X) = E[X^2] - \mu^2$, where $\mu = E[X]$

(e) If c and d are constants, then $\mathrm{Var}(c\,X + d) = c^2\,\mathrm{Var}(X)$

Proof. (c) The constant a can be viewed as a random variable X with one state $x = a$ and p.m.f. $f(x) = 1$ if $x = a$ and 0 otherwise. Thus,

$$E[a] = a \cdot P[X = a] = a \cdot 1 = a$$

(d) Since μ and μ^2 are constants and $E[X] = \mu$,

$$
\begin{aligned}
\mathrm{Var}(X) &= E[(X - \mu)^2] \\
&= E[X^2 - 2\mu X + \mu^2] \\
&= E[X^2 - 2\mu E[X] + \mu^2] \\
&= E[X^2] - \mu^2
\end{aligned}
$$

(e) First, the mean of the random variable $cX + d$ is $c\mu + d$, by parts (b) and (c) of this theorem. Therefore,

$$
\begin{aligned}
\mathrm{Var}(c\,X + d) &= E[\{(c\,X + d) - (c\,\mu + d)\}^2] \\
&= E[\{c\,X - c\,\mu\}^2] \\
&= c^2\,E[\{X - \mu\}^2] = c^2\,\mathrm{Var}(X)
\end{aligned}
$$

This theorem is a very important result which deserves some commentary. Properties (a) and (b) of Theorem 1 make expectation pleasant to work with: it is a linear operator on random variables, meaning that the expected value of a sum (or difference) of random variables is the sum (or difference) of the expected values, and constant coefficients factor out. The intuitive interpretation of part (c), as indicated in the proof, is that a random variable that is constant must also have that constant as its average value. Part (d) is a convenient computational formula for the variance that expresses it in terms of the second moment about 0. For example if the p.m.f. of a random variable X is $f(1) = 1/8, f(2) = 1/3, f(3) = 1/4, f(4) = 7/24$, then

$$\mu = E[X] = \tfrac{1}{8}\cdot 1 + \tfrac{1}{3}\cdot 2 + \tfrac{1}{4}\cdot 3 + \tfrac{7}{24}\cdot 4 = \tfrac{65}{24}$$

$$E[X^2] = \tfrac{1}{8}\cdot 1^2 + \tfrac{1}{3}\cdot 2^2 + \tfrac{1}{4}\cdot 3^2 + \tfrac{7}{24}\cdot 4^2 = \tfrac{201}{24}$$

$$\mathrm{Var}(X) = \tfrac{201}{24} - \left(\tfrac{65}{24}\right)^2 = \tfrac{599}{576}$$

The last computation is a bit easier than computing the variance directly from the definition $\mathrm{Var}(X) = E[(X - 65/24)^2]$, though it is not a major issue in light of the availability of technology. The formula in part (d) is actually more useful on some occasions when for various reasons we know the variance and the mean and want to find the second moment. Finally with regard to the theorem, part (e) shows that the variance is not linear as the mean is: constant coefficients factor out as squares, which happens because variance is defined as an expected square. Also, the variance measures spread rather than central tendency, so that the addition of a constant like d in part (e) of the theorem does not affect the variance.

Two Special Distributions

The rest of this chapter is primarily a catalog of the most common discrete distributions, their moments, and their applications. We will begin this catalog here by looking at two simple distributions that arise easily from work we have already done on random sampling: the discrete uniform distribution and the hypergeometric distribution.

Example 6 When one item is sampled at random from a set of consecutive integers $\{a, a + 1, ..., b\}$, since there are $(b - a + 1)$ equally likely states, the probability mass function of the selected item is clearly

$$f(x) = \tfrac{1}{b-a+1} \text{ if } x \in \{a, a + 1, ..., b\} \tag{10}$$

which is called the *discrete uniform distribution* on $\{a, a + 1, ..., b\}$.

Many examples of this kind of random experiment spring to mind, from draft lotteries to dice rolling and other games of chance. Because states are equally likely, the probability that the sampled item X falls into a set B is simply the number of states in B divided by the number of states in the whole space.

Usually this distribution is used as a building block in other kinds of problems however. There is a famous story, perhaps apocryphal, about German tanks in World War II, which apparently were numbered sequentially from 1 to the highest number. Allied intelligence had observed a sample of tank numbers, and was interested in using the sample to estimate how many tanks the Germans had.

To formulate a mathematical model, suppose that tank numbers $X_1, X_2, ... , X_n$ are observed in the sample, suppose also that events involving different X_i's are mutually independent (we will have a good definition of independent random variables later), and suppose that each X_i has the discrete uniform distribution on $\{1, 2, ..., \theta\}$, where θ is the total number of tanks in existence. Note that the c.d.f. $F(x) = P[X \le x]$ of this distribution would have the value (x/θ) at each state x

(why?). We can then proceed just as in Example 4 to obtain the c.d.f. and mass function of $Y = \max\{X_1, X_2, ..., X_n\}$, the largest tank number observed in the sample. (You should fill in the details.)

$$G(y) = P[Y \leq y] = \left(\frac{y}{\theta}\right)^n$$

$$g(y) = P[Y = y] = \left(\frac{y}{\theta}\right)^n - \left(\frac{y-1}{\theta}\right)^n, \quad y \in \{1, 2, ..., \theta\}$$

One sensible way of proceeding from here is to have *Mathematica* compute the expected value of this distribution for a fixed sample size n and various values of θ, and choose as our estimate of θ the one for which the expected value of Y comes closest to the highest tank number actually observed in the sample. Suppose for the sake of the example that 40 tanks are observed and the highest tank number in the sample is 125. The first function below is the p.m.f. of the maximum tank number in the sample, as a function of the sample size n and the highest tank number in the population of tanks θ. The second function gives the expected value of the highest tank number in the sample, as a function of n and θ.

```
g[y_, n_, θ_] := (y / θ)ⁿ - ((y - 1) / θ)ⁿ
```

```
ExpectedHighTank[n_, θ_] :=
  NSum[y * g[y, n, θ], {y, 1, θ}]
```

The graph in Figure 6 shows the graph of the expected value in terms of θ for a sample size of $n = 40$.

```
ListPlot[Table[
    {θ, ExpectedHighTank[40, θ]}, {θ, 125, 130}],
  AxesLabel -> {"θ", "Expected High Tank"},
  PlotStyle → PointSize[0.02],
  DefaultFont → {"Times-Roman", 8}];
```

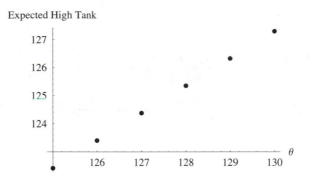

Figure 6 - Expected value of highest tank number as a function of θ

The two values of θ of 127 and 128 have the closest expected high tank number to 125. By evaluating ExpectedHighTank at each of these choices, you will find that θ = 128 is the closest estimate. ∎

Activity 5 Show that the expected value of the discrete uniform distribution on $\{1, 2, \ldots, n\}$ is $(n + 1)/2$, and the variance is $(n^2 - 1)/12$.

Example 7 Suppose that each member of a finite population can be classified as either having a characteristic or not. There are numerous circumstances of this kind: in a batch of manufactured objects each of them is either good or defective, in a deck of cards each card is either a face card or not, etc. The particular application that we will look at in this example is the classical capture-recapture model in ecology, in which a group of animals is captured, tagged, and then returned to mix with others of their species. Thus, each member in the population is either tagged or not.

Suppose that there are N animals in total, M of which are tagged. A new random sample of size n is taken in a batch without replacement. Let the random variable X be the number of tagged animals in the new sample. Reasoning as in Section 1.4, the probability that there are x tagged animals in the sample, and hence $n - x$ untagged animals, is

$$P[X = x] = \frac{\binom{M}{x}\binom{N-M}{n-x}}{\binom{N}{n}} \quad \text{, for integers } x \text{ such that } 0 \leq x \leq M \text{ and} \tag{11}$$

$$0 \leq n - x \leq N - M$$

The p.m.f. in formula (11) is called the *hypergeometric* p.m.f. with parameters N, M, and n. The restrictions on the state x guarantee that we must sample at least 0 tagged and untagged animals, and we cannot sample more than are available of each type.

For instance, if $N = 20$ animals, $M = 10$ of which were tagged, the probability of at least 4 tagged animals in a sample of size 5 is

$$\frac{\texttt{Binomial[10, 4]} * \texttt{Binomial[10, 1]}}{\texttt{Binomial[20, 5]}} + \\ \frac{\texttt{Binomial[10, 5]} * \texttt{Binomial[10, 0]}}{\texttt{Binomial[20, 5]}}$$

$$\frac{49}{323}$$

Let us try to compute the expected value of the hypergeometric distribution with general parameters N, M, and n, and then apply the result to the animal tagging context. First we need to check that the assumption that the sample was taken in a batch can be changed to the assumption that the sample was taken in sequence, without changing the distribution of the number of tagged animals. By basic combinatorics, if the sample was taken in sequence,

$$P[X = x] = \frac{\#\text{ sequences with exactly } x \text{ tagged}}{\#\text{ sequences possible}}$$

$$= \frac{(\#\text{ orderings of a batch of } n)\cdot(\#\text{ batches with } x \text{ tagged and } n-x \text{ untagged})}{\#\text{ sequences possible}}$$

$$= \frac{n!\cdot\binom{M}{x}\binom{N-M}{n-x}}{N\cdot(N-1)\cdot\,\cdots\,\cdot(N-n+1)} = \frac{\binom{M}{x}\binom{N-M}{n-x}}{\binom{N}{n}}$$

Since the last quotient is the same as formula (11), we may dispense with the assumption of batch sampling. This allows us to write X, the number of tagged animals in the sample, as

$$X = X_1 + X_2 + \cdots + X_n$$

where

$$X_i = \begin{cases} 1 & \text{if the } i^{\text{th}} \text{ animal in the sample is tagged} \\ 0 & \text{otherwise} \end{cases}$$

By linearity of expectation, $E[X] = \sum_{i=1}^{n} E[X_i]$. But each X_i has the same probability distribution, which gives weight $P[X_i = 1] = M/N$ to state 1 and weight $1 - M/N$ to state 0. Thus, the expected value of the hypergeometric random variable X is

$$E[X] = \sum_{i=1}^{n}(1 \cdot \frac{M}{N} + 0 \cdot (1 - \frac{M}{N})) = \frac{nM}{N} \tag{12}$$

The mean number of tagged animals in our sample of 5 is therefore $5(10)/20 = 2.5$. We will have to wait to compute the variance of the hypergeometric distribution until we know more about dependence between random variables. ∎

Mathematica's Statistics`DiscreteDistributions` package, which is loaded automatically when you load KnoxProb`Utilities`, contains objects that represent the distributions which we will meet in this chapter, in addition to which are ways of using these objects to find mass function values and cumulative distribution function values, and to produce simulated observations of random variables. The two named distributions that we know so far are written in *Mathematica* as:

DiscreteUniformDistribution[n]
(i.e., states $\{1, 2, \ldots, n\}$)

HypergeometricDistribution[n, M, N]
(i.e., sample size *n*, *M* with characteristic, *N* total)

When you want a value of the p.m.f. of a named distribution (including its parameter arguments) at a state *x* you issue the command

PDF[distribution, *x*]

(The "D" in "PDF" stands for "density", a word that will be motivated in Chapter 3.) For example, the earlier computation of the probability of at least 4 tagged animals could also have been done as follows. To save writing we give a temporary name to the distribution, and then we call for the sum of the p.m.f. values:

```
dist1 = HypergeometricDistribution[5, 10, 20];
PDF[dist1, 4] + PDF[dist1, 5]
```

$$\frac{49}{323}$$

For cumulative probabilities the function

CDF[distribution, *x*]

gives the value of the c.d.f. $F(x) = P[X \le x]$. For example, we can get a list of c.d.f. values at the states of the uniform distribution on $\{1, 2, \ldots, 10\}$ as follows:

```
dist2 = DiscreteUniformDistribution[10];
Table[N[CDF[dist2, i]], {i, 1, 10}]
```

{0.1, 0.2, 0.3, 0.4, 0.5, 0.6, 0.7, 0.8, 0.9, 1.}

Notice that the c.d.f. increments by exactly 1/10 at each state.

Here we display a histogram of the hypergeometric distribution with parameters $n = 5$, $M = 10$, $N = 20$ from the animal tagging example. We give a shorthand name to the distribution, write out the list of states longhand in a *Mathematica* list, create the list of probability masses by using Table and the PDF command, and finally call on ProbabilityHistogram.

```
dist1 = HypergeometricDistribution[5, 10, 20];
statelist = {0, 1, 2, 3, 4, 5};
problist = Table[PDF[dist1, x], {x, 0, 5}];
ProbabilityHistogram[statelist, problist,
    DefaultFont → {"Times-Roman", 8}];
```

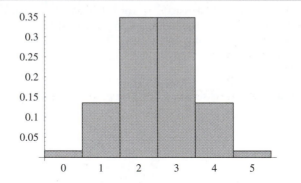

Figure 7 - Histogram of the hypergeometric(5,10,20) distribution

Finally, the Random and RandomArray commands apply to the discrete distributions in this package. Here are 10 simulated observations from the above hypergeometric distribution.

```
RandomArray[dist1, 10]
```

{2, 4, 1, 2, 3, 3, 3, 2, 2, 2}

> **Activity 6** How would you go about simulating observations from a discrete uniform distribution with states $\{a, a + 1, \ldots, b\}$? Try it.

We will see more distributions as we go through this chapter. You may also be interested in looking at other utility functions in the DiscreteDistributions package, such as Quantile, which is essentially the inverse of the c.d.f., and Mean which returns the mean of a given distribution.

Mathematica for Section 2.1

Command	Location
SeedRandom[seed]	kernel
Histogram[data, numrecs]	KnoxProb` Utilities`
DrawIntegerSample[popsize, n]	KnoxProb` Utilities`
PlotStepFunction[F, domain, jumplist]	KnoxProb` Utilities`
ProbabilityHistogram[statelist, problist]	KnoxProb` Utilities`
DiscreteUniformDistribution[n]	Statistics` DiscreteDistributions`
HypergeometricDistribution[n, M, N]	Statistics` DiscreteDistributions`
PDF[dist, x]	Statistics` DiscreteDistributions`
CDF[dist, x]	Statistics` DiscreteDistributions`
Random[dist]	Statistics` DiscreteDistributions`
RandomArray[dist, n]	Statistics` DiscreteDistributions`
SimulateGrades[n]	Section 2.1
ExpectedHighTank[n, θ]	Section 2.1

Exercises 2.1

1. Below are three possible probability mass functions for a random variable whose state space is $\{1, 2, 3, 4, 5\}$. Which, if any, is a valid p.m.f.? For any that are valid, give the associated c.d.f.

state	1	2	3	4	5
$f_1(x)$	.01	.42	.23	.15	.10
$f_2(x)$	.20	.30	.20	.15	.20
$f_3(x)$	.16	.31	.18	.10	.25

2. Two fair dice are rolled in succession. Find the p.m.f. of (a) the sum of the two; (b) the maximum of the two.

3. (*Mathematica*) Simulate 1000 rolls of 2 dice, produce a histogram of the sum of the two up faces, and compare the empirical distribution to the theoretical distribution in part(a) of Exercise 2. Do the same for the maximum of the two dice and compare to the theoretical distribution in Exercise 2(b).

4. Devise an example of a random variable that has the following c.d.f.

$$F(x) = \begin{cases} 0 & \text{if } x < 0 \\ 1/8 & \text{if } 0 \le x < 1 \\ 4/8 & \text{if } 1 \le x < 2 \\ 7/8 & \text{if } 2 \le x < 3 \\ 1 & \text{if } x \ge 3 \end{cases}$$

5. Find the c.d.f., and then the p.m.f., of the random variable X which is the minimum of a sample of five integers taken in sequence and with replacement from $\{1, 2, ..., 20\}$.

6. Consider a probability distribution on the set of integers $\mathbb{Z}$:

$$f(x) = \begin{cases} \frac{2}{9} \cdot \left(\frac{1}{3}\right)^{|x|-1} & \text{if } x = \pm 1, \pm 2, \pm 3, \, ... \\ \frac{1}{3} & \text{if } x = 0 \end{cases}$$

Verify that this is a valid p.m.f., and show that its c.d.f. F satisfies the conditions

$$\lim_{x \to -\infty} F(x) = 0 \, ; \quad \lim_{x \to +\infty} F(x) = 1$$

7. (*Mathematica*) Find the p.m.f. of the random variable X which is the largest item in a random sample of size 5 from $\{1, 2, ..., 20\}$ taken in a batch and without replacement. Verify using *Mathematica* that your function is a valid p.m.f.

8. Why is $E[X - \mu] = 0$ for every random variable X with finite mean μ? Why is the variance of any random variable non-negative? Is there any situation in which the variance can be zero?

9. Find the mean, variance, and third moment about the mean for the probability distribution $f(1.5) = .3$, $f(2) = .2$, $f(2.5) = .2$, $f(3.5) = .3$.

10. Compare the variances of the two distributions with probability mass functions $f(0) = .25$, $f(1) = .5$, $f(2) = .25$ and $g(0) = .125$, $g(1) = .75$, $g(2) = .125$. Draw pictures of the two p.m.f.'s that account for the differing variances.

11. Compare the third moments about the mean $E[(X - \mu)^3]$ for the two distributions with probability histograms below.

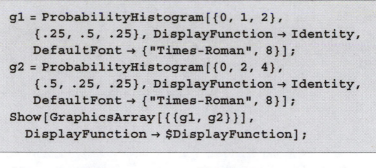

```
g1 = ProbabilityHistogram[{0, 1, 2},
     {.25, .5, .25}, DisplayFunction → Identity,
     DefaultFont → {"Times-Roman", 8}];
g2 = ProbabilityHistogram[{0, 2, 4},
     {.5, .25, .25}, DisplayFunction → Identity,
     DefaultFont → {"Times-Roman", 8}];
Show[GraphicsArray[{{g1, g2}}],
   DisplayFunction → $DisplayFunction];
```

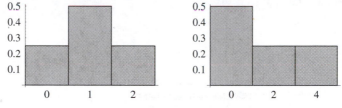

Exercise 11

12. For a random variable with state space {0, 1, 2, ...} and c.d.f. F, show that

$$E[X] = \sum_{n=0}^{\infty} (1 - F(n))$$

(Hint: Start with the right side and write $1 - F(n)$ as a sum.)

13. Derive a computational formula for $E[(X - \mu)^3]$ in terms of moments about 0.

14. (*Mathematica*) A discrete probability distribution on the non-negative integers that we will study later is the *Poisson distribution*. It has a parameter called μ, and is referred to in *Mathematica* as PoissonDistribution[mu]. Simulate a large random sample from the Poisson distribution with parameter 2, and use the empirical distribution to estimate the mean and variance. Repeat, changing the parameter to 5. Can you make a conjecture about how the mean and variance relate to the parameter μ? (Note: Instead of counting frequencies of integer states by hand, you may want to use one of the utility commands in *Mathematica*'s Statistics`DataManipulation` package, which is loaded automatically with KnoxProb`Utilities`. For instance, CategoryCounts[datalist,{{0},{1},{2},{3}, ... , {n}}] for a given numerical value of n would return the number of items in the datalist that are equal to each of the integers in the second argument of the function.)

15.   Some authors object to presenting the formula $E[g(X)] = \Sigma\ g(x) \cdot f(x)$ as a definition of the expected value of a function of X. Rather, they prefer to say that $Y = g(X)$ defines a new random variable Y, which has its own p.m.f. $f_y(y)$, and expectation $E[Y] = \Sigma\ y \cdot f_y(y)$. Give an argument that the two versions of expectation are the same in the simplest case where g is a 1-1 function.

16.   Find the mean and variance of the discrete uniform distribution on the set of states $\{-n, -n + 1, ..., 0, 1, ..., n - 1, n\}$.

17.   If a discrete random variable X has mean 6.5 and second moment about 0 equal to 50, find the mean and variance of $Y = 2X - 6$.

18.   Here is a sneak preview of the proof of linearity of expectation for two discrete random variables X and Y.   First, X and Y have *joint probability mass function* $f(x, y)$ if for all state pairs (x, y),

$$P[X = x, Y = y] = f(x, y)$$

The expectation of a function of X and Y is defined analogously to the one variable case:

$$E[g(X, Y)] = \sum_{x} \sum_{y} g(x, y) \cdot f(x, y)$$

Use these definitions to show that $E[aX + bY] = a\,E[X] + b\,E[Y]$.

19.   (*Mathematica*) In the context of the animal capture-recapture model of Example 7, suppose that $M = 10$ animals are tagged, a sample of size 8 gave 3 tagged animals, and the total population size N is unknown. What is the most likely value for N? (Compute the appropriate probabilities in *Mathematica* for various possible N.)

20.   (*Mathematica*) In a batch of 100 lightbulbs, 7 have defective filaments and will last only a very short time. A sample of size 10 is taken in a batch and without replacement. Find the mean number of defective bulbs in the sample exactly, and use *Mathematica* to numerically calculate the variance of the number of defective bulbs.

21.   (*Mathematica*) Draw probability histograms of the hypergeometric distribution with $N = 100$ for the cases: (a) $M = 50$ and $n = 10, 20, 30$;   (b) $n = 20$ and $M = 40$, 50, 60. Comment on the centers of the histograms.

22.   (*Mathematica*) Simulate (at least) 4 random samples of size 24, in sequence and with replacement, from the discrete uniform distribution on $\{1, 2, 3, 4\}$ and plot histograms of the samples. Comment on the variability of the samples. Increase the sample size to 100, and comment.

2.2 Bernoulli and Binomial Random Variables

One of the simplest discrete distributions is called the *Bernoulli distribution*. This arises in dichotomous situations, for example a flipped coin is either heads or tails, a computer bit is either set to 1 or 0, a sampled item from a population either has a certain characteristic or not, etc. With no perjorative intent we use the conventional language "success" and "failure" for the two possible experimental outcomes. A probability p is attached to the success outcome and $q = 1 - p$ is attached to the failure outcome. Using the computer bit model as our template, a random variable X is said to have the *Bernoulli distribution* if its state space is $\{0, 1\}$, and its p.m.f. is as follows:

$$\begin{aligned} f(1) = P[X = 1] = p \\ f(0) = P[X = 0] = q = 1 - p \end{aligned} \tag{1}$$

The mean and variance of the Bernoulli distribution are easy to find.

$$\mu = E[X] = 1(p) + 0(1 - p) = p \tag{2}$$

$$\sigma^2 = E[(X - p)^2] = (1 - p)^2 \, p + (0 - p)^2 \, (1 - p) = p(1 - p) \tag{3}$$

Activity 1 Recompute the variance of the Bernoulli distribution using the computational formula for variance $E[X^2] - \mu^2$.

Example 1 In the game of roulette, the roulette wheel has 38 slots into which a ball can fall when the wheel is spun. The slots are numbered 00, 0, 1, 2, ... , 36, and the numbers 0 and 00 are colored green while half of the rest are colored black and the other half red. A gambler can bet $1 on a color such as red, and will win $1 if that color comes up and lose the dollar that was bet otherwise. Consider a $1 bet on red. The gambler's winnings can then be described as a random variable $W = 2X - 1$, where X is 1 if the ball lands in a red slot and $X = 0$ if not. Clearly X has the Bernoulli distribution. The probability of a success, that is red, is $p = 18/38$ if all slots are equally likely. By linearity of expectation and formula (2), the expected winnings are:

$$E[W] = E[2X - 1] = 2E[X] - 1 = 2(\tfrac{18}{38}) - 1 = -\tfrac{2}{38}$$

Since the expected value is negative, this is not a good game for the gambler. (What do you think are the expected winnings in 500 such bets?) The winnings are variable though, specifically by the properties of variance discussed in Section 2.1 together with formula (3),

$$Var(W) = Var(2X - 1) = 4\,Var(X) = 4(\tfrac{18}{38})(\tfrac{20}{38}) = \tfrac{360}{361}$$

Since the variance and standard deviation are high relative to the mean, for the single $1 bet there is a fair likelihood of coming out ahead (exactly 18/38 in fact). We will see later that this is not the case for 500 bets. ∎

Mathematica's syntax for the Bernoulli distribution is

BernoulliDistribution[p]

The functions PDF, CDF, and Random (and RandomArray) can be applied to it in the way described at the end of the last section. Here for example is a short program to generate a desired number of replications of the experiment of spinning the roulette wheel and observing whether the color is red (R) or some other color (O). We just generate Bernoulli 0-1 observations with success parameter $p = 18/38$, and encode 0's as O and 1's as R.

```
Needs["Statistics`DiscreteDistributions`"]
```

```
SimulateRoulette[numreps_] :=
    Table[
   If[Random[BernoulliDistribution[18 / 38]] == 1,
    "R", "O"], {numreps}]
```

```
SimulateRoulette[20]
```

{O, R, O, R, R, R, R, R, O, O, O, R, R, O, R, R, R, R, R, R}

Activity 2 Write a *Mathematica* command to simulate 500 $1 bets on red in roulette, and return the net winnings. Run the command 20 times. How often are the net winnings positive?

The Bernoulli distribution is most useful as a building block for the more important *binomial distribution*, which we now describe. Instead of a single performance of a dichotomous success-failure experiment, consider n independent repetitions (usually called *trials*) of the experiment. Let the random variable X be defined as the total number of successes in the n trials. Then X has possible states $\{0, 1, 2, ..., n\}$. To find the probability mass function of X, consider a particular outcome for which the number of successes is exactly k. For concreteness, suppose that there are $n = 7$ trials, and we are trying to find the probability that the number of successes is $k = 4$. One such outcome is SSFFSFS, where the four successes occur on trials 1, 2, 5, and 7. By the independence of the trials and the fact that success occurs with probability p on any one trial, the probability of this particular outcome is

$$p \cdot p \cdot (1-p) \cdot (1-p) \cdot p \cdot (1-p) \cdot p = p^4 (1-p)^3$$

By commutativity of multiplication, it is easy to see that any such outcome with four successes and 3 failures has the same probability. In general the probability of any particular n trial sequence with exactly k successes is $p^k (1-p)^{n-k}$. How many different n trial sequences are there that have exactly k successes? A sequence is determined uniquely by sampling k positions from among n in which the S symbols are located. By combinatorics, there are $\binom{n}{k}$ such sequences. Thus the overall probability of exactly k successes in n Bernoulli trials is the following, which defines the *binomial probability mass function*.

$$f(k) = P[X = k] = \binom{n}{k} p^k (1-p)^{n-k} \ , k = 0, 1, 2, ..., n \qquad (4)$$

As a shorthand to indicate that a random variable X has a particular distribution, we use the symbol " $\sim$ ", and we also use abbreviations to stand for some of the most common distributions and their parameters. Here, to say that X has the binomial distribution with n trials and success probability p, we write

$$X \sim b(n, p)$$

The Bernoulli distribution is the special case of the binomial distribution in which there is only $n = 1$ trial. Notice in (4) that in this case the state space is $\{0, 1\}$, and the p.m.f. formula reduces to

$$f(k) = \binom{1}{k} p^k (1-p)^k = p^k (1-p)^{1-k}, \ k = 0, 1$$

which is an equivalent form to (1).

Mathematica's binomial distribution object is

$$BinomialDistribution[n, p]$$

where the meanings of the arguments are the same as described above. We will show the use of this distribution object in the example below.

Example 2 In the roulette problem, let us try to compute the probability that the gambler who bets $1 on red 500 times comes out ahead. This event happens if and only if there are strictly more wins than losses among the 500 bets. So we want to compute $P[X \geq 251]$, or what is the same thing, $1 - P[X \leq 250]$, where X is the number of reds that come up. Since X has the binomial distribution with $n = 500$ and $p = 18/38$, we can compute:

```
N[
    1 - CDF[BinomialDistribution[500, 18 / 38], 250]]
```

0.110664

So our gambler is only around 11% likely to profit on his venture. Another way to have made this computation is to add the terms of the binomial p.m.f. from 251 to 500:

```
NSum[PDF[BinomialDistribution[500, 18 / 38], k],
    {k, 251, 500}]
```

0.110664

(If you execute this command you will find that it takes a good deal longer than the one above using the CDF, because *Mathematica* has some special algorithms to approximate sums of special forms like the binomial CDF.) ∎

To recap a bit, you can identify a binomial experiment from the following properties: it consists of n independent trials, and on each trial either a success (probability p) or a failure (probability $q = 1 - p$) can occur. The binomial p.m.f. in formula (4) applies to the random variable that counts the total number of successes. Shifting the focus over to sampling for a moment, the trials could be repeated samples of one object from a population, replacing the sampled object as sampling proceeds, and observing whether the object has a certain special characteristic (a success) or not (a failure). The success probability would be the proportion of objects in the population that have the characteristic. One of the exercises at the end of this section invites you to compare the probability that k sample members have the characteristic under the model just described to the probability of the same event assuming sampling in a batch without replacement.

Activity 3 Use *Mathematica* to compute the probability that the gambler who bets 500 times on red in roulette comes out at least $10 ahead at the end. If the gambler plays 1000 times, what is the probability that he will not show a positive profit?

In Figure 8(a) is a histogram of 500 simulated values from the binomial distribution with $n = 10$ trials and $p = 1/2$. Figure 8(b) is a histogram of the probability mass function itself, which shows a similar symmetry about state 5, similar spread, and similar shape.

```
Needs["KnoxProb`Utilities`"];
datalist = RandomArray[
   BinomialDistribution[10, .5], 500];
pmf[k_] := PDF[BinomialDistribution[10, .5], k];
states = {0, 1, 2, 3, 4, 5, 6, 7, 8, 9, 10};
probs = Table[pmf[k], {k, 0, 10}];
Show[GraphicsArray[
       {Histogram[datalist,
      11, Distribution → Discrete,
      DisplayFunction → Identity],
          ProbabilityHistogram[states,
      probs, DisplayFunction → Identity,
      DefaultFont → {"Times-Roman", 8}]}],
   DisplayFunction → $DisplayFunction];
```

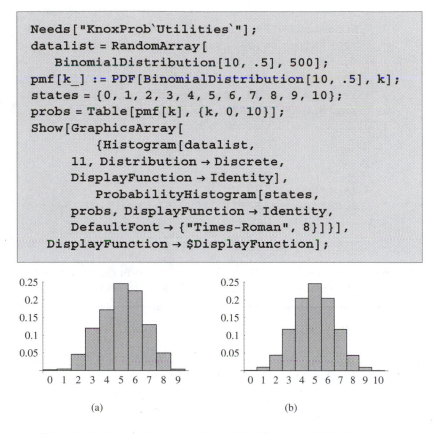

(a)   (b)

Figure 8 - (a) simulated histogram; (b) probability histogram of b(10,.5)

The binomial mass function is always symmetric about its center when $p = 1/2$ (see Exercise 9). In the cases $p > 1/2$ and $p < 1/2$, respectively, the distribution is skewed left and right. (Convince yourself of these facts analytically by looking at the formula for the p.m.f., and intuitively by thinking about the binomial experiment itself. Use *Mathematica* to plot histograms of a few examples.)

We now turn to the moments of the binomial distribution. Think about these questions: If you flip 4 fair coins, how many heads would you expect to get? What if you flipped 6 fair coins? Analyze how you are coming up with your guesses, that is, of what simpler things is your final answer composed? What does this mean about the expected value of the number of heads in an odd number of flips, say 7?

After the last paragraph you probably can guess the formula for the mean of the binomial distribution. Compare your hypothesis with the following brute force calculations of means in *Mathematica*, done by summing the probability of a state, times the state, over all states.

```
BinomialPMF[x_, n_, p_] :=
  PDF[BinomialDistribution[n, p], x]
```

```
(* case n = 20, p = .2 *)
Sum[BinomialPMF[k, 20, .2] * k, {k, 0, 20}]
```

4.

```
(* case n = 30, p = .8 *)
Sum[BinomialPMF[k, 30, .8] * k, {k, 0, 30}]
```

24.

The binomial mean is of course $n \cdot p$, which we prove in the next theorem. We also include the formula for the variance of the binomial distribution.

Theorem 1. If $X \sim b(n, p)$, then (a) $E[X] = n \cdot p$; (b) $Var(X) = n \cdot p \cdot (1 - p)$

Proof. (a) We may factor n and p out of the sum, delete the $k = 0$ term (which equals 0), and cancel the common factors of k to get

$$E[X] = \Sigma_{k=0}^{n} k \cdot \binom{n}{k} p^k (1 - p)^{n-k}$$

$$= n \cdot p \cdot \Sigma_{k=1}^{n} \frac{(n-1)!}{(k-1)!\,(n-k)!} p^{k-1} (1 - p)^{n-k}$$

Then we can change variables in the sum to $l = k - 1$, which reduces the sum to that of all terms of the $b(n - 1, p)$ probability mass function. Since this sum equals 1, the desired formula $n \cdot p$ for the mean results.

(b) We use the indirect strategy of computing $E[X(X - 1)]$ first, because this sum can be simplified very similarly to the sum above for $E[X]$.

$$E[X(X - 1)] = \Sigma_{k=0}^{n} k \cdot (k - 1) \cdot \binom{n}{k} p^k (1 - p)^{n-k}$$

$$= n \cdot (n - 1) \cdot p^2 \cdot \Sigma_{k=2}^{n} \frac{(n-2)!}{(k-2)!\,(n-k)!} p^{k-2} (1 - p)^{n-k}$$

The sum on the right adds all terms of the $b(n-2, p)$ p.m.f., hence it equals 1. Therefore, $E[X(X-1)] = E[X^2] - E[X] = n \cdot (n-1) \cdot p^2$. Finally,

$$\mathrm{Var}(X) = E[X^2] - (E[X])^2 = E[X^2] - E[X] + E[X] - (E[X])^2 =$$
$$n \cdot (n-1) \cdot p^2 + np - (np)^2$$

which simplifies to the desired formula.

Example 3   Many attempts have been made to estimate the number of planets in the universe on which there is some form of life, based on some very rough estimates of proportions of stars that have planetary systems, average numbers of planets in those systems, the probability that a planet is inhabitable, etc. These estimates use random variables that are assumed to be modeled by binomial distributions. At the time of this writing, scientists are estimating that 2% of stars have planetary systems. Suppose that only 1% of those systems have a planet (assume conservatively that there is at most one such planet per system) that is inhabitable, and that only 1% of inhabitable planets have life. For a galaxy of, say 100 billion $=10^{11}$ stars, let us find the expected number of stars with a planet that bears life.

Consider each star as a trial, and define a success as a star with a system that includes a planet that has life. Then there are $n = 10^{11}$ trials, and by the multiplication rule for conditional probability, the success probability per star is

$$p = P[\text{life}] = P[\text{has system and has inhabitable planet and life}] =$$
$$(.02)(.01)(.01) = 2 \times 10^{-6}$$

Therefore the expected value of the number of planets with life is $np = 2 \times 10^5 = 200,000$. Incidentally, the standard deviation of the number of planets with life would be

$$\sigma = \mathbf{N}\left[\sqrt{10^{11} \times (2 \times 10^{-6})\,(1 - 2 \times 10^{-6})}\,\right]$$

447.213

We will see later that most of the probability weight for any distribution falls within 3 standard deviations of the mean of the distribution. Since 3σ is around 1300 which is a great deal less than the mean of 200,000, this means that it is very likely that the number of planets with life is in the vicinity of 200,000 under the assumptions in this example. ∎

Mathematica for Section 2.2

Command	Location
BernoulliDistribution[p]	Statistics` DiscreteDistributions`
BinomialDistribution[n, p]	Statistics` DiscreteDistributions`
Random[dist]	Statistics` DiscreteDistributions`
RandomArray[dist, n]	Statistics` DiscreteDistributions`
PDF[dist, x]	Statistics` DiscreteDistributions`
CDF[dist, x]	Statistics` DiscreteDistributions`
Histogram[datalist, numrecs]	KnoxProb` Utilities`
ProbabilityHistogram[states, probs]	KnoxProb` Utilities`
SimulateRoulette[numreps]	Section 2.2

Exercises 2.2

1. Devise two examples other than the ones given at the start of the section of random phenomena that might be modeled using Bernoulli random variables.

2. Explain how a random variable X with the $b(n, p)$ distribution can be thought of as a sum of Bernoulli(p) random variables.

3. Find the third moment about the mean for the Bernoulli distribution with parameter p.

4. Suppose that the number of people among n who experience drowsiness when taking a certain blood pressure medication has the binomial distribution with success probability $p = .4$. Find the probability that no more than 3 among 10 such recipients of the medication become drowsy.

5. Find the second moment about 0 for the $b(n, p)$ distribution.

6. (*Mathematica*) Suppose that in front of a hotel in a large city, successive 20 second intervals constitute independent trials, in which either a single cab will arrive (with probability .2), or no cabs will arrive (with probability .8). Find the probability that at least 30 cabs arrive in an hour. What is the expected number of cabs to arrive in an hour?

7. (*Mathematica*) The 500 Standard & Poor's companies are often used to measure the performance of the stock market. One rough index simply counts the numbers of "advances", i.e., those companies whose stocks either stayed the same in value or rose. The term "declines" is used for those companies whose stocks fell in value. If during a certain bull market period 70% of the Standard & Poor's stocks are advances on average, what is the probability that on a particular day there are at least 300 advances? What are the expected value and variance of the number of advances?

8. For what value of the parameter p is the variance of the binomial distribution largest? (Assume that n is fixed.)

9. Prove that when $p = .5$, the p.m.f. of the binomial distribution is symmetric about p.

10. (*Mathematica*) Write a command in *Mathematica* to simulate 500 at bats of a baseball hitter whose theoretical average is .300 (that is, his probability of a hit on a given at bat is .3). Use the command to simulate several such 500 at bat seasons, making note of the empirical proportion of hits. What is the expected value of the number of hits, and what is the standard deviation of the batting average random variable (i.e., hits/at bats)?

11. Use linearity of expectation to devise an alternative method of proving the formula $E[X] = np$ where $X \sim b(n, p)$. (Hint: See Exercise 2.)

12. (*Mathematica*) Unscrupulous promoters have booked 55 people on a bus trip to a shopping mall, knowing that the bus will only hold 52. They are counting on the fact that history indicates that on average 10% of the people who book reservations do not show up for the trip. The fine print on the reservations promises a double-your-money back refund of $100 if for any reason a reserved seat is not available for the ticket holder on the day of the trip. What is the probability that the promoters will have to issue at least one refund? What is the expected total refund that will be paid?

13. (*Mathematica*) A population of 100 people contains 20 who are regular consumers of a certain soft drink. For a product survey, samples of size 10 are taken under two procedures: (1) sampling in a batch without replacement; (2) sampling in sequence with replacement. Compute and compare the probability mass functions of the number of people in the sample who are regular consumers of this soft drink for the two procedures. Compute relative differences between corresponding probability masses for the two procedures for each state 0, 1, 2, ... , 10. (The *relative difference* between positive numbers a and b is the absolute difference $|a - b|$ divided by the smaller of a and b.)

14. A generalization of the binomial experiment consists of n independent trials, each of which can result in one of k outcomes. Outcome i has probability p_i, and we must have that the sum of all p_i for $i = 1$ to k equals 1. This so-called *multino-*

mial experiment therefore involves k counting random variables X_1, X_2, ... , X_k, where X_i is the number of trials among the n that resulted in outcome i. For a multinomial experiment with 10 trials and 4 possible outcomes whose probabilities are 1/4, 1/2, 1/8, and 1/8 find an expression for the *joint probability mass function* of X_1, X_2, X_3, and X_4 defined by

$$P[X_1 = x_1 \cap X_2 = x_2 \cap X_3 = x_3 \cap X_4 = x_4]$$

2.3 Geometric and Negative Binomial Random Variables

In the last section we learned about binomial experiments, which are closed-ended and sequential. There was a fixed number of trials, each of which could result only in two possible outcomes called success and failure. We now consider the case of an open-ended experiment which continues until some fixed number r of successes is reached. Consider for example an electrical switch which endures most of its strain when it changes from the off to the on state. There is no appreciable wear on the switch, but it has some low probability of switch failure each time it is switched on, which does not change with use. Suppose that we call a "success" a malfunction of the switch and a "failure" a switch-on without malfunction. Then the lifetime of the switch, measured in terms of number of times it is switched on without malfunction, is the number of "failures" required to reach the first "success" event. A random variable X which counts the number of Bernoulli trials strictly before the first success occurs (i.e., the number of failures) has what is called the *geometric distribution*. A natural generalization is a random variable which counts the number of failures prior to the rth success where r is a positive integer. The distribution of this random variable is known as the *negative binomial distribution*.

Activity 1 Devise other examples of experimental situations where the geometric or negative binomial distribution would be a natural model.

It is fairly easy to derive the p.m.f.'s of these two distributions. Consider first the geometric experiment which repeats dichotomous trials with success probability p until the first success occurs. The event that the number of failures is exactly k is the event that there are k failures in a row, each of probability $q = 1 - p$, followed by a success, which has probability p. By the independence of the trials, the number X of failures prior to the first success has the following *geometric p.m.f. with parameter p* (abbr. geometric(p)):

$$f(k) = P[X = k] = (1 - p)^k \cdot p, \quad k = 0, 1, 2, 3, \ldots \tag{1}$$

You can think about the negative binomial experiment in a similar way, although now the possible arrangements of successes and failures in the trials prior to the last one must be considered. In order for there to be exactly k failures prior to the rth success, the experiment must terminate at trial $k + r$, and there must be exactly $r - 1$ successes during the first $k + r - 1$ trials, and a success on trial $k + r$. The first $r - 1$ successes may occur in any positions among the first $k + r - 1$ trials. Therefore the probability of this part of the event is given by a binomial probability with $k + r - 1$ trials, $r - 1$ successes and success probability p. This binomial probability is multiplied by the probability p of success on trial $k + r$. Therefore the *negative binomial p.m.f. with parameters r and p* is

$$\begin{aligned} f(k) = P[X = k] &= \binom{k + r - 1}{r - 1} p^{r-1} \, q^k \cdot p \\ &= \binom{k + r - 1}{r - 1} p^r (1 - p)^k, \quad k = 0, 1, 2, \ldots \end{aligned} \tag{2}$$

where the random variable X counts the number of failures prior to the rth success.

The *Mathematica* distribution objects for these two distributions are

GeometricDistribution[p] and NegativeBinomialDistribution[r, p]

They are contained in the Statistics`DiscreteDistributions` package, which is loaded below when the KnoxProb`Utilities` package is loaded. For example, the shape of the geometric(.5) p.m.f. is shown in Figure 9(a). A simulation of 500 observed values of a geometric(.5) random variable yields the histogram of the empirical distribution in Figure 9(b). Both graphs illustrate the exponential decrease of probability mass suggested by formula (1) as the state grows.

```
Needs["KnoxProb`Utilities`"];
```

```
f[x_] := PDF[GeometricDistribution[.5], x];
geomprobs = Table[f[x], {x, 0, 8}];
states = {0, 1, 2, 3, 4, 5, 6, 7, 8};
g1 = ProbabilityHistogram[states,
    geomprobs, DisplayFunction → Identity,
    DefaultFont → {"Times-Roman", 8}];
geomdatalist = RandomArray[
    GeometricDistribution[.5], 500];
g2 = Histogram[geomdatalist, 9, Distribution →
    Discrete, DisplayFunction → Identity];
Show[GraphicsArray[{g1, g2}], PlotRange -> All,
    DisplayFunction → $DisplayFunction];
```

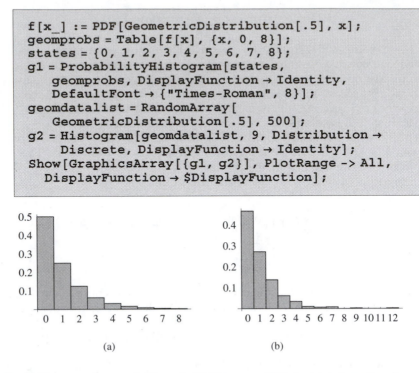

(a)   (b)

Figure 9 - (a) Histogram of geometric(.5) p.m.f.; (b) Histogram of 500 simulated observations from geometric(.5)

Activity 2   Show that the geometric p.m.f. in formula (1) defines a valid probability mass function.

Example 1 A telemarketer must call a list of cardholders of a certain lender to attempt to sell them credit card insurance, which pays off the debt in the event that the cardholder dies or becomes disabled and loses the ability to pay. Past experience indicates that only 20% of those called will agree to buy the insurance. The telemarketer would like to make 20 sales in a particular week. Find the probability that it will take at least 100 phone calls to achieve this. Comment on the variability of the distribution of the number of unsuccessful calls.

Each attempted phone call constitutes a Bernoulli trial, and if the success event is the event that the cardholder buys the insurance, then the success probability is given to be $p = .20$. Because there are to be at least 100 calls total, and 20 of them are to be successes, we are asking for the probability of the event that there will be at least 80 unsuccessful calls prior to the 20th success. The random variable X which is the number of unsuccessful calls has the negative binomial distribution with parameters $r = 20$ and $p = .2$. Therefore, we must compute:

$$P[X \geq 80] = 1 - P[X \leq 79]$$

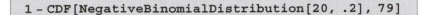

```
1 - CDF[NegativeBinomialDistribution[20, .2], 79]
```

0.480021

Thus, our telemarketer can be about 48% sure that he will make at least 100 calls. Let us look at a connected dot plot of the probability masses to get an idea of how variable the number of unsuccessful calls *X* is. (We will talk about the variance of the distribution later.)

```
f[x_] :=
    PDF[NegativeBinomialDistribution[20, .2], x];
pmf = Table[{x, f[x]}, {x, 20, 140}];
ListPlot[pmf, PlotJoined -> True,
    DefaultFont → {"Times-Roman", 8}];
```

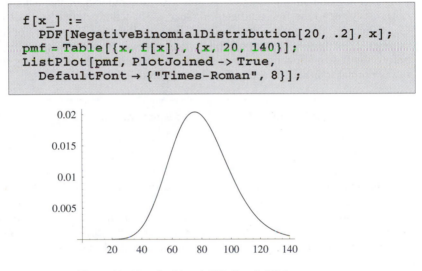

Figure 10 - Negative binomial(20,.2) probabilities

The distribution of *X* is nearly symmetric about a point that seems to be just less than 80, with a slight right skew. It appears that at least 2/3 of the probability weight is between 60 and 100, and there does not seem to be much weight at all to the right of 120. This tells the telemarketer that on repeated sales campaigns of this kind, most of the time between 60 and 100 unsuccessful calls will be made, and very rarely will he have to make more than 120 unsuccessful calls. ∎

Activity 3 Another kind of question that can be asked about the situation in Example 1 is: for what number c is it 95% likely that the number of unsuccessful calls X is less than or equal to c? Use *Mathematica* to do trial and error evaluation of the c.d.f. in search of this c.

Example 2 A snack machine dispenses bags of chips and candy. One of its ten dispenser slots is jammed and non-functioning. Assume that successive machine customers make a selection randomly. What is the probability that the first malfunction occurs strictly after the 20th customer? What is the probability that the third malfunction occurs on the 50th customer's selection?

Since one among ten slots is defective, the number of normal transactions (failures) prior to the first malfunction (success) has the geometric(.1) distribution. Consequently, the probability that the first malfunction happens for customer number 21 or higher is the probability that at least 20 normal transactions occur before the first malfunction, which is the complement of the probability that 19 or fewer normal transactions occur. This is computed as follows.

```
1 - CDF[GeometricDistribution[.1], 19]
```

0.121577

So it is highly probable, around 88%, that the first malfunction happens by the time the 20th customer uses the machine.

Similarly the number of good transactions until the third malfunction has the negative binomial distribution with parameters $r = 3$ and $p = .1$. For the second question we want the probability that there are exactly 47 good transactions, which is:

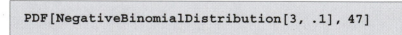

```
PDF[NegativeBinomialDistribution[3, .1], 47]
```

0.00831391

We have an opportunity to take this example in a different direction, one which indicates the power of probabilistic reasoning in drawing conclusions from data. Suppose we do not know how many among the ten dispenser slots are jammed, but we do know that malfunctions occurred on the 4th, 10th, 18th, and 24th customers among the first 24. Can the pattern of problems indicate a most likely guess for the number of jammed slots?

Because there exists at least one defect and not all slots are bad, our choices for the number of defective slots are 1, 2, 3, 4, 5, 6, 7, 8, and 9, under which the probabilities of a malfunction are, respectively, $p = .1, .2, .3, .4, .5, .6, .7, .8,$ and .9. The number of normal transactions prior to the first malfunction clearly has the geometric(p) distribution, but we can also imagine restarting a geometric experiment each time a malfunction occurs, keeping track of the number of normal transactions prior to the next malfunction. We have four such experiments, yielding independent geometric(p) random variables X_1, X_2, X_3, X_4. The observed values of the X's, that is, the number of failures between successes, are 3, 5, 7, and 5. Among the candidate p values, which gives the highest probability to this list of observations? As a function of p, the probability of this list of observed values is

$$P[X_1 = 3, X_2 = 5, X_3 = 7, X_4 = 5] = p(1 - p)^3 p(1 - p)^5 p(1 - p)^7 p(1 - p)^5$$

Summing exponents to simplify, we try to maximize this function of p over the set of possible p values above.

```
prob[p_] := p⁴ (1 - p)²⁰
```

```
Table[{p, prob[p]}, {p, .1, .9, .1}] // TableForm
```

0.1	0.0000121577
0.2	0.0000184467
0.3	6.46317×10^{-6}
0.4	9.35977×10^{-7}
0.5	5.96046×10^{-8}
0.6	1.42497×10^{-9}
0.7	8.37177×10^{-12}
0.8	4.29497×10^{-15}
0.9	6.561×10^{-21}

From the table, we see that the p value which makes the observed data likeliest is the second one, $p = .2$; in other words the data are most consistent with the assumption that exactly two slots are jammed. ∎

Activity 4 Redo the last part of Example 2, assuming that the stream of malfunctions came in on customers 1, 3, 6, and 8. Find the value of p that makes this stream most likely. Do it again, assuming that the malfunctions were on customers 2, 7, 10, and 17.

We close the section by deriving the first two moments of these distributions.

Theorem 1. (a) The mean and variance of the geometric(p) distribution are

$$\mu = \tfrac{1-p}{p}, \ \sigma^2 = \tfrac{1-p}{p^2} \tag{3}$$

(b) The mean and variance of the negative binomial(r, p) distribution are

$$\mu = \tfrac{r(1-p)}{p}, \ \sigma^2 = \tfrac{r(1-p)}{p^2} \tag{4}$$

Proof. (a) By the definition of expected value,

$$
\begin{aligned}
E[X] &= \Sigma_{k=0}^{\infty} k \cdot (1 - p)^k \cdot p \\
&= p(1-p)\Sigma_{k=0}^{\infty} k \cdot (1-p)^{k-1}
\end{aligned}
$$

The infinite series is of the form $\sum\limits_{k=0}^{\infty} k \cdot x^{k-1}$ for $x = 1 - p$. This series is the derivative of the series $\sum\limits_{k=0}^{\infty} x^k$ with respect to x, and the latter series has the closed form $1/(1-x)$. Hence

$$\Sigma_{k=0}^{\infty} k \cdot x^{k-1} = \tfrac{d}{dx}\left(\tfrac{1}{1-x}\right) = \tfrac{1}{(1-x)^2}$$

Evaluating at $x = 1 - p$,

$$\mu = E[X] = p(1-p) \cdot \tfrac{1}{(1-(1-p))^2} = \tfrac{1-p}{p}$$

The variance of the geometric distribution is the subject of Exercise 7.

(b) The number of failures X until the rth success can be thought of as the sum of the number of failures X_1 until the first success, plus the number of failures X_2 between the first and second successes, etc., out to X_r, the number of failures between successes $r - 1$ and r. Each of the X_i's has the geometric(p) distribution; hence by linearity of expectation $E[X] = r \cdot E[X_1]$, which yields the result when combined with part (a).

The proof of the variance formula is best done using an important result that we have not yet covered: when random variables are independent of one another the variance of their sum is the sum of their variances. Then, reasoning as we did for the mean, $\text{Var}(X) = r \cdot \text{Var}(X_1)$, which yields the formula for the variance of the negative binomial distribution when combined with the formula in part (a) for the variance of the geometric distribution. We will prove this result on the variance of a sum in Section 2.6, but Exercise 13 gives you a head start.

Mathematica for Section 2.3

Command	Location
GeometricDistribution[p]	Statistics` DiscreteDistributions`
NegativeBinomialDistribution[r, p]	Statistics` DiscreteDistributions`
RandomArray[distribution, n]	Statistics` DiscreteDistributions`
PDF[dist, x]	Statistics` DiscreteDistributions`
CDF[dist, x]	Statistics` DiscreteDistributions`
Histogram[datalist, numrecs]	KnoxProb` Utilities`
ProbabilityHistogram[states, probs]	KnoxProb` Utilities`

Exercises 2.3

1. For the electrical switch discussed at the start of the section, how small must the switch failure probability be such that the probability that the switch will last at least 1000 trials (including the final trial where the switch breaks) is at least 90%?

2. Derive the c.d.f. of the geometric(p) distribution.

3. If X has the geometric(p) distribution, find $P[X > m + n \mid X > n]$.

4. (*Mathematica*) On a long roll of instant lottery tickets, on average one in 20 is a winner. Find the probability that the third winning ticket occurs somewhere between the 50th and 70th tickets on the roll.

5. (*Mathematica*) Simulate several random samples of 500 observations from the negative binomial distribution with parameters $r = 20$ and $p = .2$, produce associated histograms, and comment on the shape of the histograms as compared to the graph in Figure 10.

6. A certain home run hitter in baseball averages a home run about every 15 at bats. Find the probability that it will take him (a) at least 10 at bats to hit his first home run; (b) at least 25 at bats to hit his second home run; (c) at least 25 at bats to hit his second home run given that it took exactly 10 at bats to hit his first. (d) Find also the expected value and standard deviation of the number of at bats required to hit the second home run.

7. Derive the formula for the variance of the geometric(p) distribution. (Hint: it will be simpler to first find $E[X(X-1)]$.)

8. (*Mathematica*) A shoe store owner is trying to estimate the probability that customers who enter his store will purchase at least one pair of shoes. He observes one day that the 3rd, 6th, 10th, 12th, 17th, and 21st customers bought shoes. Assuming that the customers make their decisions independently, what purchase probability p maximizes the likelihood of this particular customer stream?

9. (*Mathematica*) Write a command that simulates a stream of customers as described in Exercise 8. It should take the purchase probability p as one argument, and the number of customers in total that should be simulated as another. It should return a list of coded observations like {B,N,N,B,...} indicating "buy" and "no buy".

10. Show in the case $r = 3$ that the negative binomial p.m.f. in formula (2) defines a valid probability mass function.

11. (*Mathematica*) A *random walk to the right* is a random experiment in which an object moves from one time to the next on the set of integers. At each time, as shown in the figure, either the object takes one step right to the next higher integer with probability p, or stays where it is with probability $1 - p$. If as below the object starts at position 0, and $p = .3$, what is the expected time required for it to reach position 5? What is the variance of that time? What is the probability that it takes between 15 and 30 time units to reach position 5?

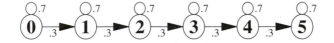

Exercise 11

12. (*Mathematica*) Study the dependence of the negative binomial distribution with parameter $r = 3$ on the success probability parameter p by comparing connected line graphs of the p.m.f. for several values of p. Be sure to comment on the key features of the graph of a p.m.f.: location, spread, and symmetry.

13. This exercise foreshadows the result to come later on the variance of a sum of independent random variables which was needed in the proof of the theorem about the variance of the negative binomial distribution. Suppose that two random variables X and Y are such that joint probabilities involving both of them factor into the product of individual probabilities, for example:

$$P[X = i, Y = j] = P[X = i]P[Y = j]$$

Let the expected value of any function $g(X, Y)$ of the two random variables be defined by the weighted average

$$E[g(X, Y)] = \sum_i \sum_j g(i, j) \, P[X = i, Y = j]$$

Write down and simplify an expression for $\text{Var}(X + Y) = E[\{(X + Y) - (\mu_x + \mu_y)\}^2]$ and show that it reduces to $\text{Var}(X) + \text{Var}(Y)$.

14. (*Mathematica*) Consider a basketball team which is estimated to have a 60% chance of winning any game, and which requires 42 wins to clinch a playoff spot. Find the expected value and standard deviation of the number of games that are required to do this. Think about the assumptions you are making when you do this problem, and comment on any possible difficulties.

2.4 Poisson Distribution

The last member of the collection of single-variable discrete distributions that we will look at is the *Poisson distribution*. It arises as a model for random variables that count the number of occurrences of an event in a given region of space or time, such as the number of users of a computer lab during a fixed time period, the number of daisies in a field, the number of dust particles in a room, the number of cars arriving to an intersection in a fixed time period, etc. Its state space is the non-negative integers $\{0, 1, 2, ...\}$, and the probability mass function that characterizes it is in formula (1) below. So, this distribution depends on a single parameter called μ, hence we refer to it as the Poisson(μ) distribution.

In the cases of the other discrete distributions that we have studied, the mass function of the associated random variable was derivable from primitive assumptions about the randomness of an experiment. The Poisson distribution is a bit different, since it comes up as an idealization of a binomial experiment whose number of trials approaches infinity while the success probability becomes vanishingly small. To see the idea, consider some event that happens sporadically on a time axis, in such a way that the chance that it happens during a short time interval is proportional to the length of the interval, and the events that it happens in each of two disjoint time intervals are independent. Without loss of generality, let the time axis be the bounded interval $[0, 1]$, and also let μ be the proportionality constant, that is, $P[\text{event happens in } [a, b]] = \mu(b - a)$. Then, if the time interval $[0, 1]$ is broken up into n disjoint, equally sized subintervals $[0, 1/n], (1/n, 2/n), ... , ((n - 1)/n, 1]$,

$$p = P[\text{event happens in interval } ((j - 1)/n, j/n]\,] = \frac{\mu}{n}$$

for all subintervals. The number of occurrences during the whole interval $[0,1]$ is the sum of the number of occurrences in the subintervals; hence this number has the binomial distribution with parameters n and $p = \mu/n$. But what do the probability masses approach as $n \longrightarrow \infty$? We will proceed to argue that they approach the following *Poisson(μ) p.m.f.*:

$$f(x) = P[X = x] = \frac{e^{-\mu} \mu^x}{x!} \ , \ x = 0, 1, 2, ... \tag{1}$$

We can write out the binomial probabilities as

$$
\begin{aligned}
\text{P[exactly } x \text{ successes]} \ &= \ \binom{n}{x} p^x (1-p)^{n-x} \\[2mm]
&= \ \frac{n!}{x!\,(n-x)!} \left(\frac{\mu}{n}\right)^x \left(1 - \frac{\mu}{n}\right)^{n-x} \\[2mm]
&= \ \frac{\mu^x}{x!} \left(1 - \frac{\mu}{n}\right)^n \frac{n(n-1)\cdots(n-x+1)}{n^x} \left(1 - \frac{\mu}{n}\right)^{-x}
\end{aligned}
\tag{2}
$$

By a standard result from calculus, the factor $(1 - \frac{\mu}{n})^n$ approaches $e^{-\mu}$ as $n \longrightarrow \infty$. In the activity below you will analyze the other factors to finish the proof that the binomial probabilities approach the Poisson probabilities as n becomes large and the success probability p becomes correspondingly small.

Activity 1 Show that the third and fourth factors in the bottom expression on the right side of (2) approach 1 as $n \longrightarrow \infty$. You will have to deal with the case $x = 0$ separately. Conclude that the binomial probability masses b$(x; n, \mu/n)$ approach the Poisson(μ) masses in formula (1) as n approaches infinity.

In summary, the Poisson distribution is an approximation for the binomial distribution with large n and small p, and the Poisson parameter relates to these by $\mu = n\,p$.

   Mathematica has an object for the Poisson distribution, called

PoissonDistribution[μ]

which is contained in the Statistics‘DiscreteDistributions‘ package, and which can be used in the usual way with the Random, PDF, and CDF functions. Let us use it to see how close the binomial distribution is to the Poisson distribution for large n and small p. We take $n = 100$, $\mu = 4$, and $p = \mu/n = 4/100$. You will be asked to extend the investigation to other values of the parameters in the exercises. First is a table of the first few values of the Poisson and binomial p.m.f.'s, followed by superimposed connected list plots.

```
Needs["Statistics`DiscreteDistributions`"];
Poissonpmf[x_] :=
  N[PDF[PoissonDistribution[4], x]];
Binomialpmf[x_] :=
  PDF[BinomialDistribution[100, .04], x];
TableForm[Join[{{"x", "Poisson", "binomial"}},
  Table[{x, Poissonpmf[x], Binomialpmf[x]},
    {x, 0, 12}]]]
```

x	Poisson	binomial
0	0.0183156	0.0168703
1	0.0732626	0.070293
2	0.146525	0.144979
3	0.195367	0.197333
4	0.195367	0.199388
5	0.156293	0.159511
6	0.104196	0.105233
7	0.0595404	0.0588803
8	0.0297702	0.0285201
9	0.0132312	0.0121475
10	0.00529248	0.00460591
11	0.00192454	0.0015702
12	0.000641512	0.000485235

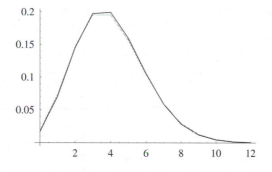

Figure 11 - List plots of b(100,.04) and Poisson(4) p.m.f.'s

The connected line graphs hardly differ at all, and in fact the largest absolute difference between probability masses is only around .004 which takes place at $x = 4$.

Activity 2   Show that the Poisson(μ) probabilities in formula (1) do define a valid probability mass function. Try to obtain a closed form for the c.d.f. of the distribution. Can you do it? (In Exercise 2 you are asked to use *Mathematica* to form a table of the c.d.f. for several values of μ.)

Example 1   We will suppose for this example that the number of ant colonies in a small field has the Poisson distribution with parameter $\mu = 16$. This is reasonable if the field can be broken up into a large number of equally sized pieces, each of which has the same small probability of containing a colony, independent of the other pieces. Then the probability that there are at least 16 colonies is

$$P[X \geq 16] = 1 - P[X \leq 15] = 1 - \Sigma_{k=0}^{15} \frac{e^{-16}\,16^k}{k!}$$

Using *Mathematica* to evaluate this probability we compute

```
N[1 - CDF[PoissonDistribution[16], 15]]
```

```
0.533255
```

Next, suppose that we no longer know μ, but an ecology expert tells us that fields have 8 or fewer ant colonies about 90% of the time. What would be a good estimate of the parameter μ? For this value of μ, $P[X \leq 8] = .90$. We may write $P[X \leq 8]$ as a function of μ as below, and then check several values of μ in search of the one that makes this probability closest to .90. I issued a preliminary Table command to narrow the search from μ values between 2 and 10 to values between 5 and 6. The Table command below shows that to the nearest tenth, $\mu = 5.4$ is the parameter value we seek. ∎

```
eightprob[mu_] :=
 CDF[PoissonDistribution[mu], 8]
```

```
Table[N[{mu, eightprob[mu]}], {mu, 5, 6, .1}]
```

```
{{5., 0.931906}, {5.1, 0.925182},
 {5.2, 0.918065}, {5.3, 0.910554}, {5.4, 0.90265},
 {5.5, 0.894357}, {5.6, 0.885678}, {5.7, 0.876618},
 {5.8, 0.867186}, {5.9, 0.857389}, {6., 0.847237}}
```

The parameter μ of the Poisson distribution gives complete information about its first two moments, as the next theorem shows.

Theorem 1. If $X \sim \text{Poisson}(\mu)$, then $E[X] = \mu$ and $\text{Var}(X) = \mu$.

Proof. First, for the mean,

$$
\begin{aligned}
E[X] &= \Sigma_{k=0}^{\infty} k \cdot \frac{e^{-\mu} \mu^k}{k!} \\
&= e^{-\mu} \cdot \mu \cdot \Sigma_{k=1}^{\infty} \frac{\mu^{k-1}}{(k-1)!} \\
&= e^{-\mu} \cdot \mu \cdot e^{\mu} \\
&= \mu
\end{aligned}
$$

The first line is just the definition of expected value. In the second line we note that the $k = 0$ term is just equal to 0 and can be dropped, after which we cancel k with the k in $k!$ in the bottom, and then remove two common factors $e^{-\mu} \cdot \mu$. Changing variables to say $j = k - 1$ we see that the series in line 2 is just the Taylor series expansion of e^{μ}.

To set up the computation of the variance, we first compute $E[X(X - 1)]$ similarly to the above computation.

$$
\begin{aligned}
E[X(X - 1)] &= \Sigma_{k=0}^{\infty} k \cdot (k - 1) \frac{e^{-\mu} \mu^k}{k!} \\
&= e^{-\mu} \cdot \mu^2 \cdot \Sigma_{k=2}^{\infty} \frac{\mu^{k-2}}{(k-2)!} \\
&= e^{-\mu} \cdot \mu^2 \cdot e^{\mu} \\
&= \mu^2
\end{aligned}
$$

You are asked to complete the computation in the next activity.

Activity 3 By expanding the expression $E[X(X-1)]$ in the proof of Theorem 1 and using the computational formula for the variance, finish the proof.

Since μ is both the mean and the variance of the Poisson(μ) distribution, as μ grows we would expect to see the probability weight shifting to the right and spreading out. In Figure 12 we see a connected list plot of the probability mass function for three Poisson distributions with parameters 2 (bold, leftmost), 5 (middle), and 8 (light, rightmost), and this dependence on μ is evident. We also see increasing symmetry as μ increases. (In the notebook, feel free to open up the closed input cell for the figure and edit it to try some other values of μ.)

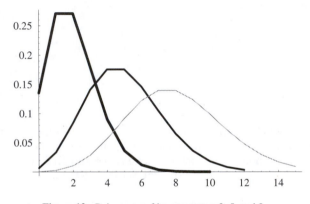

Figure 12 - Poisson p.m.f.'s, parameters 2, 5, and 8

Poisson Processes

Probably the most important instance of the Poisson distribution is in the study of what is called the *Poisson process*. A Poisson process records the cumulative number of occurrences of some phenomenon as time increases. Therefore such a process is not just a single random variable, but a family of random variables (X_t) indexed by a variable t usually thought of as time. We interpret X_t as the number of occurrences of the phenomenon in $[0, t]$.

Since occurrences happen singly and randomly, and X_t counts the total number of them, if for a fixed experimental outcome ω we plot $X_t(\omega)$ as a function of t we get a non-decreasing step function starting at 0, which jumps by exactly 1 at the random times $T_1(\omega)$, $T_2(\omega)$, $T_3(\omega)$, ... of occurrence of the phenomenon. Such a function, called a *sample path* of the process, is shown in Figure 13, for the case where the first three jumps occur at times $T_1 = 1.2$, $T_2 = 2.0$, $T_3 = 3.4$.

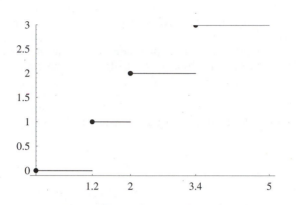

Figure 13 - Sample path of a Poisson process

By our earlier discussion of the domain of application of the Poisson distribution, it is reasonable to assume that as long as the probability of an occurrence in a short interval is proportional to the length of the interval, and the numbers of occurrences on disjoint intervals are independent, X_t should have a Poisson distribution.

What should be the parameter of this Poisson distribution? Let λ be the expected number of occurrences per unit time, i.e., $\lambda = E[X_1]$. In a time interval of length t such as $[0, t]$ we should expect $\lambda \cdot t$ occurrences. Thus, $E[X_t] = \lambda t$ and since the expected value of a Poisson random variable is the same as its parameter, it makes sense that $X_t \sim \text{Poisson}(\lambda t)$. The constant λ is called the *rate of the process*.

All of this can be put on a rigorous footing, but we will not do that here. When we study a continuous distribution called the *exponential distribution* later, we will see a constructive way to define a Poisson process, by supposing that the times between successive occurrences $T_n - T_{n-1}$ are independent and have a common exponential distribution. For now though, we will simply assume that certain situations generate a Poisson process, which means a family (X_t) of random variables for which X_t has the Poisson(λt) distribution and whose sample paths are as in Figure 13.

Activity 4 Many examples of Poisson processes revolve around service situations, where the occurrence times are arrival times of customers. Give some examples of such situations, and discuss what the underlying assumptions must be in order that a Poisson process is a good model for the cumulative number of customers as a function of time.

Example 2 Suppose that cars travelling on an expressway arrive to a toll station according to a Poisson process with rate $\lambda = 5$ per minute. Thus, we are assuming that the numbers of cars arriving in disjoint time intervals are independent, and the probability of an arrival in a short time interval is proportional to the length of the interval. Let X_t be the total number of cars in $[0, t]$. Find: (a) $P[X_2 > 8]$; (b) $P[X_1 < 5 \mid X_5 > 2]$; (c) assuming that λ is no longer known, find the largest possible arrival rate λ such that the probability of 8 or more arrivals in the first minute is no more than .1.

To answer question (a), we note that since $\lambda = 5$ and the time interval is 2 minutes, $X_2 \sim \text{Poisson}(10)$; hence by complementation of the c.d.f., the probability is

```
N[1 - CDF[PoissonDistribution[10], 8]]
```

    0.66718

It is therefore about 67% likely that more than 8 cars will come in two minutes, which is intuitively reasonable given the average arrival rate of 5 per minute.

For question (b), we have that $X_1 \sim \text{Poisson}(5)$ and $X_5 \sim \text{Poisson}(2.5)$. In order for both $\{X_5 > 2\}$ and $\{X_1 < 5\}$ to occur, either $X_5 = 3$ or $X_5 = 4$, in which case the number of arrivals in $[.5, 1]$ is determined. Thus,

$$P[X_1 < 5 \mid X_5 > 2] = \frac{P[X_1 < 5 \cap X_5 > 2]}{P[X_5 > 2]}$$

$$= \frac{P[X_5 = 3, X_1 - X_5 \le 1] + P[X_5 = 4, X_1 - X_5 = 0]}{P[X_5 > 2]}$$

Now the random variable $X_1 - X_5$ in the numerator of the last expression is the number of arrivals in the time interval $(.5, 1]$, which is independent of X_5, the number of arrivals in $[0, .5]$. The intersection probabilities in the numerator therefore factor into the product of the individual event probabilities. But also, we may think of restarting the Poisson process at time .5, after which it still satisfies the Poisson process hypotheses so that $X_1 - X_5$ has the same distribution as X_5, which is Poisson(2.5). By these observations, we are left to compute

$$P[X_1 < 5 \mid X_5 > 2]$$

$$= \frac{P[X_5 = 3] \, P[X_5 \le 1] + P[X_5 = 4] \, P[X_5 = 0]}{P[X_5 > 2]}$$

In *Mathematica*, this is the following:

```
f[x_] := PDF[PoissonDistribution[2.5], x]
F[x_] := CDF[PoissonDistribution[2.5], x]
  f[3] * F[1] + f[4] * f[0]
  ─────────────────────────
           1 - F[2]
```

0.158664

For part (c), we now know only that $X_1 \sim$ Poisson($\lambda \cdot 1$). The probability of 8 or more arrivals is the same as $1 - P[X_1 \leq 7]$, and we would like this probability to be less than or equal to .1. We can define the target probability as a function of λ, and then compute it for several values of λ. A preliminary Table command narrows the search to values of λ in [4, 5]. The Table command below indicates that the desired rate λ is around 4.6. ■

```
prob[lambda_] :=
  1 - CDF[PoissonDistribution[lambda], 7];
Table[{lambda, N[prob[lambda]]},
  {lambda, 4, 5, .1}]
```

```
{{4, 0.0511336}, {4.1, 0.0573121},
 {4.2, 0.0639433}, {4.3, 0.0710317}, {4.4, 0.0785794},
 {4.5, 0.0865865}, {4.6, 0.095051}, {4.7, 0.103969},
 {4.8, 0.113334}, {4.9, 0.123138}, {5., 0.133372}}
```

Mathematica for Section 2.4

Command	Location
PoissonDistribution[μ]	Statistics` DiscreteDistributions`
BinomialDistribution[n, p]	Statistics` DiscreteDistributions`
PDF[dist, x]	Statistics` DiscreteDistributions`
CDF[dist, x]	Statistics` DiscreteDistributions`

Exercises 2.4

1. Assume that the number of starfish X in an ocean region has the Poisson(20) distribution. Find (a) $P[X = 20]$; (b) $P[X > 20]$; (c) $P[\mu - \sigma < X < \mu + \sigma]$, where μ and σ are the mean and standard deviation of the distribution of X.

2. (*Mathematica*) Use *Mathematica* to form a table of cumulative probabilities $F(x) = P[X \le x]$ for each of the four Poisson distributions with parameters $\mu = 1, 2, 3$, and 4. For all distributions, begin at $x = 0$ and end at $x = 10$.

3. (*Mathematica*) Suppose that the sample 3, 2, 4, 3, 4, 5, 6, 2 comes from a Poisson(μ) distribution. What value of μ makes this sample most likely to have occurred? Compare this value of μ to the sample mean $\overline{X}$.

4. Find the third moment about 0 for the Poisson(μ) distribution.

5. If $X \sim$ Poisson(μ), find an expression for the expectation:
$$E[X(X - 1)(X - 2) \cdots (X - k)].$$

6. (*Mathematica*) Write a function of n, p, and k that finds the largest absolute difference between the values of the $b(n,p)$ probability mass function and the Poisson(np) mass function, among states $x = 0, 1, 2, \dots, k$. Use it to study the change in the maximum absolute difference for fixed $p = 1/10$ as values of n increase from 20 to 100 in multiples of 10. Similarly, fix $n = 200$ and study the maximum absolute difference as p takes on the values 1/2, 1/3, 1/4, 1/5, 1/6, 1/7, 1/8, 1/9, 1/10.

7. (*Mathematica*) Animate connected list plots of the Poisson probability mass function using parameters $\mu = 1, 2, 3, \dots, 8$ for values of x in $\{0, 1, 2, \dots, 14\}$. Describe what happens to the mass function as μ increases.

8. If the number of defects in a length of drywall has the Poisson distribution, and it is estimated that the proportion with no defects is .3, estimate the proportion with either 1 or 2 defects.

9. If (X_t) is a Poisson process with rate 2/min., find the probability that X_3 exceeds its mean by at least 2 standard deviations.

10. Suppose that customer arrivals to a small jewelry store during a particular period in the day form a Poisson process with rate 12/hr. (a) Find the probability that there will be more than 12 customers in a particular hour; (b) Given that there are more than 12 customers in the first hour, find the probability that there will be more than 15 customers in that hour.

11. If outside line accesses from within a local phone network form a Poisson process with rate 1/min., find the joint probability that the cumulative number of accesses is equal to 2 at time 1 minute, 3 at time 2 minutes, and 5 at time 3 minutes.

12. Find a general expression for $P[X_s = j \cap X_{t+s} = k]$ for a Poisson process (X_t) of rate λ.

13. (a) If T_1 is the first occurrence time of a Poisson process of rate λ, find $P[T_1 > s]$.
(b) As in part (a), find $P[T_1 > s+t \mid T_1 > s]$.

14. (*Mathematica*) Suppose that emergency 911 calls come in to a station according to a Poisson process with unknown rate λ per hour. Find the value of λ that is most likely to have resulted in a stream of calls at times: .1, .3, .6, 1.0, 1.1, 1.4, 1.8, 2.0, 2.2, 2.3, 2.8, 3.6.
(Hint: consider the numbers of calls in successive intervals of length .5: [0, .5], (.5, 1.0], ...)

15. (*Mathematica*) Assume that cars pull up to a drive-in window according to a Poisson process of rate λ per minute. What is the smallest value of λ such that the probability of 4 or more cars in the first minute is at least .8?

2.5 Joint, Marginal, and Conditional Distributions

In this section we would like to take the ideas of conditional probability and independence that were introduced in Sections 1.5 and 1.6 into the domain of random variables and their distributions. This transition should be a very straightforward one.

Earlier we studied a contingency table for the Peotone airport problem, and a similar example here should motivate the ideas well. Suppose that a researcher in cognitive psychology is conducting an experiment on memory. She exposes to several subjects four numbers between 1 and 100 on flash cards, and after fifteen minutes she asks the subjects to recall the numbers. The experimenter records how many of the numbers the subjects correctly recalled. Then the same subjects are presented four more numbers between 1 and 100 orally. After another fifteen minutes, the experimenter again records how many of them each subject got right. At the end of the experiment, she classifies each subject as to how many of the numbers that were shown visually and how many that were spoken were correctly remembered. Assume that the frequencies are in the table below. I have computed the total number of subjects in each row and column for your convenience.

Y = visual correct	0	1	2	3	4	row sum
0	4	6	10	8	3	31
1	2	5	7	12	6	32
2	1	4	10	15	6	36
3	2	6	9	14	10	41
4	1	3	5	11	11	31
col sum	10	24	41	60	36	171

(X = oral correct labels the rows 0–4.)

Consider the experiment of picking a subject randomly from this group of 171 subjects. Let X and Y, respectively, denote the oral and visual scores of the subject. Then for instance

$$P[X = 0, Y = 0] = \tfrac{4}{171}, \ P[X = 0, Y = 1] = \tfrac{6}{171}, \ P[X = 0, Y = 2] = \tfrac{10}{171},$$

etc. The complete list of the probabilities

$$f(x, y) = P[X = x, Y = y] \tag{1}$$

over all possible values (x, y) is called the *joint probability mass function* of the random variables X and Y. Joint mass functions for more than two discrete random variables are defined similarly (see Exercise 14).

Activity 1.   Referring to the memory experiment, what is $P[X = 0]$? $P[X = 1]$? $P[X = 2]$? $P[X = 3]$? $P[X = 4]$? What do these probabilities add to?

The activity above suggests another kind of distribution that is embedded in the joint distribution described by f. Working now with Y using the column sums we find:

$$P[Y = 0] = \tfrac{10}{171}; \ P[Y = 1] = \tfrac{24}{171}; \ P[Y = 2] = \tfrac{41}{171}; \ P[Y = 3] = \tfrac{60}{171};$$
$$P[Y = 4] = \tfrac{36}{171}$$

It is easy to check that these probabilities sum to 1; hence the function $q(y) = P[Y = y]$ formed in this way is a valid probability mass function. Notice that to compute each of these probabilities, we add up joint probabilities over all x values for the fixed y of interest, for example,

$$P[Y = 0] = P[X = 0, Y = 0] + P[X = 1, Y = 0] + P[X = 2, Y = 0]$$
$$+ P[X = 3, Y = 0] + P[X = 4, Y = 0]$$
$$= \tfrac{4}{171} + \tfrac{2}{171} + \tfrac{1}{171} + \tfrac{2}{171} + \tfrac{1}{171} = \tfrac{10}{171}$$

The Law of Total Probability justifies this procedure. In general, the *marginal probability mass function* of Y is

$$q(y) = P[Y = y] = \sum_x P[X = x, Y = y] = \sum_x f(x, y) \qquad (2)$$

where f is the joint p.m.f. of X and Y. Similarly, the *marginal probability mass function* of X is obtained by adding the joint p.m.f. over all values of y:

$$p(x) = P[X = x] = \sum_y P[X = x, Y = y] = \sum_y f(x, y) \qquad (3)$$

The idea of a *conditional distribution of one discrete random variable given another* is analogous to the discrete conditional probability of one event given another. Using the memory experiment again as our model, what is the conditional probability that a subject gets 2 visual numbers correct given that the subject got 3 oral numbers correct? The condition on oral numbers restricts the sample space to subjects in the line numbered 3 in the table, which has 41 subjects. Among them, 9 got 2 visual numbers correct. Hence,

$$P[Y = 2 \mid X = 3] = \frac{9}{41}$$

It should be easy to see from this example that for jointly distributed discrete random variables, $P[Y = y \mid X = x]$ is a well defined conditional probability of one event given another. This leads us to the definition of the *conditional probability mass function of Y given X = x*:

$$q(y \mid x) = P[Y = y \mid X = x] = \frac{P[X = x, Y = y]}{P[X = x]} = \frac{f(x,y)}{p(x)} \qquad (4)$$

Similarly, the *conditional probability mass function of X given Y = y* is

$$p(x \mid y) = P[X = x \mid Y = y] = \frac{P[X = x, Y = y]}{P[Y = y]} = \frac{f(x,y)}{q(y)} \qquad (5)$$

Formulas (4) and (5) tie together the three kinds of distributions: joint, marginal, and conditional.

The concept of independence also carries over readily to two or more discrete random variables. In the case of two discrete random variables X and Y, we say that they are *independent* of each other if and only if for all subsets A and B of their respective state spaces,

$$P[X \in A, Y \in B] = P[X \in A] \cdot P[Y \in B] \qquad (6)$$

(Exercise 15 gives the analogous definition of independence of more than two random variables.) Alternatively, we could define independence by the condition

$$P[Y \in B \mid X \in A] = P[Y \in B] \qquad (7)$$

for all such sets A and B, with the added proviso that $P[X \in A] \neq 0$. (How does (7) imply the factorization criterion (6), and how is it implied by (6)?)

For example, from the memory experiment table, $P[Y = 4 \mid X = 4] = 11/31$, since there are 31 subjects such that $X = 4$, and 11 of them remembered 4 numbers that were visually presented. However, $P[Y = 4]$ is only 36/171, so that the chance that 4 visual numbers are remembered is increased by the occurrence of the event that 4 auditory numbers were remembered. Even this one violation of condition (7) is enough to show that the random variables X and Y are dependent (that is, not independent). But in general, do you really have to check all possible subsets A of the X state space ($2^5 = 32$ of them here) with all possible subsets B of the Y state space (again 32 of them) in order to verify independence? Fortunately the answer is no, as the following important activity shows.

Activity 2 Show that X and Y are independent if and only if their joint p.m.f. $f(x, y)$ factors into the product of the marginal p.m.f.'s $p(x) \cdot q(y)$.

Thus, in our example instead of checking $32 \cdot 32 = 1024$ possible combinations of sets A and B, we need only check the factorization of f for $5 \cdot 5 = 25$ possible combinations of states x and y. Here is a *Mathematica* interaction that compares joint probabilities to the products of the marginals for our example. The variable *jointpmf* is a list of joint probabilities formatted in row-ordered matrix form similar to our frequency table, in which entry i, j equals $f(i, j)$. The other two variables *xmarginal* and *ymarginal* are lists of probabilities corresponding to the two marginals. The output table is also formatted like our frequency table, with two entries in each cell, one for $f(i, j)$ and one for $p(i) q(j)$. (We define a utility function NPlaces in the closed cell below the next one, to cut the number of digits displayed in the table to 3.)

```
jointpmf = {{4, 6, 10, 8, 3},
    {2, 5, 7, 12, 6}, {1, 4, 10, 15, 6},
    {2, 6, 9, 14, 10}, {1, 3, 5, 11, 11}} / 171;
xmarginal = {31, 32, 36, 41, 31} / 171;
ymarginal = {10, 24, 41, 60, 36} / 171;
```

```
TableForm[
  Table[{{{NPlaces[jointpmf[[i, j]], 3], NPlaces[
      xmarginal[[i]] * ymarginal[[j]], 3]}}},
    {i, 1, 5}, {j, 1, 5}]]
```

0.023	0.035	0.058	0.047	0.018
0.011	0.025	0.043	0.064	0.038
0.012	0.029	0.041	0.07	0.035
0.011	0.026	0.045	0.066	0.039
0.006	0.023	0.058	0.088	0.035
0.012	0.03	0.05	0.074	0.044
0.012	0.035	0.053	0.082	0.058
0.014	0.034	0.057	0.084	0.05
0.006	0.018	0.029	0.064	0.064
0.011	0.025	0.043	0.064	0.038

Observe that many of the factorizations that would be desired for independence do approximately hold. But there are several, such as $X = 0$, $Y = 4$ where one probability differs by a factor of two from the other. Even considering our data to be randomly sampled data from some universe that is subject to chance variability, it does not seem likely that visual and oral memory are independent of each other.

Independence for more than two random variables is explored in the exercises. The basic idea is that several random variables are independent if and only if any subcollection of them satisfies factorization of intersection probabilities as in formula (6). This also follows if and only if the joint p.m.f. factors into the product of the marginals.

We will now look at a series of examples to elaborate on the ideas of joint, marginal, and conditional distributions and independence.

Example 1 Suppose that a mall department store has two entrances, one on the east and the other on the west end of the store. Customers enter from the east according to a Poisson process with rate 2 per minute, and they enter from the west by a Poisson process with rate 1.5 per minute. Furthermore, the two Poisson processes are independent. Compute the joint distribution of the numbers N_e and N_w of customers who have entered from the two directions by time t, and compute the distribution of the total number of customers N who have entered by time t.

The problem description indicates that we should assume that $N_e \sim$ Poisson($2t$), $N_w \sim$ Poisson($1.5t$), and N_e and N_w are independent random variables. By the result of Activity 2, the joint p.m.f. of these two random variables is the product of their marginals:

$$f(x,y) = p(x) \cdot q(y) = \frac{e^{-2t}(2t)^x}{x!} \cdot \frac{e^{-1.5t}(1.5t)^y}{y!}$$

Now let $N = N_e + N_w$ be the total number of customers by time t. Then its distribution is, by the law of total probability,

$$
\begin{aligned}
g(n) = \mathrm{P}[N = n] &= \textstyle\sum_{i=0}^{n} \mathrm{P}[N_e = i, N_w = n - i] \\
&= \textstyle\sum_{i=0}^{n} \frac{(e^{-2t}\,(2\,t))^i}{i!} \cdot \frac{(e^{-1.5t}\,(1.5\,t))^{n-i}}{(n-i)!} \\
&= \frac{e^{-3.5t}}{n!} \textstyle\sum_{i=0}^{n} \frac{n!}{i!\,(n-i)!} (2\,t)^i\,(1.5\,t)^{n-i} \\
&= \frac{e^{-3.5t}}{n!} (2\,t + 1.5\,t)^n \\
&= \frac{e^{-3.5t}\,(3.5\,t)^n}{n!}
\end{aligned}
$$

Note that the fourth line of the derivation follows from the binomial theorem. This shows that N, the total number of arrivals, has the Poisson($3.5t$) distribution. We have an indication, though not a full proof, that the sum of two independent Poisson processes is also a Poisson process, whose rate is the sum of the rates of the two components.

Let us try another problem related to this one: what is the conditional distribution of the total number of customers by time $t + s$, given that there have been n customers by time t?

Write the total number of customers by time t as N_t. We want to compute for all $m \geq n$,

$$
\mathrm{P}[N_{t+s} = m \mid N_t = n] = \frac{\mathrm{P}[N_{t+s} = m, N_t = n]}{\mathrm{P}[N_t = n]}
$$

Now in order for both $N_{t+s} = m$ and $N_t = n$ to occur, there must be exactly n arrivals during $[0, t]$ and $m - n$ arrivals during $(t, t + s]$. Since these time intervals are disjoint, by the properties of the Poisson process the probability of the intersection of the two events must factor. Therefore,

$$
\begin{aligned}
\mathrm{P}[N_{t+s} = m \mid N_t = n] &= \frac{\mathrm{P}[N_{t+s} - N_t = m-n, N_t = n]}{\mathrm{P}[N_t = n]} \\
&= \frac{\mathrm{P}[N_{t+s} - N_t = m-n]\,\mathrm{P}[N_t = n]}{\mathrm{P}[N_t = n]} \\
&= \mathrm{P}[N_{t+s} - N_t = m - n] \\
&= \frac{e^{-3.5s}\,(3.5\,s)^{m-n}}{(m-n)!}, \quad m \geq n \quad \blacksquare
\end{aligned}
$$

Example 2 This example concerns a generalization of the hypergeometric distribution that we studied earlier. Suppose that the faculty at Old VineCovered College is sharply divided along political lines, with 45 of the 100 faculty members who are political liberals, 30 who are centrists, and 25 who are political conservatives. The Dean of the College is appointing a committee to revise the general education curriculum by selecting 8 faculty members at random, in a batch and without replacement, from the 100. Let us find the joint distribution of the numbers of liberals, centrists, and conservatives on the committee, the marginal distributions of each political group, and the probability that the liberals will have a majority.

In order to have, say i liberals, j centrists, and k conservatives on the committee, where $i + j + k$ is required to be 8, we must select i liberals in a batch without replacement from the 45 liberals, and similarly j centrists from the 30 and k conservatives from the 25. By combinatorics, since there are $\binom{100}{8}$ possible equally likely committees,

$$
\begin{aligned}
f(i, j, k) &= \text{P}[i \text{ liberals, } j \text{ centrists, } k \text{ conservatives}] \\
&= \frac{\binom{45}{i}\binom{30}{j}\binom{25}{k}}{\binom{100}{8}}, \quad i, j, k \in \{0, 1, \ldots, 8\}, \ i + j + k = 8
\end{aligned}
$$

Now the marginals may be found by using the definition in (2) of marginal p.m.f.'s and summing out over the other random variable states. Try this in the activity that follows this example. But we can also work by combinatorial reasoning. In order to have exactly i liberals, we must select them from the subgroup of 45, and then select any other $8 - i$ people from the 55 non-liberals to fill out the rest of the committee. Therefore, the marginal p.m.f. of the number of liberals on the committee is

$$
p(i) = \text{P}[i \text{ liberals}] = \frac{\binom{45}{i}\binom{55}{8-i}}{\binom{100}{8}} \tag{8}
$$

Similarly, the other marginals are

$$
q(j) = \text{P}[j \text{ centrists}] = \frac{\binom{30}{j}\binom{70}{8-j}}{\binom{100}{8}},
$$

$$
r(k) = \text{P}[k \text{ conservatives}] = \frac{\binom{25}{k}\binom{75}{8-k}}{\binom{100}{8}}
$$

All of these marginal distributions are of the hypergeometric family.

By the way, it is easy to check that

$$
f(i, j, k) \neq p(i) \cdot q(j) \cdot r(k)
$$

so that the numbers of the three political types in the sample are not independent random variables. Intuitively, knowing the number of liberals, for instance, changes the distribution of the number of centrists (see also Exercise 5).

Now the probability that the liberals will have a majority is the probability that there are at least five liberals on the committee. It is easiest to use the marginal distribution in formula (8) for the number of liberals, and then to let *Mathematica* do the computation:

$$\text{P[liberals have majority]} = \Sigma_{i=5}^{8} \frac{\binom{45}{i}\binom{55}{8-i}}{\binom{100}{8}}$$

```
Needs["Statistics`DiscreteDistributions`"]
```

```
N[1 - CDF[
    HypergeometricDistribution[8, 45, 100], 4]]
```

0.251815

We see that the probability of a liberal majority occurring by chance is only about 1/4. ∎

Activity 3 In the above example, find $p(i)$, the marginal p.m.f. of the number of liberals by setting $k = 8 - i - j$ in the joint p.m.f. and then adding the joint probabilities over the possible j values.

Example 3 If two random variables X and Y are independent and we observe instances of them $(X_1, Y_1), (X_2, Y_2), \ldots, (X_n, Y_n)$, what patterns would we expect to find? If they are dependent, how do those patterns change?

We will answer these questions by simulating the list of n pairs. For concreteness, let X have the discrete p.m.f. $p(1) = 1/4, p(2) = 1/4, p(3) = 1/2$ and let Y have the discrete p.m.f. $q(1) = 5/16, q(2) = 5/16, q(3) = 3/8$. The program below simulates two numbers at a time from *Mathematica*'s random number generator and converts them to observations of X and Y, respectively. If the random number generator works as advertised, the values of X and Y that are simulated should be (or appear to be) independent. We will then tabulate the number of times the pairs fell into each of the 9 categories $(1, 1), (1, 2), \ldots, (3, 3)$ and study the pattern.

The first two commands simulate individual observations from the distributions described in the last paragraph, the third uses them to simulate a list of pairs, and the last tabulates frequencies of the nine states and presents them in a table.

```
SimX[] := Module[{rand},
               rand = Random[];
               If[rand < 1/4, 1,
                      If[
   1/4 ≤ rand < 1/2, 2, 3]]]
```

```
SimY[] := Module[{rand},
               rand = Random[];
               If[rand < 5/16, 1,
                      If[
   5/16 ≤ rand < 10/16, 2, 3]]]
```

```
SimXYPairs[numpairs_] :=
 Table[{SimX[], SimY[]}, {numpairs}]
```

```
XYPairFrequencies[numpairs_] :=
   Module[
 {simlist, freqtable, nextpair, x, y},
      simlist = SimXYPairs[numpairs];
 (* generate data *)
        freqtable = Table[0, {i, 1, 3},
   {j, 1, 3}]; (* initialize table *)
   Do[nextpair = simlist[[i]];
 (* set up next x,y pair *)
            x = nextpair[[1]];
 y = nextpair[[2]];
             freqtable[[x, y]] =
   freqtable[[x, y]] + 1,
 (* increment the frequency table *)
          {i, 1, Length[simlist]}];
     (* output the table *)
     TableForm[
 {{" ", "Y=1", "Y=2", "Y=3"},
                     {"X=1",
   freqtable[[1, 1]], freqtable[[1, 2]],
   freqtable[[1, 3]]},
      {"X=2", freqtable[[2, 1]],
   freqtable[[2, 2]], freqtable[[2, 3]]},
             {"X=3", freqtable[[3, 1]],
   freqtable[[3, 2]], freqtable[[3, 3]]}}]]
```

Below is one run among many that I did. There is quite a bit of variability in category frequencies, even when the sample size is 1000, but a few patterns do emerge. In each X row, the $Y = 1$ and $Y = 2$ frequencies are about equal, and the $Y = 3$ frequency is a bit higher. Similarly, in each Y column, the $X = 1$ frequency is close to that of $X = 2$, and their frequencies are only around half that of $X = 3$. If you look back at the probabilities assigned to each state, you can see why you should have expected this. In general, if independence holds, the distribution of data into columns should not be affected by which row you are in, and the distribution of data into rows should not be affected by which column you are in.

```
SeedRandom[1654];
XYPairFrequencies[1000]
```

	Y=1	Y=2	Y=3
X=1	76	64	85
X=2	77	78	97
X=3	143	160	220

How do we simulate dependence? If X and Y are dependent, then once X is simulated to have a value of x, Y no longer has the marginal p.m.f. $q(y)$ as its distribution; rather, it has the conditional p.m.f. $q(y \mid x) = f(x, y)/p(x)$. Thus, our simulation strategy will be to simulate X first using its marginal p.m.f. $p(x)$, then simulate y using the conditional mass function for the particular simulated x.

As above, let each of X and Y take values in $\{1, 2, 3\}$, and suppose that the joint p.m.f. $f(x, y)$ is as tabulated below.

		Y			
		1	2	3	$p(x)$
	1	1/8	1/16	1/16	1/4
X	2	1/16	1/8	1/16	1/4
	3	1/8	1/8	1/4	1/2
	$q(y)$	5/16	5/16	3/8	

Notice that the joint p.m.f. has been set up so that the marginals of both X and Y are exactly the same as before. Then the three conditional p.m.f.'s of Y given X are (check them):

$$q(y|1) = \begin{cases} 1/2 & \text{if } y = 1 \\ 1/4 & \text{if } y = 2 \\ 1/4 & \text{if } y = 3 \end{cases} \qquad q(y|2) = \begin{cases} 1/4 & \text{if } y = 1 \\ 1/2 & \text{if } y = 2 \\ 1/4 & \text{if } y = 3 \end{cases}$$

$$q(y|3) = \begin{cases} 1/4 & \text{if } y = 1 \\ 1/4 & \text{if } y = 2 \\ 1/2 & \text{if } y = 3 \end{cases}$$

The earlier simulator of an X value can be used again. Below is a simulator of Y conditioned on X, followed by a revised version of the simulator of pairs. The old XYPairFrequencies command above can then be reused to produce the table.

```
SimYGivenX[x_] :=
    Module[{rand},
        rand = Random[];
        If[x == 1,
          If[rand < 1 / 2,
      1, If[1 / 2 ≤ rand < 3 / 4, 2, 3]],
       If[x == 2, If[rand < 1 / 4, 1,
       If[1 / 4 ≤ rand < 3 / 4, 2, 3]],
                      If[rand < 1 / 4, 1,
          If[1 / 4 ≤ rand < 1 / 2, 2, 3]]]]]
```

```
SimXYPairs[numpairs_] :=
    Module[{x, y, pairlist},
          pairlist = {};
          Do[x = SimX[]; y = SimYGivenX[x];
                AppendTo[pairlist, {x, y}],
    {i, 1, numpairs}];
        pairlist]
```

```
SeedRandom[20863];
XYPairFrequencies[1000]
```

	Y=1	Y=2	Y=3
X=1	120	65	51
X=2	60	128	58
X=3	139	140	239

By contrast to the independent case, the rows look markedly different from one another in terms of relative frequencies of Y values; when $X = i$, i is also the most frequent Y value. Of course the frequencies in simulated tables like this one follow the patterns of the joint mass function, but had you not seen that in advance, you would detect dependence by looking for cases where the distribution of data into columns changes depends on the row you look in, or the distribution of data into rows depends on the column (the latter also can be seen here). ∎

Activity 4 Write similar programs to simulate (X, Y) pairs in the dependent case in which the joint distribution of the two random variables is $f(1,1) = 1/6$, $f(1,2) = 2/6$, $f(2,1) = 2/6$, $f(2,2) = 1/6$. After imitating my conditional approach, try a different approach to simulate the pairs directly using the joint p.m.f. Run your programs a few times and observe the structure of the frequency table.

Example 4 For our last example we will look at a generalization of the binomial distribution called the *multinomial distribution*. Consider n repeated, independent trials in which there are several outcomes 1, 2, ... , k that are possible on each trial instead of just two. This would describe, for example, a random sample drawn in sequence and with replacement taken from a population which splits into k categories. For $i = 1, 2, ... , k$ let p_i be the probability that a single trial results in category i. We are interested in the joint distribution of the numbers X_i of category i items among the n, for categories $i = 1, 2, ... , k$.

For instance, we could be studying the on-time performance of a bus service at a particular stop, and each day the outcomes might be: early, on time, and late, say with probabilities .12, .56, and .32, respectively. What is the probability that in an eight day period the bus is early once, on time 4 times, and late 3 times? Denoting the three daily outcomes as E, O, and L, a couple of configurations that are consistent with the requirements are:

$$\text{E O O O O L L L}$$
$$\text{O E O L O L L O}$$

The probability of each such configuration is clearly $(.12)^1(.56)^4(.32)^3$, by the independence of the trials. There are $\binom{8}{1}\binom{7}{4}\binom{3}{3}$ such configurations, since a configuration is determined uniquely by a choice of 1 position from the 8 for an E, 4 positions among the remaining 7 for the O's, and 3 positions from the last 3 for the L's. We have seen a quantity like this before in Section 1.4 when we were counting the number of partitions of a set. It can be rewritten as

$$\frac{8!}{1!\,7!} \cdot \frac{7!}{4!\,3!} \cdot \frac{3!}{3!\,0!} = \frac{8!}{1!\,4!\,3!} = \binom{8}{1\ 4\ 3}$$

and it is called the *multinomial coefficient* for 8 choose 1, 4, and 3. Therefore the overall probability of 1 early arrival, 4 on time arrivals, and 3 late arrivals is

$$P[X_1 = 1, X_2 = 4, X_3 = 3] = \binom{8}{1\ 4\ 3}(.12)^1(.56)^4(.32)^3 = .108$$

In general, the *multinomial distribution* is the joint distribution of the numbers X_1, X_2, ..., X_k of items among n trials that belong to the categories 1, 2, ..., k, respectively. If the category probabilities are p_1, p_2, ... , p_k, where $\sum_i p_i = 1$, then this joint p.m.f. is

$$f(x_1, x_2, ..., x_k) = P[X_1 = x_1, X_2 = x_2, ..., X_k = x_k]$$

$$= \binom{n}{x_1 \ x_2 \ ... x_k} p_1^{x_1} p_2^{x_2} \cdots p_k^{x_k}, \ \sum_i x_i = 1 \qquad (9)$$

where the multinomial coefficient is

$$\binom{n}{x_1 \ x_2 \ ... x_k} = \frac{n!}{x_1! \ x_2! \cdots x_k!}$$

Mathematica knows about the multinomial coefficient, whose syntax is

Multinomial[x1, x2, ..., xk]

Here we compute the probability we found above in the bus example:

```
Multinomial[1, 4, 3] * .12 * (.56)^4 * (.32)^3
```

0.108278

The marginal distributions of random variables that have the joint multinomial distribution are easy to find. But as in the case of the multidimensional hypergeometric distribution, the easiest approach is not to sum the joint p.m.f. over the other variables. If we go back to first principles we can find the marginals easily. Consider X_1 for example. Each of the n trials can either result in a category 1 outcome, or some other. Thus, X_1 counts the number of successes in a binomial experiment with success probability p_1. This reasoning shows that if X_1, ... , X_k have the multinomial distribution with n trials and category probabilities p_1, ..., p_k, then each X_i has the binomial distribution with parameters n and p_i. ∎

Activity 5 If X_1, X_2, X_3, X_4, and X_5 have the multinomial distribution with $n = 20$ trials and category probabilities .1, .2, .3, .2, and .2, respectively, find the joint p.m.f. of X_1, X_2, and X_3.

Mathematica for Section 2.5

Command	Location
CDF[dist, x]	Statistics` DiscreteDistributions`
HypergeometricDistribution[n, M, N]	Statistics` DiscreteDistributions`
Random[]	kernel
SeedRandom[seed]	kernel
Multinomial[x1, x2, ..., xk]	kernel
SimX[], SimY[]	Section 2.5
SimXYPairs[numpairs]	Section 2.5
XYPairFrequencies[numpairs]	Section 2.5
SimYGivenX[x]	Section 2.5
NPlaces[number, places]	Section 2.5

Exercises 2.5

1. In the memory experiment example at the beginning of the section, find the conditional p.m.f. of Y given $X = 0$, and the conditional p.m.f. of X given $Y = 2$.

2. Suppose that X and Y have the joint distribution in the table below. Find the marginal distributions of X and Y. Are X and Y independent random variables?

		\(Y\)			
		1	2	3	4
	1	1/16	1/16	1/16	1/16
X	2	1/32	1/32	1/16	1/16
	3	1/16	1/16	1/32	1/32
	4	1/8	1/8	1/16	1/16

3. Argue that two discrete random variables X and Y cannot be independent unless their joint state space is a Cartesian product of their marginal state spaces, i.e., $E = E_x \times E_y = \{(x, y): x \in E_x, \ y \in E_y\}$.

4. Suppose that a joint p.m.f. $f(x, y)$ puts equal weight on all the integer grid points marked in the diagram below. Find: (a) the marginal distribution of X; (b) the marginal distribution of Y; (c) the conditional distribution of X given $Y = 1$; (d) the conditional distribution of Y given $X = 1$.

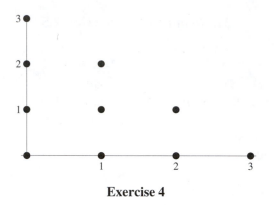

Exercise 4

5. In Example 2, find the conditional distribution of the number of centrists on the committee given that the number of liberals is 3.

6. Show that if two discrete random variables X and Y are independent, then their *joint cumulative distribution function*, defined by $F(x, y) = P[X \le x, Y \le y]$, factors into the product of the marginal c.d.f.'s $F_x(x) = P[X \le x]$ and $F_y(y) = P[Y \le y]$.

7. (*Mathematica*) Use *Mathematica* to compare joint probabilities to the product of marginal probabilities if one subject is drawn at random from the following contingency table. In it, subjects were classified according to their age group and their opinion about what should be done with a government budget surplus. Considering the data in the table to be a random sample of the American voters, do the age group and opinion variables seem independent?

	Save Social Security	Reduce National Debt	Lower Taxes	Increase Defense Spending
21 – 35	22	10	63	15
36 – 50	46	20	85	60
50 – 65	89	54	70	41
over 65	106	32	15	20

8. (*Mathematica*) Write a *Mathematica* program to simulate observations (X, Y) having the joint density in Exercise 4.

9. Show that if two discrete random variables X and Y are independent under the intersection form of the definition, then for all x such that $p(x) \ne 0$, $q(y \mid x) = q(y)$, and for all y such that $q(y) \ne 0$, $p(x \mid y) = p(x)$, where $p(x)$ and $q(y)$ are the marginal p.m.f.'s of X and Y, and $p(x \mid y)$ and $q(y \mid x)$ are the conditional p.m.f.'s.

10. Suppose that in baseball a pitcher throws a fastball (coded by 0), a curve (1), or a slider (2) with probabilities 1/2, 1/4, and 1/4, respectively. Meanwhile the hitter guesses fastball (coded by 0 again), curve (1), or slider (2) with probabilities 5/8,

1/8, and 1/4, respectively. The hitter's guess is independent of the pitcher's choice. The probabilities that the hitter will hit safely under each possible pair of pitcher pitches and hitter guesses are in the table below. Find the probability that the hitter hits safely.

		Y (hitter)		
		0	1	2
X (pitcher)	0	.4	.05	.2
	1	.1	.3	.25
	2	.25	.2	.4

11. A machine can either be in a working state ($X = 1$) or malfunctioning ($X = 0$). A diagnostic test either reports that the machine is working ($Y = 1$), or that it is malfunctioning ($Y = 0$). Assume that the marginal distribution of X is $p(1) = .9$, $p(0) = .1$, assume that the conditional distribution of Y given $X = 0$ is $q(0 \mid 0) = .95$, $q(1 \mid 0) = .05$, and assume that the conditional distribution of Y given $X = 1$ is $q(0 \mid 1) = .10$, $q(1 \mid 1) = .90$. Find the conditional distribution of X given $Y = 1$, and the conditional distribution of X given $Y = 0$.

12. Show that the conditional p.m.f. $q(y \mid x)$ is always a valid p.m.f. for each fixed x such that $p(x) \neq 0$.

13. (*Mathematica*) Write a *Mathematica* function which computes values of the multinomial probability mass function discussed in Example 4 (see formula (9)). The input parameters should be a list of number of desired occurrences of each of the states $\{x_1, x_2, ..., x_k\}$, the number of trials n, and a list of multinomial probabilities $\{p_1, p_2, ..., p_k\}$. Use it to find the probability that in 20 rolls of a fair die, you roll 4 sixes, 7 fives, 6 fours, and 1 three. (*Mathematica* note: The function Multinomial will not apply directly to a sequence enclosed in list braces such as the x's above, but if you compose Multinomial with the *Mathematica* function Sequence using the syntax: Multinomial[Sequence@@xlist], then the list braces will be stripped off before feeding the list into Multinomial. Or, just use the Factorial function to make your own version of the multinomial coefficient.)

14. The *joint probability mass function* of many discrete random variables X_1, X_2, ... , X_k is the function

$$f(x_1, x_2, ..., x_k) = P[X_1 = x_1, X_2 = x_2, ..., X_k = x_k]$$

Joint marginal distributions of subgroups of the X's can be found by summing out over all values of the states x for indices not in the subgroup. If a joint mass function $f(x_1, x_2, x_3)$ puts equal probability on all corners of the unit cube $[0,1] \times [0,1] \times [0,1]$, find the joint marginals of X_1 and X_2, X_2 and X_3, and X_1 and X_3. Find the one variable marginals of X_1, X_2, and X_3.

15. Discrete random variables X_1, X_2, ... , X_k are *mutually independent* if for any subsets B_1, B_2, ... , B_k of their respective state spaces

$$P[X_1 \in B_1, X_2 \in B_2, \dots , X_k \in B_k] = P[X_1 \in B_1] \cdot P[X_2 \in B_2] \cdots P[X_k \in B_k]$$

(a) Argue that if the entire group of random variables is mutually independent, then so is any subgroup.

(b) Show that if X_1, X_2, ... , X_k are mutually independent, then their joint p.m.f. (see Exercise 14) factors into the product of their marginal p.m.f.'s.

(c) Show that if X_1, X_2, ... , X_k are mutually independent, then

$$P[X_1 \in B_1, \ X_2 \in B_2 \ | X_3 \in B_3 \dots , \ X_k \in B_k] = P[X_1 \in B_1] \cdot P[X_2 \in B_2]$$

2.6 More on Expectation

Now that we know about joint, marginal, and conditional distributions, there are some essential matters that we need to finish involving expected value of discrete random variables. Of particular importance in this section are the results on expected value and variance of a linear combination of random variables, and the definitions of the summary measures of association between random variables called the *covariance* and *correlation*. These ideas have a great deal of impact on sampling, and its use in making inferences about the population from which the sample is taken, because they reveal important properties of the mean of a random sample.

Recall from the last section the joint probability mass function of two random variables X and Y, defined by

$$f(x, y) = P[X = x, Y = y]$$

For several random variables X_1, X_2, ... , X_n the joint p.m.f. is defined analogously:

$$f(x_1, x_2, ..., x_n) = P[X_1 = x_1, X_2 = x_2, ..., X_n = x_n] \qquad (1)$$

In the single variable case we defined the expectation of a function g of a random variable X as the weighted average of the possible states $g(x)$, weighted by their probabilities $f(x)$. The analogous definition for many random variables follows.

Definition 1. The *expected value* of a function $g(X_1, X_2, ..., X_n)$ of random variables whose joint p.m.f. is as in (1) is:

$$E[g(X_1, X_2, ..., X_n)] = \sum_{x_1} \cdots \sum_{x_n} g(x_1, x_2, ..., x_n) \cdot f(x_1, x_2, ..., x_n) \qquad (2)$$

where the multiple sum is taken over all possible states $(x_1, x_2, ..., x_n)$.

Example 1 Let X be the total number of heads in two flips of a fair coin and let Y be the total number of heads in two further flips, which we assume are independent of the first two. Then $X + Y$ is the total number of heads among all flips. Each of X and Y have the binomial distribution with parameters $n = 2$ and $p = 1/2$, and by the independence assumption,

$$
\begin{aligned}
E[X + Y] &= \sum_{x=0}^{2} \sum_{y=0}^{2} (x + y) f(x, y) \\
&= \sum_{x=0}^{2} \sum_{y=0}^{2} (x + y) \binom{2}{x} .5^2 \binom{2}{y} .5^2 \\
&= .5^4 (0 \cdot 1 \cdot 1 + 1 \cdot 1 \cdot 2 + 2 \cdot 1 \cdot 1 + 1 \cdot 2 \cdot 1 + 2 \cdot 2 \cdot 2 \\
&\qquad\quad + 3 \cdot 2 \cdot 1 + 2 \cdot 1 \cdot 1 + 3 \cdot 1 \cdot 2 + 4 \cdot 1 \cdot 1) \\
&= .5^4 \cdot 32 = 2
\end{aligned}
$$

Notice that the result is just $1 + 1$, that is, the expected value of X plus the expected value of Y. This is not a coincidence, as you will see in the next theorem. ∎

Theorem 1. If X and Y are discrete random variables with finite means, and c and d are constants, then

$$E[c \cdot X + d \cdot Y] = c \cdot E[X] + d \cdot E[Y] \qquad (3)$$

Proof. Let $f(x, y)$ be the joint p.m.f. of the two random variables and let $p(x)$ and $q(y)$ be the marginals of X and Y. By formula (2) the expectation on the left is

$$
\begin{aligned}
\mathrm{E}[c \cdot X + d \cdot Y] &= \textstyle\sum_x \sum_y (c \cdot x + d \cdot y) f(x, y) \\
&= \textstyle\sum_x c \cdot x \sum_y f(x, y) + \sum_y d \cdot y \sum_x f(x, y) \\
&= c \textstyle\sum_x x \cdot p(x) + d \cdot \sum_y y \cdot q(y) \\
&= c \cdot \mathrm{E}[X] + d \cdot \mathrm{E}[Y]
\end{aligned}
$$

It is easy to extend the property in formula (3), called *linearity of expectation*, to many random variables. (See Exercise 5.) Think about the following question, to which we will return shortly after a theorem about the variance of a linear combination of random variables.

Activity 1 What does Theorem 1 imply about the simple arithmetical average of two random variables X and Y?

Theorem 2. If X and Y are independent discrete random variables with finite variances, and c and d are constants, then

$$
\mathrm{Var}(c \cdot X + d \cdot Y) = c^2 \cdot \mathrm{Var}(X) + d^2 \cdot \mathrm{Var}(Y) \qquad (4)
$$

Proof. By Theorem 1, the mean of $c \cdot X + d \cdot Y$ is $c \cdot \mu_x + d \cdot \mu_y$, where the μ's are the individual means of X and Y. By the definition of variance,

$$
\mathrm{Var}(c \cdot X + d \cdot Y) = \mathrm{E}[((c \cdot X + d \cdot Y) - (c \cdot \mu_x + d \cdot \mu_y))^2]
$$

Group the terms involving X together, and group those involving Y to get

$$
\mathrm{Var}(c \cdot X + d \cdot Y) = \mathrm{E}[(c(X - \mu_x) + d(Y - \mu_y))^2]
$$

Expansion of the square yields

$$
\mathrm{Var}(c \cdot X + d \cdot Y) = \\
c^2 \cdot \mathrm{E}[(X - \mu_x)^2] + d^2 \cdot \mathrm{E}[(Y - \mu_y)^2] + 2 \cdot c \cdot d \cdot \mathrm{E}[(X - \mu_x)(Y - \mu_y)] \qquad (5)
$$

But the last term in the sum on the right equals zero (see Activity 2), which proves (4).

Activity 2 Use the independence assumption in the last theorem to show that $\mathrm{E}[(X - \mu_x)(Y - \mu_y)] = 0$. Generalize: show that if X and Y are independent, then $\mathrm{E}[g(X) h(Y)] = \mathrm{E}[g(X)] \mathrm{E}[h(Y)]$.

Theorem 2 also extends easily to the case of linear combinations of many random variables.

Example 2 Let X_1, X_2, ..., X_n be a random sample drawn in sequence and with replacement from some population with mean μ and variance σ^2. The *mean of the sample* is the simple arithmetical average of the sample values

$$\overline{X} = \frac{X_1 + X_2 + ... + X_n}{n} \tag{6}$$

If the sample is in a *Mathematica* list and the DiscreteDistributions package is loaded, then *Mathematica* can calculate the sample mean with the command Mean[datalist] as follows.

```
Needs["Statistics`DiscreteDistributions`"]
```

```
sample = {3, 2, 4, 6, 5};
Mean[sample]
```

4

In the study of statistics, one of the most fundamental ideas is that because a random sample is subject to chance influences, so is a statistic based on the sample such as $\overline{X}$. So, $\overline{X}$ is a random variable, and you can ask what its mean and variance are. The previous theorems allow us to answer the question. By linearity of expectation,

$$\mathrm{E}[\overline{X}] = \mathrm{E}[\tfrac{X_1+X_2+...+X_n}{n}] = \tfrac{1}{n}(\mathrm{E}[X_1] + \mathrm{E}[X_2] + \cdots + \mathrm{E}[X_n]) = \tfrac{1}{n}n\mu = \mu \tag{7}$$

So $\overline{X}$ has the same mean value as each sample value X_i. It is in this sense an accurate estimator of μ. Also, by Theorem 2,

$$\mathrm{Var}(\overline{X}) = \mathrm{Var}(\tfrac{X_1+X_2+...+X_n}{n}) = \tfrac{1}{n^2}(\mathrm{Var}(X_1) + \cdots + \mathrm{Var}(X_n))$$
$$= \tfrac{1}{n^2}n\sigma^2 = \tfrac{\sigma^2}{n} \tag{8}$$

As the sample size n grows, the variance of $\overline{X}$ becomes small, and in this sense it becomes a more precise estimator of μ as n increases.

Let us use *Mathematica* to see these properties in action. First, we will build a simulator of a list of sample means from a given discrete distribution. The arguments of the next command are the number of sample means we want to simulate, the distribution from which we are simulating, and the size of each random sample. The RandomArray function is used to create a sample of the given size, then Mean is applied to the sample to produce a sample mean.

```
SimulateSampleMeans[nummeans_,
    distribution_, sampsize_] :=
      Table[Mean[RandomArray[
      distribution, sampsize]], {nummeans}]
```

Now we will simulate and plot a histogram of 100 sample means of random samples of size 20 from the geometric distribution with parameter 1/2.

```
Needs["KnoxProb`Utilities`"];
SeedRandom[4532];
```

```
sample = SimulateSampleMeans[
    100, GeometricDistribution[.5], 20];
Histogram[sample, 10];
```

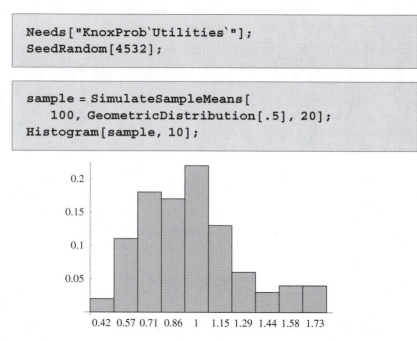

Figure 14 - Sample histogram of 100 means of samples of size 20 from geometric(.5)

This histogram is the observed distribution of the sample mean $\overline{X}$ in our simulation. Recall that the mean of the underlying geometric distribution is $\frac{1/2}{1-1/2} =$ 1, and we do see that the center of the frequency histogram is roughly at $x = 1$.

To see the effect of increasing the sample size of each random sample, let's try quadrupling the sample size from 20 to 80.

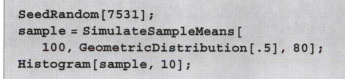

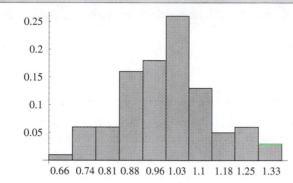

0.66 0.74 0.81 0.88 0.96 1.03 1.1 1.18 1.25 1.33

Figure 15 - Sample histogram of 100 means of samples of size 80 from geometric(.5)

Again the center point is around 1, but whereas the range of observed values in the case that the sample size was 20 extended from around .3 to 2, now it only extends from around .6 to 1.4, about half as wide. The theory explains this phenomenon. According to Theorem 2, since the variance of the geometric(.5) distribution is $\frac{1-1/2}{1/2^2} = 2$, the standard deviation of $\overline{X}$ in the first experiment is $\sigma/\sqrt{n} = \sqrt{2}/\sqrt{20}$, and in the second experiment it is $\sqrt{2}\,/\,\sqrt{80} = \sqrt{2}\,/(2 \cdot \sqrt{20}\,)$, which is exactly one half of the standard deviation in the first experiment. ∎

Activity 3 Try simulating 100 sample means of samples of size 10 from the Poisson(4) distribution. Predict the histogram you will see before actually plotting it. Then try samples of size 90. What happens to the standard deviation of the sample mean?

The assumption of independence in Theorem 2 on the variance of a linear combination is a rather severe one. (Note that in Theorem 1 on the mean, we did not need to assume that the random variables in the combination were independent.) As we know, when sampling occurs without replacement, successive sampled values are not independent. And yet, it would be interesting to know about the variability of the sample mean. Can anything be said in the dependent case about the variance of the combination?

First we need a detour to introduce the concepts of covariance and correlation.

Definition 2. If X and Y are discrete random variables with means μ_x and μ_y, then the *covariance* between X and Y is

$$\sigma_{xy} = \text{Cov}(X, Y) = \text{E}[(X - \mu_x)(Y - \mu_y)] \qquad (9)$$

Furthermore, if X and Y have finite standard deviations σ_x and σ_y, then the *correlation* between X and Y is

$$\rho_{xy} = \text{Corr}(X, Y) = \frac{\text{Cov}(X, Y)}{\sigma_x \sigma_y} \qquad (10)$$

Example 3 To get an idea of how these two expectations measure dependence between X and Y, consider the two simple discrete distributions in Figure 16.

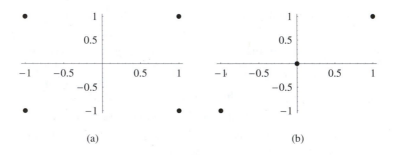

(a)   (b)

Figure 16 - Two discrete distributions

In part (a) of the figure, we put equal probability weight of 1/4 on each of the corners $(-1,-1),(-1,1)$, $(1,1)$, and $(1,-1)$. It is easy to check that both μ_x and μ_y are zero. The covariance for the distribution in (a) is therefore

$$\begin{aligned}
\sigma_{xy} &= \text{E}[X\,Y] \\
&= \tfrac{1}{4}(-1)(-1) + \tfrac{1}{4}(-1)(1) + \tfrac{1}{4}(1)(1) + \tfrac{1}{4}(1)(-1) = 0
\end{aligned}$$

and so the correlation is 0 as well. If you look at the defining formula for covariance, you see that it will be large and positive when Y tends to exceed its mean at the same time X exceeds its mean. But for this joint distribution, when X exceeds its mean (i.e., when it has the value 1), Y still will either equal 1 or -1 with equal probability, which is responsible for the lack of correlation. However in Figure 16(b), suppose we put equal probability of 1/3 on each of the states $(-1,-1)$, $(0,0)$, and $(1,1)$. Again, it is easily checked that both μ_x and μ_y are zero. This time the covariance is

$$\sigma_{xy} = \frac{1}{3}(-1)(-1) + \frac{1}{3}(0)(0) + \frac{1}{3}(1)(1) = \frac{2}{3}$$

The random variables X and Y actually have the same marginal distribution (check this) which puts a probability weight of 1/3 on each of -1, 0, and 1, and so their common variance is also easy to compute as $\sigma^2 = 2/3$. Thus, the correlation between X and Y is

$$\rho = \frac{\sigma_{xy}}{\sigma_x \sigma_y} = \frac{2/3}{\sqrt{2/3}\sqrt{2/3}} = 1$$

For the distribution in Figure 16(b), X and Y are perfectly (positively) correlated. (Notice that when X is bigger than its mean of 0, namely when $X = 1$, Y is certain to be greater than its mean; in fact it can only have the value 1.) It turns out that this is the most extreme case. In Exercise 12 you are led through a proof of the important fact that:

> If ρ is the correlation between random variables X and Y, then $|\rho| \leq 1$ (11)

and in addition it is true that ρ equals positive 1 or negative 1 if and only if, with certainty, Y is a linear function of X. ■

Remember also the result of Activity 2. If X and Y happen to be independent, then $E[(X - \mu_x)(Y - \mu_y)] = E[X - \mu_x]E[Y - \mu_y] = 0$. This case represents the other end of the spectrum. Statistical independence implies that the covariance (and correlation) are zero. Perfect linear dependence implies that the correlation is 1 (or -1 if Y is a decreasing linear function of X).

Another set of results in the exercises (Exercise 11) about covariance and correlation is worth noting. If two random variables are both measured in a different system of units, that is if X is transformed to $aX + b$ and Y is transformed to $cY + d$, where a and c are positive, then

$$\text{ɔv}(aX + b, cY + d) = a \cdot c \cdot \text{Cov}(X, Y), \quad \text{Corr}(aX + b, cY + d) = \text{Corr}(X, \quad (12)$$

In particular, the covariance changes under the change of units, but the correlation does not. This makes the correlation a more standardized measure of dependence than the covariance.

Finally, expansion of the expectation that defines the covariance yields the useful simplifying formula:

$$\begin{aligned}
\text{Cov}(X, Y) &= E[(X - \mu_x)(Y - \mu_y)] \\
&= E[X \cdot Y] - \mu_x E[Y] - \mu_y E[X] + \mu_x \mu_y \\
&= E[X \cdot Y] - \mu_x \mu_y
\end{aligned} \qquad (13)$$

Example 4   Recall the psychology experiment on memory from the beginning of Section 2.5. If once again we let X be the number of orally presented numbers that were remembered, and Y be the number of visually presented numbers that were remembered, then we can compute the covariance and correlation of X and Y. The arithmetic is long, so we will make good use of *Mathematica*. First we define the marginal distributions of the two random variables, which are the marginal totals divided by the overall sample size. We also introduce the list of states for the two random variables

```
xmarginal = {31, 32, 36, 41, 31} / 171;
ymarginal = {10, 24, 41, 60, 36} / 171;
xstates = {0, 1, 2, 3, 4};
ystates = {0, 1, 2, 3, 4};
```

The marginal means are the dot products of the marginal distributions and the state lists:

```
{mux, muy} =
  {xmarginal.xstates, ymarginal.ystates}
```

$$\left\{ \frac{39}{19}, \frac{430}{171} \right\}$$

The variances can be found by taking the dot product of the marginal mass lists with the square of the state lists minus the means:

```
{sigsqx, sigsqy} =
  {xmarginal.((xstates - mux) * (xstates - mux)),
     ymarginal.((ystates - muy) * (ystates - muy))}
```

$$\left\{ \frac{2030}{1083}, \frac{38084}{29241} \right\}$$

By computational formula (13), it now suffices to subtract the product of marginal means from the expected product. For the latter, we need the full dot product of the joint probabilities with the product of X and Y states, which is below.

```
jointmassfn =
  {4, 6, 10, 8, 3, 2, 5, 7, 12, 6, 1, 4, 10, 15,
    6, 2, 6, 9, 14, 10, 1, 3, 5, 11, 11} / 171;
stateprods = {0, 0, 0, 0, 0, 0, 1, 2, 3, 4, 0,
    2, 4, 6, 8, 0, 3, 6, 9, 12, 0, 4, 8, 12, 16};
expectedprod = jointmassfn.stateprods
```

$$\frac{943}{171}$$

Hence the covariance and correlation are

```
covar = N[expectedprod - mux * muy]
```

0.353032

```
corr = N[ ─────────covar─────────── ]
          √sigsqx * √sigsqy
```

0.225946

These two random variables are therefore rather weakly correlated. ∎

Example 5 If (X_t) is a Poisson process with rate λ, find $\text{Cov}(X_t, X_{t+s})$.

We know that since the value of the process at time t has the Poisson distribution with parameter λt, $E[X_t] = \lambda t$ and $E[X_{t+s}] = \lambda(t+s)$. By formula (13), it remains to compute $E[X_t \cdot X_{t+s}]$. But also we note that $X_{t+s} - X_t$ is independent of X_t and has the Poisson(λs) distribution; hence

$$
\begin{aligned}
E[X_t \cdot X_{t+s}] &= E[X_t \cdot (X_{t+s} - X_t + X_t)] \\
&= E[X_t] \cdot E[X_{t+s} - X_t] + E[X_t^2] \\
&= E[X_t] \cdot E[X_{t+s} - X_t] + (\text{Var}(X_t) + E[X_t]^2) \\
&= \lambda t \cdot \lambda s + \lambda t + (\lambda t)^2
\end{aligned}
$$

Therefore,

$$\text{Cov}(X_t, X_{t+s}) = \lambda t \cdot \lambda s + \lambda t + (\lambda t)^2 - (\lambda t)(\lambda(t+s)) = \lambda t$$

after a little simplification. Oddly enough this does not depend on s, the length of the time interval. (Can the same be said for the correlation?) ∎

It is time to turn back to the distributional properties of the sample mean in the case of sampling without replacement. To set up this study, look at the variance of a linear combination of two random variables which are not assumed to be independent. By formula (5),

$$\text{Var}(c \cdot X + d \cdot Y) = c^2 \cdot \text{Var}(X) + d^2 \cdot \text{Var}(Y) + 2\,c \cdot d \cdot \text{Cov}(X, Y) \qquad (14)$$

Extend this result to three random variables by doing the next activity.

Activity 4 Find an expression analogous to (14) for $\text{Var}(c \cdot X + d \cdot Y + e \cdot Z)$.

The proof of Theorem 2 can be generalized (see Exercise 6) to give the formula

$$\text{Var}(\textstyle\sum_{i=1}^{n} c_i X_i) = \sum_{i=1}^{n} c_i^2 \,\text{Var}(X_i) + 2 \sum_{i=1}^{n} \sum_{j=1}^{i-1} c_i c_j \,\text{Cov}(X_i, X_j) \qquad (15)$$

In particular, for the sample mean $\overline{X} = \frac{1}{n}\sum_{i=1}^{n} X_i$ each coefficient c_i equals $1/n$, and if the X_i's are identically distributed with variance σ^2 then we obtain

$$
\begin{aligned}
\text{Var}(\overline{X}) &= \sum_{i=1}^{n} \frac{1}{n^2} \sigma^2 + 2 \sum_{i=1}^{n} \sum_{j=1}^{i-1} \frac{1}{n^2} \,\text{Cov}(X_i, X_j) \\
&= \frac{\sigma^2}{n} + \frac{2}{n^2} \sum_{i=1}^{n} \sum_{j=1}^{i-1} \text{Cov}(X_i, X_j)
\end{aligned}
\qquad (16)
$$

Since σ^2/n is the variance of $\overline{X}$ in the independent case, we can see that if the pairs (X_i, X_j) are all positively correlated (i.e., their covariance is positive), then the variance of $\overline{X}$ in the dependent case is greater than the variance in the independent case, and if the pairs are all negatively correlated the dependent variance is less than the independent variance.

Example 6 Suppose that a population can be categorized into type 0 individuals and type 1 individuals. This categorization may be done on virtually any basis: male vs. female, at risk for diabetes vs. not, preferring Coke vs. Pepsi, etc. We draw a random sample of size n in sequence of these 0's and 1's: $X_1, X_2, \ldots, X_n$. Notice that in this case, since the X's only take on the values 0 and 1, the sample mean $\overline{X}$ is also the sample proportion of 1's, which should be a good estimator of the population proportion of 1's. This is because each X_i has the Bernoulli(p) distribution, whose mean is p and whose variance is $p(1 - p)$ (formulas (2) and (3) of Section 2.2). Thus, by formula (7),

$$E[\overline{X}] = p$$

and so $\overline{X}$ is an accurate estimate of p. If the sample is drawn with replacement, then the X's will be independent, hence formula (8) gives in the independent case

$$\text{Var}(\overline{X}) = \frac{p(1-p)}{n}$$

Therefore $\overline{X}$ becomes an increasingly precise estimator of p as the sample size n increases.

Now let's look at the case of sampling in sequence without replacement. Suppose our population has finite size M, and L individuals in it are of type 1, hence $p = L/M$. Each X_i is marginally Bernoulli(p), and so the equation $\text{E}[\overline{X}] = p$ still holds. But the sample values are now correlated. To compute the covariance between X_i and X_j, note that the probability that both of them equal 1 is

$$\frac{L}{M} \cdot \frac{L-1}{M-1}$$

since we require $X_i = 1$, and once that is given, there are $L - 1$ type 1's remaining to select from the $M - 1$ population members who are left which can make $X_j = 1$. Hence

$$\text{E}[X_i \cdot X_j] = 1 \cdot \text{P}[X_i = 1, X_j = 1] = \frac{L}{M} \cdot \frac{L-1}{M-1}$$

and by computational formula (13),

$$\text{Cov}\,(X_i, X_j) = \frac{L}{M} \cdot \frac{L-1}{M-1} - \frac{L}{M} \cdot \frac{L}{M} = \frac{L}{M} \cdot \frac{L-M}{M(M-1)}$$

after simplification. This can be rewritten as

$$\text{Cov}\,(X_i, X_j) = -\frac{p(1-p)}{M-1}$$

if we use the facts that $p = L/M$ and $1 - p = (M - L)/M$. Notice that we are in one of the cases cited above, where all of the X pairs are negatively correlated. Finally, substitution into (16) gives

$$
\begin{aligned}
\text{Var}\,(\overline{X}) &= \frac{p(1-p)}{n} - \frac{2}{n^2} \sum_{i=1}^{n} \sum_{j=1}^{i-1} \frac{p(1-p)}{M-1} \\
&= \frac{p(1-p)}{n} - \frac{2}{n^2} \frac{(n-1)n}{2} \frac{p(1-p)}{M-1} \qquad (17) \\
&= \frac{p(1-p)}{n} \left(1 - \frac{n-1}{M-1}\right)
\end{aligned}
$$

One interesting thing about formula (17) is that as the sample size n approaches the population size M, the variance of $\overline{X}$ decreases to zero, which is intuitively reasonable since sampling is done without replacement, and the more of the population we sample, the surer we are of $\overline{X}$ as an estimator of p.

In Figure 17 is a *Mathematica* plot of the variance of $\overline{X}$ as a function of n for fixed $M = 1000$ and $L = 500$ (hence $p = 1/2$). The thin curve is the variance of the sample mean in the independent case, shown for comparison. You can open up the closed cell in the electronic version of the text, and try a few different plot domains to get a complete look at the graphs. Here we see that for sample sizes in the range of 100 to 500, both variances are getting small, but the variance for the non-replacement case is less than that for the replacement case, making it a more precise estimator of p. ∎

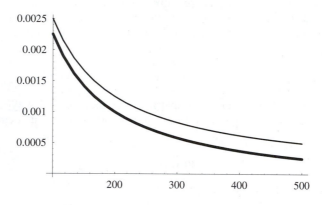

Figure 17 - Variance of the sample mean as a function of sample size, with (thin) and without (thick) replacement

We would like to introduce one more notion involving expectation before leaving this section. In Section 2.5 we developed the idea of the conditional distribution of one random variable given the observed value of another. To this distribution is associated an expectation as well, according to the following definition.

Definition 3. Let X and Y be discrete random variables with joint p.m.f. $f(x, y)$, and let $q(y \,|\, x)$ be the conditional p.m.f. of Y given $X = x$. Then the *conditional expectation* of a function g of Y given $X = x$ is

$$E[g(Y) \,|\, X = x] = \sum_y g(y)\, q(y \,|\, x) \qquad (18)$$

Thus, the conditional expectation is not really a new quantity; it is just expectation as defined earlier (the weighted average of the values $g(y)$) except that the conditional p.m.f. of Y is used as the weighting function instead of the marginal p.m.f. of Y. It follows that the conditional expectation may well depend on the observed X value x.

Definition 3 can be extended slightly to include the case where g is a function of both X and Y; simply replace the random variable X by its observed value x as follows:

$$E[g(X, Y) \,|\, X = x] = E[g(x, Y) \,|\, X = x] = \sum_y g(x, y)\, q(y \,|\, x) \qquad (19)$$

The activity below lists an important property of conditional expectation.

Activity 5 Use the definition of conditional expectation to show that if X and Y are independent, then $E[Y \,|\, X = x] = E[Y]$.

In Exercise 16 you will derive a special case of another important property. To set it up, observe that formula (18) implies that $E[g(Y) \,|\, X = x]$ is a function of x; call it $h(x)$ say. We write the composition $h(X)$ as $E[g(Y) \,|\, X]$, thinking of it as the expectation of Y given the random variable X. The function $h(X)$ is random, and its expectation simplifies this way:

$$E[E[g(Y) \,|\, X]] \;=\; E[g(Y)] \qquad (20)$$

This property appears similar to, and is in fact referred to as, the *law of total probability for expectation*; we can find the expected value of a function $g(Y)$ by conditioning on X, then "un-conditioning", which here means taking the expected value with respect to X.

Example 7 Recall Example 1.5-3, in which a Xerox machine could be in one of four states of deterioration labeled 1, 2, 3, and 4 on each day. The matrix of conditional probabilities of machine states tomorrow given today is reproduced here for your convenience.

		tomorrow		
	1	2	3	4
1	3/4	1/8	1/8	0
today 2	0	3/4	1/8	1/8
3	0	0	3/4	1/4
4	0	0	0	1

Now that we know about random variables and conditional distributions, we can introduce X_1, X_2, and X_3 to represent the random states on Monday through Wednesday, respectively. Now let us suppose that the state numbers are proportional to the daily running costs when the machine is in that state. Thus, we can find for example the expected total cost through Wednesday, given that $X_1 = 1$ was the initial state on Monday. The conditional distribution of X_2 given $X_1 = 1$ is, from the table,

$$P[X_2 = 1 \mid X_1 = 1] = 3/4, \quad P[X_2 = 2 \mid X_1 = 1] = 1/8,$$
$$P[X_2 = 3 \mid X_1 = 1] = 1/8$$

Also, you can check that by the law of total probability, the conditional distribution of X_3 given $X_1 = 1$ is (see Figure 13 of Chapter 1)

$$P[X_3 = 1 \mid X_1 = 1] = \frac{9}{16}, \quad P[X_3 = 2 \mid X_1 = 1] = \frac{6}{32},$$
$$P[X_3 = 3 \mid X_1 = 1] = \frac{13}{64}, \quad P[X_3 = 4 \mid X_1 = 1] = \frac{3}{64}$$

Therefore the expected total cost from Monday through Wednesday given that the machine was in state 1 on Monday is

$$
\begin{aligned}
E[X_1 + X_2 + X_3 \mid X_1 = 1] &= 1 + E[X_2 \mid X_1 = 1] + E[X_3 \mid X_1 = 1] \\
&= 1 + (1 \cdot \tfrac{3}{4} + 2 \cdot \tfrac{1}{8} + 3 \cdot \tfrac{1}{8}) \\
&\quad + (1 \cdot \tfrac{9}{16} + 2 \cdot \tfrac{6}{32} + 3 \cdot \tfrac{13}{64} + 4 \cdot \tfrac{3}{64}) \\
&= \tfrac{273}{64} = 4.265625 \qquad \blacksquare
\end{aligned}
$$

Mathematica for Section 2.6

Command	Location
Mean[datalist]	Statistics` DiscreteDistributions`
RandomArray[distribution, sampsize]	Statistics` DiscreteDistributions`
GeometricDistribution[p]	Statistics` DiscreteDistributions`
Histogram[sample, numrectangles]	KnoxProb` Utilities`
SeedRandom[seed]	kernel

Exercises 2.6

1. Rederive the results for the mean and variance of the b(n, p) distribution using another method, namely by observing that the number of successes Y is equal to the sum of random variables $X_1 + X_2 + \cdots + X_n$ where $X_i = 1$ or 0, respectively, according to whether the ith trial is a success or failure.

2. Suppose that the pair of random variables (X, Y) has a distribution that puts equal probability weight on each of the points in the integer triangular grid shown in the figure. Find (a) $E[X + Y]$; (b) $E[2X - Y]$.

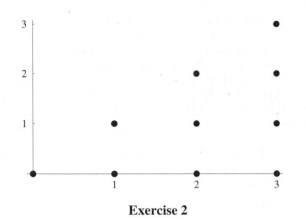

Exercise 2

3. If X_1, X_2, X_3, X_4, and X_5 have the multinomial distribution with parameters $n = 20$, $p_1 = 1/5$, $p_2 = 1/10$, $p_3 = 1/10$, $p_4 = 2/5$, and $p_5 = 1/5$, find $E[X_1 + X_2 + X_3]$. Without computation, what is $E[X_1 + X_2 + X_3 + X_4 + X_5]$?

4. If (X_t) is a Poisson process of van arrivals with rate 2 per hour and 6 individuals in each van, find the expected value and variance of the number of individuals arriving in either the first or third hours.

5. Extend the linearity of expectation (Theorem 1) to linear combinations of n random variables X_1, X_2, ... , X_n.

6. Prove formula (15), the general expression for the variance of a linear combination $\mathrm{Var}(\sum_i c_i X_i)$. To what does the formula reduce when the X_i's are independent?

7. (*Mathematica*) Write *Mathematica* commands which take as input a finite list $\{x_1, x_2, ..., x_m\}$ of states of a random variable X, a list $\{y_1, y_2, ..., y_n\}$ of states of a random variable Y, and a matrix $\{\{p11, p12, ..., p1n\}, ..., \{pm1, pm2, ..., pmn\}\}$ of joint probabilities of states (i.e., $pij = P[X = i, Y = j]$), and return: (a) the mean of X; (b) the mean of Y; (c) the variance of X; (d) the variance of Y; (e) the covariance of X and Y; (f) the correlation of X and Y.

8. Find the covariance and correlation between the random variables X and Y in Exercise 2.

9. If X and Y, respectively, are the numbers of successes and failures in a binomial experiment with n trials and success probability p, find the covariance and correlation between X and Y.

10. (*Mathematica*) We saw in the section that the probability distribution of the sample mean $\overline{X}$ of a random sample of size n clusters more and more tightly around the distributional mean μ as $n \longrightarrow \infty$. Here let us look at a different kind of conver-

gence, studying a time series of the sample mean as a function of n, $\overline{X}(n) = (X_1 + X_2 + \cdots + X_n)/n$ as we add more and more observations.

Write a *Mathematica* command that successively simulates a desired number of observations from a discrete distribution, updating $\overline{X}$ each time m new observations have been generated (where m is also to be an input parameter). The command should display a connected list plot of that list of means. Run it several times for several distributions and larger and larger values of n. Report on what the sequence of means seems to do.

11. Show that if a, b, c, and d are constants with $a, c > 0$, and if X and Y are random variables, then

$$\text{Cov}(aX + b, cY + d) = a \cdot c \cdot \text{Cov}(X, Y)$$
$$\text{Corr}(aX + b, cY + d) = \text{Corr}(X, Y)$$

12. Show that the correlation ρ between two random variables X and Y must always be less than or equal to 1 in magnitude, and furthermore, show that if X and Y are perfectly linearly related, then $|\rho| = 1$. (The latter implication is almost an equivalence but we don't have the tools to see that as yet. To show that $|\rho| \leq 1$, carry out the following plan. First reduce to the case where X and Y have mean 0 and variance 1 by considering the random variables $(X - \mu_x)/\sigma_x$ and $(Y - \mu_y)/\sigma_y$. Then look at the discriminant of the non-negative valued quadratic function of t: $E[(W + tZ)^2]$, where W and Z are random variables with mean 0 and variance 1.)

13. (*Mathematica*) If the population in Example 6 is of size 1000, and sampling is done without replacement, use a graph to find out how large must the sample be to guarantee that the standard deviation of $\overline{X}$ is no more than .03. (Hint: you do not know a priori what p is, but at most how large can $p(1 - p)$ be?)

14. (*Mathematica*) Consider the problem of simulating random samples taken from the finite population $\{1, 2, \ldots, 500\}$, and computing their sample means. Plot histograms of the sample means of 50 random samples of size 100 in both the cases of sampling with replacement and sampling without replacement. Comment on what you see, and how it relates to the theoretical results of this section.

15. As in Example 7, compute the expected total cost through Wednesday given that the initial state on Monday was 2.

16. Show that if X and Y are discrete random variables, $E[E[Y \mid X]] = E[Y]$.

17. Find $E[Y \mid X = 1]$ and $E[Y \mid X = 2]$ for the random variables X and Y whose joint distribution is given in Exercise 2.5-2.

18. Compute $h(x) = E[Y^2 \mid X = x]$ for each possible x, if X and Y have the joint distribution in Exercise 2. Use this, together with formula (20), to find $E[Y^2]$.

CHAPTER 3
CONTINUOUS PROBABILITY

3.1 From the Finite to the (Very) Infinite

In this chapter we begin the transition from discrete probability, in which sample spaces have either finitely many or countably many outcomes, to continuous probability, in which there is a continuum of possible outcomes. This area of study continues in the next chapter when we review some of the most useful probability distributions and the properties of continuous random variables. We will see many interesting examples and applications there; but here we will focus more on the underlying concepts.

Calculus and set theory are the tools for adapting what we have done so far to the more complicated continuous setting. Remember that when the sample space of a random experiment is finite or countable, to construct a probability it is enough to make an assignment $P[\omega]$ of positive probability to each outcome $\omega \in \Omega$. To be valid, these outcome probabilities must be non-negative and must sum to 1. In continuous probability, we are usually interested in sample spaces $\Omega \subset \mathbb{R}$ which are intervals. But the problem is that we cannot assign strictly positive probability to all points in an interval $[a, b]$ without violating the condition that the sample space has probability 1. To see this, observe that in order for $P[[a, b]]$ to be less than or equal to 1, we can have at most 1 point in $[a, b]$ of probability in $(1/2, 1]$, at most 2 points of probability in $(1/3, 1/2]$, at most 3 points of probability in $(1/4, 1/3]$, etc. The set of points in $[a, b]$ of strictly positive probability is the disjoint union as n ranges from 2 to ∞ of the sets of points whose probability is in $(1/n, 1/(n-1)]$. Since the set of points of positive probability is therefore a countable union of finite sets, it is at most countable. In particular it follows that the set of points of positive probability on Ω can contain no intervals.

Another problem concerns what we mean by events. A typical sampling experiment in discrete probability is to sample two items, in sequence and with replacement, from the universe $\{a, b, c\}$. The sample space for this experiment is

$$\Omega = \{(a, a), (a, b), (a, c), (b, a), (b, b), (b, c), (c, a), (c, b), (c, c)\}$$

which has 9 elements. Each of the $2^9 = 512$ subsets of the sample space is an event. The analogous problem in the continuous domain of sampling two real numbers has all of $\mathbb{R}^2$ as its sample space. The set of all subsets of $\mathbb{R}^2$ is a large breed of infinity indeed, and it also includes some pathological sets whose properties are incompatible with the way we would like to formulate continuous probability. So in the rest of this book we will restrict our attention to a smaller, more manageable class of events: the so-called *Borel sets*, which are described next.

Let the sample space be $\Omega = \mathbb{R}$. (The construction for higher dimensional sample spaces is analogous.) Certainly intervals should qualify as events. For definiteness, let us use intervals of the form $(a, b]$ as our most primitive building blocks, where a is allowed to be $-\infty$. Notice that the intersection of any two of these is either empty or is another interval of this form. We are accustomed to taking unions, intersections, and complements of events and expecting the resulting set to be an event. This is the idea behind the following definition.

Definition 1. A σ-*algebra* $\mathcal{H}$ of a set Ω is a family of subsets of Ω which is closed under countable union and complementation; i.e., if $A_1, A_2, A_3, \ldots \in \mathcal{H}$ then $\bigcup_i A_i \in \mathcal{H}$, and if $A \in \mathcal{H}$ then $A^c \in \mathcal{H}$.

The definition does not specifically refer to countable intersections, but the next activity covers that case.

Activity 1 Use DeMorgan's laws to argue that if $\mathcal{H}$ is a σ-algebra then $\mathcal{H}$ is also closed under countable intersection.

Exercise 2 asks you to show that an arbitrary intersection of σ-algebras is also a σ-algebra. Hence if we start with a generating family of sets such as the collection of all intervals $\mathcal{L} = \{(a, b]\}$ we can talk about the family of all σ-algebras $\mathcal{H}$ that contain $\mathcal{L}$. The intersection of all those σ-algebras is again a σ-algebra, and it must be the smallest σ-algebra containing $\mathcal{L}$. The sets in this σ-algebra are the Borel sets.

Definition 2. The *Borel subsets* of the real line are the sets comprising the smallest σ-algebra containing all intervals $(a, b]$.

Example 1 Let us look more concretely at examples of members of the σ-algebra of Borel sets on the real line. All primitive half-open, half-closed intervals such as $(9, 11]$ and $(-\infty, 3]$ are Borel sets. Since the family of Borel sets is closed under complementation, open rays such as (b, ∞), which are complementary to closed rays $(-\infty, b]$ are Borel sets. Singleton sets $\{b\}$ are also Borel sets, because they can be expressed as countable intersections of intervals $(a_n, b]$ where the sequence of left endpoints a_n increases to b. Similarly, a closed ray of the form $[a, \infty)$ is a Borel set, because it is expressible as a countable intersection of intervals (a_n, ∞) where the left endpoints a_n increase to a. Exercise 4 asks you to show that intervals of all other possible forms (a, b), $[a, b]$, $[a, b)$, for a and b finite, and $(-\infty, b)$ are Borel sets as well. Also, any countable union of intervals of any form is a Borel set. But we will rarely see any set more complicated than a finite, disjoint union of intervals, which is certainly a Borel set by closure under union. In higher dimensions, the σ-algebra of Borel sets is the smallest σ-algebra containing all generalized rectangles $[a_1, b_1) \times [a_2, b_2) \times \cdots$, which is sufficiently rich to include most events of interest. ∎

So, in cases where the sample space Ω is the real line, the family of events $\mathcal{H}$ will be the collection of Borel subsets of Ω. If the sample space is a subset of $\mathbb{R}$, then the events are the Borel sets contained in that subset.

Next we must look at how to characterize probability of events. Whatever we do should be consistent with the definition of a probability measure in Section 1.2, which requires: (i) $P[\Omega] = 1$; (ii) For all $A \in \mathcal{H}$, $P[A] \geq 0$; and (iii) If $A_1, A_2, \cdots$ is a sequence of pairwise disjoint events then $P[A_1 \cup A_2 \cup \cdots] = \sum_i P[A_i]$. In this case, Propositions 1-5 in Section 1.2, which give the main properties of probability, still hold without further proof, because they were proved assuming only axioms (i)-(iii) of the definition, not the finiteness of Ω. We will discover that many of the other properties of probability, random variables, and distributions will carry over as well.

Much as probabilities of events in the discrete case are found by *summing* values of a function that gives probability *weight* to each outcome, in the continuous case we will *integrate* a function that gives probability *density* to each outcome. To motivate the idea, consider a discrete distribution that gives positive probability weight to points 0, Δx, $2\Delta x$, $3\Delta x$, etc. which have a constant spacing of length Δx. A scaled probability histogram can be drawn, by setting the bar height for interval x to $x + \Delta x$ equal to the probability weight at x divided by the length of the interval Δx. If we do this, then the bar area, which is the bar height times Δx, is equal to the probability of outcome x. For instance, look at Figure 1, in which the probability measure on $\Omega = \{0, .1, .2, .3, .4, .5, .6\}$ with probability weights .1, .2, .3, .2, .1, .05, and .05, respectively, is plotted on a scaled histogram with $\Delta x = .1$, and heights 1, 2, 3, 2, 1, .5, and .5. The bar heights are the probability weights divided by Δx, or what is the same thing, multiplied by 10. We think of the bar height, which is probability weight per unit length, as a probability density. For that reason, let us call such a scaled histogram a *discrete density histogram*.

```
Needs["KnoxProb`Utilities`"]
```

```
ProbabilityHistogram[{0, .1, .2, .3, .4, .5, .6},
    {1, 2, 3, 2, 1, .5, .5},
    DefaultFont → {"Times-Roman", 8}];
```

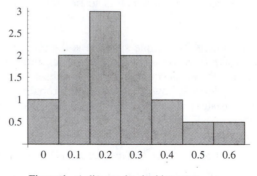

Figure 1 - A discrete density histogram

In the sample space above, outcomes lie between 0 and .6, and they are measured only to the nearest one-tenth. In continuous probability we begin with the (idealized) model that many measured variables, such as physical dimensions, measurement errors, and economic quantities, are continuous in nature. If we had infinitely precise instruments, we could measure them with arbitrary precision and their possible values would lie on an interval of the real line. One way of thinking about the situation is that a discrete density histogram like the one above is only a discrete approximation of the true function that gives density of probability at every point. To visualize the approximation, the command

DensityHistogram[densityfn, n, a, b]

in the KnoxProb'Utilities' package is used below. It takes as arguments the actual continuous density function, an integer *n* determining the number of points in the discrete approximation (the interval [*a*, *b*] will be broken into $n + 1$ subintervals), and the endpoints *a* and *b* which bound the sample space. The command outputs the graph of the density together with the graph of the associated discrete density histogram. It would be instructive to open up the Utilities package to see the code, which is fairly straightforward.

In Figure 2 are two such discrete approximations to a probability distribution whose density is $f(x) = \frac{3}{4}(1 - x^2)$ on the interval [−1, 1], using values $n = 5$ and $n = 10$. In the electronic version of this section you may run the Table command below the figure, and select and animate the graphics it generates to get a visual perspective on the convergence of the discrete density histograms to a continuous density function. As we measure the outcome to higher and higher precision, the discrete density histogram settles down to a smooth curve, which is the graph of the density function.

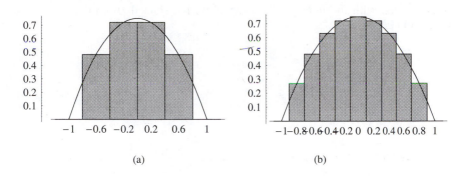

Figure 2 - Approximation of continuous density by discrete density histograms (a) $n = 5$; (b) $n = 10$

```
Table[DensityHistogram[f, n,
    -1, 1, NumDigits → 1], {n, 5, 25, 5}]
```

As noted earlier, for a particular bar on a discrete density histogram, the area of that bar is the probability of the outcome associated with that bar. Hence the total area of several bars is the total probability associated with their outcomes. This observation, together with the convergence illustrated by Figure 2, suggests a good way of modeling the assignment of probability to events in the continuous case.

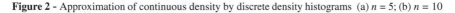

Definition 3. A probability measure on $\Omega \subset \mathbb{R}$ is said to have a *probability density function (p.d.f.)* $f(x)$ if, for any event E,

$$P[E] = \int_E f(x)\,dx \tag{1}$$

So the probability density function allows us to find probabilities of events (Borel sets) by integration, that is, by finding the area under the density curve corresponding to the event on the x-axis. For our building block events $(a, b]$ we have

$$P[\,(a, b]\,] = \int_a^b f(x)\,dx \tag{2}$$

It can be shown in more advanced work that the integral in (1) is well-defined on the family of Borel sets which we are taking as our events.

For instance, for the parabolic density function from Figure 2 the probability associated with the interval (0,.5], which is the shaded area under the density function in Figure 3, is:

$$\int_0^{.5} f[x] \, dx$$

```
0.34375
```

To display the area, we have used the command

PlotContsProb[densityfn, domain, between]

in KnoxProb`Utilities`, which takes the density function, a plot domain, and another pair called *between* which designates the endpoints of the shaded interval.

```
PlotContsProb[f[x], {x, -1, 1}, {0, .5},
   Ticks → {{-1, 0, .5, 1}, Automatic},
   AxesOrigin → {0, 0}, PlotRange → All,
   DefaultFont → {"Times-Roman", 8}];
```

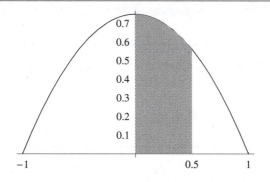

Figure 3 - Shaded area = P[(0, .5]]

Activity 2 Why is $\int_a^b f(x) \, dx$ also equal to P[(a, b)], P[$[a, b]$], and P[$[a, b)$]?

At the start of the section we argued that a probability measure can only put strictly positive weight on at most countably many points. In the density model of continuous probability expressed by Definition 4, the probability weight of <u>every</u>

singleton point $\{a\}$ is 0, since $\int_a^a f(x)\,dx = 0$. This is not a logical problem, because $f(x)$ does not give the probability weight of x. Rather, f is a density function; hence the value of $f(x)$ gives the rate at which probability is accumulating at x.

We are obliged to check whether P defined by (1) is a valid probability measure. The following conditions on f are sufficient:

(i) $f(x) \geq 0$ for all $x \in \Omega$;
(ii) $\int_\Omega f(x)\,dx = 1$

Condition (ii) clearly implies the first axiom of probability $P[\Omega] = 1$. Condition (i) implies that $P[E] = \int_E f(x)\,dx \geq 0$ for all events E, which is the second axiom. The third axiom of probability, that is, the additivity over countable disjoint unions, is actually tricky to show in general, because of the fact that the Borel events can be quite complicated. We will omit the proof. But at least for unions of finitely many disjoint intervals the additivity property of the integral, shown in calculus, yields the conclusion of the third axiom without difficulty, as follows.

$$P[(a_1, b_1] \cup (a_2, b_2] \cup [(a_3, b_3]]$$
$$= \int_{(a_1,b_1]\cup(a_2,b_2]\cup(a_3,b_3]} f(x)\,dx$$
$$= \int_{a_1}^{b_1} f(x)\,dx + \int_{a_2}^{b_2} f(x)\,dx + \int_{a_3}^{b_3} f(x)\,dx$$
$$= P[(a_1, b_1]] + P[(a_2, b_2]] + P[(a_3, b_3]]$$

The parabolic density function from Figure 2 does satisfy the appropriate density conditions. Clearly $f(x) = .75\,(1 - x^2) \geq 0$ on the sample space $[-1, 1]$. Also,

$$\int_{-1}^1 f(x)\,dx = \int_{-1}^1 .75\,(1 - x^2)\,dx = .75\left(x - \frac{x^3}{3}\right)\Big|_{-1}^1 = .75\left(\frac{2}{3} - \frac{-2}{3}\right) = 1$$

There is a huge variety of valid density functions; for instance, for a suitably chosen positive constant c, you can check that the function

$$f(x) = c \cdot \frac{2 + \sin(\sqrt{x})}{e^{x/6}}$$

is a valid density on $\Omega = [0, \pi^2]$. Only a few density functions arise very often in applications though, including the one in the following example.

Example 2 Let $\Omega = [0, 8]$ and let a probability be defined on Ω using the density function

$$f(x) = \begin{cases} 1/8 & \text{if } x \in \Omega \\ 0 & \text{otherwise} \end{cases}$$

Then for example,

$$P[\,(1,\,3)\,] = \int_1^3 \tfrac{1}{8}\,dx = \tfrac{1}{8}\,x\,\big|_1^3 = \tfrac{2}{8}$$

$$P[\,[0,\,1)\cup(6,\,7)\,] = \int_0^1 \tfrac{1}{8}\,dx + \int_6^7 \tfrac{1}{8}\,dx = \tfrac{2}{8}$$

$$P[\,[0,\,2]^c\,] = P[\,(2,\,8]\,] = \int_2^8 \tfrac{1}{8}\,dx = \tfrac{6}{8}$$

$$P[\,\{2\}\,] = \int_2^2 \tfrac{1}{8}\,dx = 0$$

This f is called the *continuous uniform density* on [0, 8]. Constant density is attached to each point, which gives no point favoritism over any other. This probability model is therefore appropriate for the random selection of one real number from the interval [0, 8]. Another way of thinking of this probability distribution is listed in the next activity. ■

Activity 3 For the density in Example 2, check that the probability of any interval is its length divided by 8. Hence, sets of equal length have equal probability.

Mathematica for Section 3.1

Command	Location
ProbabilityHistogram[statelist, problist]	KnoxProb` Utilities`
DensityHistogram[densityfn, n, a, b]	KnoxProb` Utilities`
PlotContsProb[density, domain, between]	KnoxProb` Utilities`

Exercises 3.1

1. Explain why the set of all integers is a Borel set. Explain why any at most countable set of points on the real axis is a Borel set.

2. Prove that an arbitrary intersection (not necessarily a countable one) of σ-algebras is a σ-algebra. Conclude that the intersection of all σ-algebras containing a collection of sets $\mathcal{L}$ is the smallest σ-algebra containing $\mathcal{L}$.

3. In the case of a finite sample space Ω, argue that the set of all subsets of Ω forms a σ-algebra. What is the smallest σ-algebra containing a singleton set $\{x\}$, for $x \in \Omega$?

4. Show that all of the intervals (a, b), $[a, b]$, $(-\infty, b)$, $[a, b)$ are Borel sets for any real numbers $a < b$.

5. (*Mathematica*) Consider the density function $f(x) = (x-1)^2$, $x \in \left[0, 1 + \sqrt[3]{2}\right]$. Use the DensityHistogram command to compare the graph of f to the histogram in the cases when $n = 10, 20, 30, 40, 50$. Why may it not turn out that the discretized probabilities form a valid probability measure on the discretized set of states?

6. Find a constant c such that $f(x) = c/x^2$ is a valid probability density function on $\Omega = [1, \infty)$. Then find $P[\, (2, 4) \cup (8, \infty)\,]$.

7. Show that $f(x) = \frac{1}{\pi(1+x^2)}$ is a valid probability density function on $\Omega = \mathbb{R}$. Then find $P[\, [0, \infty)\,]$ and $P[\, [-1, 1]\,]$.

8. Find a constant c such that $f(x) = c \log(x)$ is a valid density on $\Omega = [1, e]$, and then compute $P[\, [2, e]\,]$ and $P[\, [1, 2]\,]$.

9. (*Mathematica*) Plot a graph of the density function $f(x) = c\, x^4\, e^{-3x}$ on $[0, \infty)$ for the appropriate choice of c. Compute $P[\, (3, 4]^c\,]$.

3.2 Continuous Random Variables and Distributions

Next we would like to study the notion of random variables in the continuous setting. Random variables, as we know, attach a numerical value to the outcome of a random phenomenon. In the discrete case we used the probability mass function to characterize the likelihood $f(x) = P[X = x]$ that a random variable X takes on each of its possible states x. Then for subsets B of the state space E,

$$P[X \in B] = \sum_{x \in B} f(x) = \sum_{x \in B} P[X = x]$$

The continuous case will come out similarly, with an integral replacing the discrete sum. But why are we concerned with continuous random variables? There are many interesting experiments which give rise to measurement variables that are inherently continuous, that is, they take values in intervals on the line rather than at finitely many discrete points. For instance, the pressure on an airplane wing, the amount of yield in a chemical reaction, the measurement error in an experimental determination of the speed of light, and the time until the first customer arrives to a store are just a few among many examples of continuous variables. To make predictions about them, and to understand their central tendency and spread, we must have a way of modeling their probability laws.

We have already set up the machinery to characterize the probability distribution of a continuous random variable in Section 3.1: the probability density function is the key. But now we consider density functions on the state spaces E of random variables rather than on the sample spaces Ω. From here on, Ω will be pushed to the background.

Definition 1.   A random variable X with state space $E \subset \mathbb{R}$ is said to have the *probability density function* (abbr. p.d.f.) f if, for all Borel subsets B of E,

$$P[X \in B] = \int_B f(x)\,dx \tag{1}$$

Here, as in the last section, to be a valid density function f must be non-negative and $\int_E f(x)\,dx = 1$. Figure 4 gives the geometric meaning of formula (1). The probability that X will fall into set B is the area under the density curve on the subset B of the real axis.

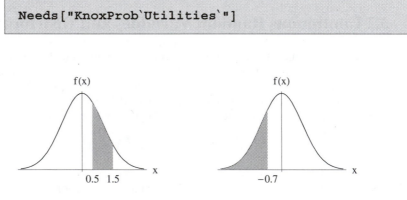

Needs["KnoxProb`Utilities`"]

Figure 4 - Shaded area is $P[X \in [.5,\ 1.5]]$ **Figure 5** - Shaded area is $F(-.7) = P[X \le -.7]$

The *cumulative distribution function* of a continuous random variable will be prominently featured from here on. The definition of the c.d.f. F is the same as it was before, though now it is computed as an integral:

$$F(x) = P[X \le x] = \int_{-\infty}^{x} f(t)\,dt \qquad (2)$$

It totals all probability weight to the left of and including point x (see Figure 5). By the Fundamental Theorem of Calculus, differentiating both sides of (2) with respect to x gives the relationship

$$F'(x) = f(x) \qquad (3)$$

Thus, the c.d.f. also characterizes the probability distribution of a continuous random variable, since if it is given then the density function is determined.

Activity 1   From calculus, the differential equation $F'(x) = f(x)$ only determines F up to an additive constant, given f. In our setting, what further conditions are present which guarantee that each density f has exactly one c.d.f.?

For example, if f is the continuous uniform density on [0, 4], then

$$f(x) = \begin{cases} 1/4 & \text{if } x \in [0,\,4] \\ 0 & \text{otherwise} \end{cases}$$

$$F(x) = \int_0^x \tfrac{1}{4}\,dt = \tfrac{1}{4}\,x \text{ if } x \in [0,\,4]$$

For $x < 0$, $F(x) = P[X \le x] = 0$, and for $x > 4$, $F(x) = P[X \le x] = 1$, since X must take a value in [0, 4]. The graphs of the density and c.d.f. are shown in Figure 6. Note according to (3) that the density at each point x is the slope of the c.d.f. graph at that x.

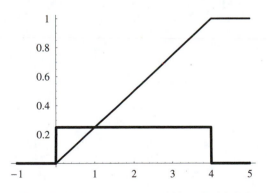

Figure 6 - Uniform(0,4) density (bold), and c.d.f. (light)

Mathematica's standard packageStatistics`ContinuousDistributions`, loaded by default when you load KnoxProb`Utilities`, contains definitions of many of the most common continuous distributions that we will use in this chapter and the next. The only one that we are acquainted with at this point is

Uniform Distribution[a, b]

which is the continuous uniform distribution with density $f(x) = 1/(b-a)$ on $[a, b]$. As in the discrete case, the functions

PDF[distribution, x]   and   CDF[distribution, x]

are available when you load this package. PDF returns the density function, CDF returns the cumulative distribution function, and x is the desired argument of each function.

For example, for the uniform distribution on $[0, 4]$ we can define the density and c.d.f. as follows:

```
f[x_] := PDF[UniformDistribution[0, 4], x];
F[x_] := CDF[UniformDistribution[0, 4], x];
```

Then we can calculate a probability like $P[X \in (1, 3]]$ in either of two ways in *Mathematica*; by integrating the density from 1 to 3, or by computing $P[1 < X \le 3] = P[X \le 3] - P[X \le 1] = F(3) - F(1)$.

```
  ⌠3
  ⎮  f[x] dx
  ⌡1

F[3] - F[1]
```

$\dfrac{1}{2}$

$\dfrac{1}{2}$

Example 1 If we simulate large random samples from a continuous uniform distribution, will a histogram of the data show the same flatness as the density function?

Consider for instance the uniform distribution on [0, 10]. Let us simulate 100 values from this distribution using the RandomArray command, and plot a histogram of the result with 10 rectangles. Two replications of this experiment are shown in Figure 7.

```
SeedRandom[63471]
```

```
uniflist1 =
    RandomArray[UniformDistribution[0, 10], 100];
uniflist2 = RandomArray[
    UniformDistribution[0, 10], 100];
Show[GraphicsArray[
        {Histogram[uniflist1, 10, Endpoints → {0, 10},
            DisplayFunction → Identity, NumDigits → 1],
        Histogram[uniflist2, 10,
            Endpoints → {0, 10}, NumDigits → 1,
            DisplayFunction → Identity]}],
    DisplayFunction → $DisplayFunction];
```

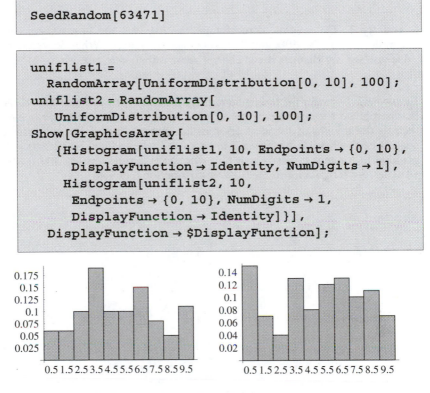

Figure 7 - Two simulations of 100 uniform(0,10) observations

Because the distribution of probability is uniform, we expect to see 1/10th of the observations in each category, but in repeated runs it is typical to see a variation of relative frequencies ranging from a low of around .05 to a high of around .15. The moral of the story is that you should not expect histograms of random samples to look exactly like the density functions, although if there are drastic departures of data from the density, one should suspect that the density is not appropriate for the population being sampled from. Do the next Activity to study the effect of increasing the sample size.

Activity 2   Do several more simulations like the one above, changing the sample size to 200, 500, and then 1000 sample values. What do you observe? Do you see any changes if you change the distribution to uniform(0, 1)?

Example 2 Consider the triangular density function sketched in Figure 8. (It could be appropriate for example to model the distribution of a random variable X that records the amount of roundoff error in the computation of interest correct to the nearest cent, where X is measured in cents and takes values between negative and positive half a cent.) Verify that this function is a valid density and find a piece-wise defined formula for its cumulative distribution function.

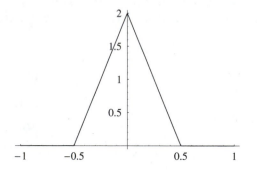

Figure 8 - A triangular density function

From the graph, f is clearly non-negative on its state space $[-.5, .5]$ and its formula is

$$f(x) = \begin{cases} 2 + 4x & \text{if } -.5 \leq x \leq 0 \\ 2 - 4x & \text{if } 0 < x \leq .5 \\ 0 & \text{otherwise} \end{cases}$$

The area under f corresponding to all of the state space is the area of the triangle, which is $(1/2) \cdot 1 \cdot 2 = 1$. Therefore f is a valid density function. For values of x between $-.5$ and 0 the c.d.f. is

$$F(x) = \int_{-.5}^{x} 2 + 4t\,dt = 2t + 2t^2 \,|_{-.5}^{x} = 2x^2 + 2x + \tfrac{1}{2}$$

Notice that $F(0) = 1/2$, which is appropriate because the density is symmetric about 0, and therefore half of its overall probability of 1 lies to the left of $x = 0$. For values of x between 0 and .5,

$$F(x) = \tfrac{1}{2} + \int_{0}^{x} 2 - 4t\,dt = \tfrac{1}{2} + (2t - 2t^2 \,|_{0}^{x}) = -2x^2 + 2x + \tfrac{1}{2}$$

The final form for F is

$$F(x) = \begin{cases} 0 & \text{if} \quad x < -.5 \\ 2x^2 + 2x + \tfrac{1}{2} & \text{if} \quad -.5 \le x < 0 \\ -2x^2 + 2x + \tfrac{1}{2} & \text{if} \quad 0 \le x < .5 \\ 1 & \text{if} \quad x \ge .5 \end{cases}$$

In the interest rounding example then, the probability that the rounding error is between $-.2$ and $.2$ is $F(.2) - F(-.2)$. This turns out to be $.82 - .18 = .64$ after calculation. ∎

Example 3 The *Weibull distribution* is a familiar one in reliability theory, because it is used as a model for the distribution of lifetimes. It depends on two parameters called α and β, which control the graph of its density function (see Exercise 6). The state space of a Weibull random variable X is the half line $[0, \infty)$, and the formula for the c.d.f. is

$$F(x) = P[X \le x] = 1 - e^{-(x/\beta)^{\alpha}}, \quad x \ge 0 \tag{4}$$

Mathematica's syntax for this distribution is

WeibullDistribution$[\alpha, \beta]$

Below we construct the density and c.d.f. functions, and plot the two together in the case $\alpha = 2$, $\beta = 3$. Again as in formula (3), the density graph is clearly the derivative graph of the c.d.f.

```
g[x_, α_, β_] :=
  PDF[WeibullDistribution[α, β], x];
G[x_, α_, β_] :=
  CDF[WeibullDistribution[α, β], x];
Plot[{g[x, 2, 3], G[x, 2, 3]},
  {x, 0, 8}, PlotStyle ->
    {Thickness[.01], Thickness[.007]},
  DefaultFont → {"Times-Roman", 8}];
```

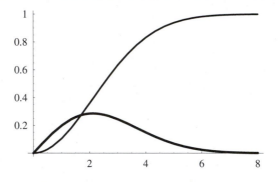

Figure 9 - Weibull(2, 3) p.d.f. (bold), c.d.f. (light)

If the lifetimes (in thousands of miles) of auto shock absorbers of a particular kind have a Weibull distribution with parameters $\alpha = 2$, $\beta = 30$, then for example the proportion of them that last at least 30,000 miles is $P[X \geq 30] = 1 - F(30)$, which is about 37%, as shown below.

```
F[x_] := CDF[WeibullDistribution[2, 30], x];
N[1 - F[30]]
```

```
0.367879
```

Activity 3 Find the equation of the Weibull density function in general. Use *Mathematica* to plot it in the case where $\alpha = 2$, $\beta = 30$ as in the last example, and write a short paragraph about the shape and location of the density, and what the graph implies about the distribution of shock absorber lifetimes.

Joint, Marginal, and Conditional Distributions

In this subsection we carry over the ideas of joint, marginal, and conditional probability distributions from discrete to continuous probability. The adaptation is straightforward. In addition, we take another look at independence of random variables.

To characterize the joint distribution of two continuous random variables X and Y, we combine the ideas of the single variable density function and the two variable discrete probability mass function.

Definition 2. Random variables X and Y are said to have *joint probability density function* $f(x, y)$ if for any Borel subset B of $\mathbb{R}^2$,

$$P[(X, Y) \in B] = \int_B \int f(x, y) \, dx \, dy \qquad (5)$$

The most familiar case of (5) occurs when B is a rectangular set, $B = [a, b] \times [c, d]$. Then (5) reduces to:

$$P[a \le X \le b, \ c \le Y \le d] = \int_a^b \int_c^d f(x, y) \, dy \, dx \qquad (6)$$

For more complicated Borel subsets of $\mathbb{R}^2$, the techniques of multivariable calculus can be brought to bear to find the proper limits of integration. One final comment before we do an example computation: the extension of (5) and (6) to more than two jointly distributed random variables is simple. For n random variables $X_1, X_2, \ldots, X_n$, the density f becomes a function of the n-tuple $(x_1, x_2, \ldots, x_n)$ and the integral is an n-fold iterated integral. We will not deal with this situation very often, except in the case where the random variables are independent, where the integral becomes a product of single integrals.

Example 4 Consider the experiment of randomly sampling a pair of numbers X and Y from the continuous uniform $(0, 1)$ distribution. Suppose we record the pair (U, V), where $U = \min(X, Y)$ and $V = \max(X, Y)$. First let us look at a simulation of many replications of this experiment. Then we will pose a theoretical model for the joint distribution of U and V and ask some questions about them.

The next *Mathematica* command is straightforward. Given the distribution to be sampled from and the number of pairs to produce, it repeatedly simulates an X, then a Y, and then appends the min and max of the two to a list, which is plotted.

```
SimPairs[dist_, numpairs_] :=
    Module[{X, Y, UVlist},
        UVlist = {};
        Do[X = Random[dist];
                Y = Random[dist];
            AppendTo[
    UVlist, {Min[X, Y], Max[X, Y]}],
                {i, 1, numpairs}];
            ListPlot[UVlist, AspectRatio -> 1,
    DefaultFont → {"Times-Roman", 8},
    PlotStyle → {PointSize[.02]}]]
```

Here is a sample run of SimPairs; in the e-version of the text you should try running it again a few times, and increasing the number of pairs to see that this run is typical.

```
SimPairs[UniformDistribution[0, 1], 500];
```

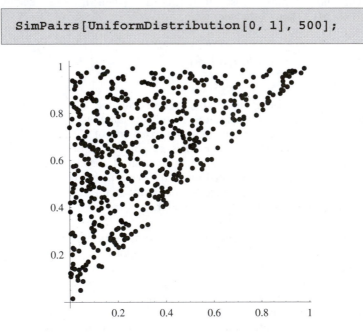

Figure 10 - 500 simulated sorted uniform(0, 1) pairs

Notice that points seem to be randomly spread out within the triangle whose vertices are $(0, 0)$, $(0, 1)$, and $(1, 1)$, which is described by the inequalities $0 \le u \le v \le 1$. No particular region appears to be favored over any other. A moment's thought about how U and V are defined shows that indeed both can take values in the interval [0, 1], but the minimum U must be less than or equal to the maximum V; hence the state space of the pair (U, V) is the aforementioned triangle. But both of the data points from which this pair originated were uniformly distributed on [0, 1], which combined with the empirical evidence of Figure 10 suggests to us that a good model for the joint distribution of U and V would be a constant density over the triangle. In order for that density to integrate to 1, its value must be 1 divided by the area of the triangle, i.e., $1/(1/2) = 2$. So we model the joint density of U and V as

$$f(u, v) = 2 \text{ if } 0 \le u \le v \le 1$$

and $f = 0$ otherwise. (We will deal with the problem of finding the distribution of transformed random variables analytically later in the book.)

We can ask, for example, for the probability that the larger of the two sampled values is at least 1/2. This is written as $P[V \ge .5]$ and is computed by integrating the joint density f over the shaded region in Figure 11. The easiest way to do this is to integrate first on u, with fixed v between .5 and 1, and u ranging from 0 to v. (You could also integrate on v first, but that would require splitting the integral on u into two parts, since the boundaries for v change according to whether u is less than .5 or not.)

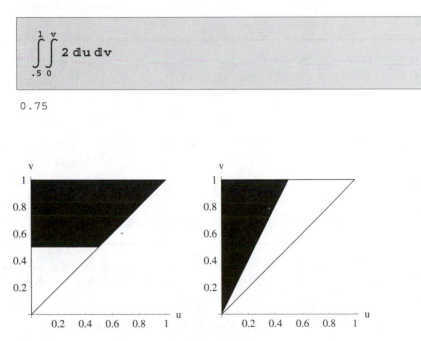

$$\int_{.5}^{1} \int_{0}^{v} 2 \, du \, dv$$

0.75

Figure 11 - Region of integration for $P[V \ge .5]$ **Figure 12** - Region of integration for $P[V > 2U]$

As another example, let us find the probability that the maximum exceeds twice the minimum, i.e., $P[V > 2U]$. The boundary line $v = 2u$ for the inequality

cuts through the state space as shown in Figure 12, and the solution of the inequality is the shaded area. This time the limits of integration are easy regardless of the order you use. Here we integrate v first, from $2u$ to 1, then u from 0 to .5:

$$\int_0^{.5} \int_{2u}^{1} 2 \, dv \, du$$

0.5

∎

Activity 4   Give geometric rather than calculus based arguments to calculate the two probabilities in Example 4.

Marginal probability density functions are defined in the continuous case much as marginal p.m.f.'s are in the discrete case. Instead of summing the joint mass function, we integrate the joint density over one of the variables.

Definition 3.   Let X and Y be continuous random variables with joint density $f(x, y)$. The *marginal density function* of X is

$$p(x) = \int f(x, y) \, dy \tag{7}$$

where the integral is taken over all Y states that are possible for the given x. Similarly the *marginal density* of Y is

$$q(y) = \int f(x, y) \, dx \tag{8}$$

where the integral is taken over all X states that are possible for the given y.

Notice that the marginal densities completely characterize the individual probability distributions of the two random variables. For example, the probability that X lies in a Borel subset B of its state space is

$$P[X \in B] = P[X \in B, Y \in E_Y] = \int_B \int_{E_Y} f(x, y) \, dy \, dx = \int_B p(x) \, dx$$

where E_Y is the state space of Y. You can do a similar derivation for Y. So, $p(x)$ and $q(y)$ are the probability density functions of X and Y in the one-variable sense.

In Example 4, for instance, where the joint density $f(u, v)$ was constantly equal to 2 on the triangle, the marginal density of the smaller sampled value U is

$$p(u) = \int_u^1 2 \, dv = 2(1 - u), \quad 0 \le u \le 1$$

and the marginal density of the larger sampled value V is

$$q(v) = \int_0^v 2 \, du = 2v, \quad 0 \le v \le 1$$

Example 5 Let X and Y have a joint density of the form

$$f(x, y) = c(x^2 + y^2), \quad x, y \in [0, 1]$$

Find the two marginal densities, find $P[X > 1/2, Y > 1/2]$, and find both $P[X > 1/2]$ and $P[Y > 1/2]$.

First, the coefficient c must be found such that the double integral of f over the rectangle $[0, 1] \times [0, 1]$ is 1, so that f is a valid joint density. We have

$$1 = \int_0^1 \int_0^1 c(x^2 + y^2) \, dy \, dx = c \int_0^1 \int_0^1 x^2 + y^2 \, dy \, dx$$

hence c is the reciprocal of the last double integral. We compute this integral below in *Mathematica*.

```
⌠1 ⌠1
⎮  ⎮  (x² + y²) dx dy
⌡0 ⌡0
```

$$\frac{2}{3}$$

Therefore $c = 3/2$. Below is a graph of the joint density function. If we repeatedly sampled from this joint density, the data points that we would see would be more numerous as we move farther away from the origin.

```
f[x_, y_] := 1.5 * (x² + y²);
Plot3D[f[x, y], {x, 0, 1}, {y, 0, 1},
    DefaultFont → {"Times-Roman", 8},
    AxesLabel → {"x", "y", "f(x,y)"},
    ColorOutput → GrayLevel];
```

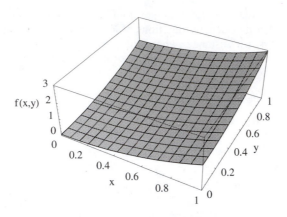

Figure 13 - Joint density $f(x, y) = \frac{3}{2} (x^2 + y^2)$

The marginal density of X is

$$p(x) = \frac{3}{2} \int_0^1 x^2 + y^2 \, d y = \frac{3}{2} \left(x^2 \, y + \frac{y^3}{3} \big|_0^1 \right) = \frac{3}{2} (x^2 + \frac{1}{3}), \, x \in [0, 1]$$

and by symmetry the marginal of Y has the same form,

$$q(y) = \frac{3}{2} (y^2 + \frac{1}{3}), \, y \in [0, 1]$$

We compute the requested probabilities in *Mathematica* below, by appropriately integrating the joint and marginal densities.

```
(* P[X > 1/2, Y > 1/2] = *)

3
--- *  ∫   ∫   (x² + y²) dx dy
2     1/2  1/2
```

$$\frac{7}{16}$$

```
(* P[X > 1/2] = *)    3        1
                     --- *  ∫   (x² + 1 / 3) dx
                      2    1/2
```

$$\frac{11}{16}$$

The probability that $Y > 1/2$ is also 11/16, since X and Y have the same distribution. Notice that since $(11/16)(11/16) \neq 7/16$, the joint probability that both random variables exceed 1/2 does not factor into the product $P[X > 1/2] \cdot P[Y > 1/2]$. Thus, the event that X exceeds 1/2 is not independent of the event that Y exceeds 1/2. We will explore independence of two continuous random variables more thoroughly in a few more paragraphs. ∎

In the case where there are more than two jointly distributed random variables, the marginal distribution of each of them is obtained by integrating the joint density over all possible values of all random variables in the group other than the one you are computing the marginal p.d.f. for. For example, if X, Y, and Z have joint density $f(x, y, z)$ then the marginal density of Y would be

$$q(y) = \int \int f(x, y, z) \, dx \, dz$$

Activity 5 Carefully formulate the notion of a joint marginal distribution of a subgroup of a larger group of jointly distributed random variables. Write an integral expression for the joint marginal density of X and Y when random variables X, Y, Z, and W have a joint density function f.

There is an analogue of the idea of a conditional distribution of one random variable given another in the world of continuous random variables too. Remember that for discrete random variables, the expression

$$q(y \mid x) = P[Y = y \mid X = x] = \frac{P[Y = y, X = x]}{P[X = x]} = \frac{f(x,y)}{p(x)}$$

made perfect sense from basic definitions of probability. When X and Y are continuous random variables with joint density $f(x, y)$, the ratio of probabilities in the third term above no longer makes sense, because both numerator and denominator are 0. But the ratio of joint density function to marginal density function in the fourth term does make sense, and we will take that as our definition of the conditional density, by analogy with the discrete case. (Actually, in Exercise 21 you will be led to make a better argument for the legitimacy of this definition by taking a conditional c.d.f. as the starting point.)

Definition 4. Let X and Y be continuous random variables with joint density $f(x, y)$. Denote by $p(x)$ and $q(y)$ the marginal density functions of X and Y, respectively. Then the *conditional density of Y given X = x* is given by

$$q(y \mid x) = \frac{f(x, y)}{p(x)} \qquad (9)$$

and the *conditional density of X given Y = y* is

$$p(x \mid y) = \frac{f(x, y)}{q(y)} \qquad (10)$$

Example 6 Consider again the random variables of Example 5. The joint density is

$$f(x, y) = \tfrac{3}{2}(x^2 + y^2), \quad x, y \in [0, 1]$$

and the two marginal densities are

$$p(x) = \tfrac{3}{2}(x^2 + \tfrac{1}{3}), x \in [0, 1] \quad \text{and} \quad q(y) = \tfrac{3}{2}(y^2 + \tfrac{1}{3}), y \in [0, 1]$$

Therefore, by formula (9) the conditional density of y given x is

$$q(y \mid x) = \frac{f(x,y)}{p(x)} = \frac{(x^2+y^2)}{(x^2+\tfrac{1}{3})}, y \in [0, 1]$$

The density of Y does depend on what x value is observed. For example, here is a plot of the marginal density of Y, together with three of the conditionals, with $X = 1/4$, $X = 1/2$, and $X = 3/4$.

```
q[y_] := 3/2 (y^2 + 1 / 3)
```

```
qygivenx[y_, x_] := (x^2 + y^2) / (x^2 + 1 / 3)
```

```
Plot[{q[y], qygivenx[y, 1 / 4], qygivenx[y, 1 / 2],
    qygivenx[y, 3 / 4]}, {y, 0, 1},
  PlotStyle -> {Thickness[.01], Thickness[.007],
    GrayLevel[.5], Dashing[{.01, .01}]},
  DefaultFont → {"Times-Roman", 8}];
```

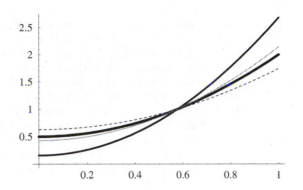

Figure 14 - Marginal of Y (bold), conditional of Y given $X = 1/4$ (thin), $X = 1/2$ (gray), $X = 3/4$ (dashed)

If we wanted to compute the probability that Y exceeds $1/2$ given an observed X value of $1/2$, we would integrate the conditional density of Y as follows:

$$P[Y > 1/2 \mid X = 1/2] = \int_{1/2}^{1} q(y \mid 1/2)\, dy$$

which we can do in *Mathematica*:

$$\int_{1/2}^{1} \frac{(1/2)^2 + y^2}{(1/2)^2 + 1/3}\, dy$$

$$\frac{5}{7}$$

∎

Activity 6 Referring to Example 6, compute $P[X \in [1/3,\, 2/3] \mid Y = 3/4]$.

The last idea that we will take up in this section is the independence of continuous random variables. Here the definition expressed by formula (6) of Section 2.5 serves the purpose again without change. We repeat it here for your convenience:

Definition 5. Two continuous random variables X and Y are *independent* of each other if and only if for all Borel subsets A and B of their respective state spaces,

$$P[X \in A,\, Y \in B] = P[X \in A] \cdot P[Y \in B] \tag{11}$$

All of the comments made in Section 2.5 about independent random variables have their analogues here. An equivalent definition of independence would be $P[Y \in B \mid X \in A] = P[Y \in B]$ for all Borel subsets A and B such that $P[X \in A]$ is not zero. X and Y are also independent if and only if the joint probability density function $f(x, y)$ factors into the product of the marginal densities. In fact, X and Y are independent if and only if the joint c.d.f. $F(x, y) = P[X \le x, Y \le y]$ factors into the product $P[X \le x]\,P[Y \le y]$. The corresponding definition of independence for many random variables would amend formula (11) in the natural way: the simultaneous probability that all of the random variables fall into sets is the product of the marginal probabilities. Again in the many variable context, independence is equivalent to both the factorization of the joint density, and the factorization of the joint c.d.f. into the product of the marginals.

Activity 7 Show that if the joint p.d.f. of two continuous random variables factors into the product of the two marginal densities, then the random variables are independent.

Independence often arises in the following context: a *random sample* taken in sequence and with replacement from a population represented by a probability density f is a sequence $X_1, X_2, ..., X_n$ of independent random variables each of which has density f. Hence by what we said in the last paragraph, the joint density of the random variables X_i is

$$g(x_1, x_2, ..., x_n) = f(x_1)\,f(x_2) \cdots f(x_n) \tag{12}$$

Thus,

$$P[X_1 \in B_1, ..., X_n \in B_n] = \int_{B_1 \times \cdots \times B_n} g(x_1, ..., x_n)\,d x_1 \cdots d x_n$$

$$= \int_{B_1} f(x_1)\,d x_1 \cdots \int_{B_n} f(x_n)\,d x_n \tag{13}$$

Example 7 What do independent continuous random variables look like? Example 3 introduced the Weibull distribution, which was governed by two parameters, α and β. Figure 9 showed the graph of the density and c.d.f. in the case $\alpha = 2, \beta = 3$. The density was asymmetrical, with a long right tail and a peak at around 2. First let us see what we should expect to observe if we simulate random samples (X, Y) of size $n = 2$ from the Weibull(2, 3) distribution. We can slightly alter the earlier command SimPairs to suit our needs. The change is that instead of tabulating the min and max of the raw data pair, we tabulate the raw data pair itself. The distinct calls to Random should make the X and Y observations in a pair independent of one another, and different pairs should also be independent.

```
SimPairs[dist_, numpairs_] :=
   Module[{X, Y, pairlist},
      pairlist = {};
         Do[X = Random[dist];
                  Y = Random[dist];
                  AppendTo[pairlist, {X, Y}],
               {i, 1, numpairs}];
            ListPlot[pairlist, AspectRatio -> 1,
      DefaultFont → {"Times-Roman", 8},
      PlotStyle → PointSize[.015]]]
```

In Figure 15 are the results of simulating 2000 such independent and identically distributed (X_i, Y_i) pairs using the SimPairs command.

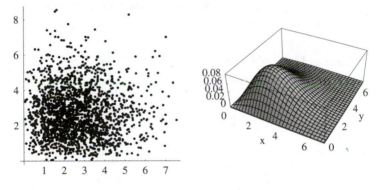

Figure 15 - 2000 simulated Weibull(2, 3) pairs   **Figure 16** - 2-dimensional Weibull(2, 3) joint density

The fact that X and Y have the same distribution shows up in the fact that the point cloud looks about the same in the orientation of Figure 15 as it would if the roles of X and Y were interchanged and the graph was stood on its side. The independence of X and Y is a little harder to see, but you can pick it out by observing that regardless of the vertical level, the dots spread out horizontally in about the same way, with a heavy concentration of points near 2, and a few points near 0 and out at large values between 6 and 8. Similarly, regardless of the horizontal position, the points spread out from bottom to top in this way. We expect this to happen because under independence, the observed value of one of the variables should not affect the distribution of the other.

The graph of the joint density in Figure 16 sheds even more light on this phenomenon. Our point cloud in Figure 15 is most dense exactly where the joint density in Figure 16 reaches its maximum, around (2, 2). Also the point cloud is sparse where the density is low, and the long right tail extending to 6 and beyond is visible in both graphs. But the joint density graph shows the independence property a little better. By examining the grid lines you can see that the relative density of points in the x direction is independent of the y value, and the relative density of points in the y direction is independent of the x value. The identical distribution of the X and Y variables is seen in Figure 16 by the fact that the cross-sections look the same regardless of whether you slice the surface perpendicularly to the x-axis or to the y-axis.

Mathematica for Section 3.2

Command	Location
PDF[dist, x]	Statistics` ContinuousDistributions`
CDF[dist, x]	Statistics` ContinuousDistributions`
UniformDistribution[a, b]	Statistics` ContinuousDistributions`
WeibullDistribution[α, β]	Statistics` ContinuousDistributions`
Random[dist]	Statistics` ContinuousDistributions`
RandomArray[dist, n]	Statistics` ContinuousDistributions`
SeedRandom[seed]	kernel
Histogram[data, numrecs]	KnoxProb` Utilities`
SimPairs[dist, numpairs]	Section 3.2

Exercises 3.2

1. Consider the function

$$f(x) = \begin{cases} C\,e^{-kx} & \text{if } x \geq 0 \\ 0 & \text{otherwise} \end{cases}$$

where C and k are positive constants. Find C such that f is a valid probability density function. Find the cumulative distribution function associated to this distribution.

2. Find the cumulative distribution function of the uniform distribution on $[a, b]$. If X has this distribution, find $P[X > \frac{a+b}{2}]$.

3. For each of the following random variables, state whether it is best modeled as discrete or continuous, and explain your choice.
(a) The error in measuring the height of a building by triangulation
(b) The number of New Year's Day babies born in New York City next year
(c) The relative humidity in Baltimore on a certain day at a certain time

4. The time T (in days) of a server malfunction in a computer system has the probability density function

$$f(t) = \begin{cases} \frac{1}{100}\, t\, e^{-t/10} & \text{if } t \geq 0 \\ 0 & \text{otherwise} \end{cases}$$

Verify that this is a valid density and compute the probability that the malfunction occurs after the tenth day.

5. The c.d.f. of a continuous random variable X is

$$F(x) = \begin{cases} 0 & \text{if } x < 1 \\ (1/4)\,x - 1/4 & \text{if } 1 \leq x < 2 \\ (1/2)\,x - 3/4 & \text{if } 2 \leq x < 3 \\ (1/4)\,x & \text{if } 3 \leq x < 4 \\ 1 & \text{if } x \geq 4 \end{cases}$$

Find a piecewise expression for the density function $f(x)$ and graph both f and F. Explain why it does not matter what value you set for f at the break points 1, 2, 3, and 4. If X has this probability distribution, find $P[1.5 \leq X \leq 3.5]$.

6. (*Mathematica*) By setting several different values for α and fixing β, graph several versions of the Weibull(α, β) density. Describe the dependence of the density on the α parameter. Do the same for the β parameter.

7. Find the location of the maximum value of the Weibull(α, β) density in terms of α and β.

8. If a random variable X has the uniform(0, 1) distribution and $Y = 2X - 1$, what possible values can Y take on? Find $P[Y \leq y]$ (i.e., the c.d.f. of Y) for all such values y. Use this to find the probability density function of Y.

9. (*Mathematica*) Suppose that the time D (in years) until death of an individual who holds a new 25 year $40,000 life insurance policy has the Weibull distribution with parameters $\alpha = 1.5$, $\beta = 20$. The policy will expire and not pay off anything if the policy holder survives beyond 25 years; otherwise if the holder dies within 25 years it pays the $40,000 face value to his heirs. What is the probability that the heirs will receive the $40,000? What is the expected value of the amount that the insurance company will pay?

10. (*Mathematica*) The value V of an investment at a certain time in the future is a random variable with probability density function

$$f(v) = \frac{1}{v} \frac{1}{\sqrt{2\pi}} e^{-(\log v - 4)^2/2}, \ v > 0$$

Find the probability that the value will be at least 40. Find the *median* v^* of the value distribution, that is, the point such that

$$P[V \leq v^*] = .5$$

11. If X, Y, and Z are three independent uniform$(0, 1)$ random variables, find $P[X + Y + Z \leq \frac{1}{2}]$.

12. For the random variables U and V of Example 4, find $P[V - U \geq \frac{1}{4}]$.

13. (*Mathematica*) If X and Y are independent uniform$(-1, 1)$ random variables, build a simulator of pairs (W, Z) where $W = X - Y$ and $Z = X + Y$. By looking at a large sample of such pairs and noting the construction of W and Z, what seems to be the state space of (W, Z)? Does a uniform model seem reasonable? Use your model to find $P[Z \geq 0]$.

14. (*Mathematica*) Two random variables X and Y have joint density

$$f(x, y) = \begin{cases} \frac{1}{2} e^{-y} & \text{if } x \in [0, 2], \ y \in [0, \infty) \\ 0 & \text{otherwise} \end{cases}$$

Sketch the graph of this joint density, and find $P[Y > X]$.

15. For the random variables X and Y of Exercise 14, find the marginal densities. Can you conclude anything about X and Y?

16. For the random variables U and V of Example 4, find the conditional densities $q(v \mid u)$ and $p(u \mid v)$ for all u and v. Compute also $P[V > \frac{3}{4} \mid U = \frac{1}{2}]$.

17. Explain why the random variables X and Y of Example 6 are not independent of each other. However, is there any case in which the marginal density $q(y)$ equals the conditional density $q(y \mid x)$?

18. (*Mathematica*) If X and Y have joint density of the form

$$f(x, y) = \frac{c}{x+y}, \quad x, y \in [1, 2]$$

find the marginal densities of X and Y, and compute $P[Y > \frac{5}{4} \mid X = \frac{3}{2}]$ and $P[X < \frac{3}{2} \mid Y = \frac{6}{5}]$.

19. If X_1, X_2, and X_3 are a random sample (in sequence and with replacement) from the uniform$(0, 3)$ distribution, find (a) $P[X_1 > 1, X_2 < 1, X_3 > 2]$; (b) $P[\min(X_1, X_2, X_3) > 1]$

20. (*Mathematica*) Does the following random sample appear to have been sampled from the distribution whose density function is

$$f(x) = \frac{27}{2} x^2 e^{-3x}, \quad x > 0 ?$$

Use graphics to explain why or why not.

6.21, 3.97, 6.34, 5.31, 6.52, 4.57, 4.03, 5.99, 4.99, 5.54, 4.69, 5.09, 4.13, 5.32, 3.16, 4.16, 4.49, 5.88, 6.23, 7.42, 3.98, 5.29, 5.22, 6.4, 5.25, 5.75, 4.4, 1.25, 5.45, 3.55, 5.83, 3.48, 3.57, 4.95, 5.06, 4.55, 3.06, 5.36, 6.6, 4.57, 6.39, 5.61, 5.72, 4.55, 3.67, 3.79, 4.84, 6.05, 5.99, 6.37, 6.58, 4.48, 5.52, 5.89, 7.19, 2.9, 4.1, 5.96, 5.56, 5.6, 4.72, 6.19, 4.14, 3.76, 4.17, 4.13, 4.05, 6.44, 5.23, 5.88, 5.4, 7.23, 2.82, 5.05, 5.57, 4.74, 4.73, 3.78, 6.0, 2.7, 6.35, 4.07, 4.98, 3.81, 4.6, 4.54, 4.39, 2.85, 5.13, 5.11, 4.83, 4.91, 4.14, 6.19, 3.64, 4.33, 4.54, 4.55, 5.33, 4.27

21. We can take as the definition of the *conditional c.d.f.* of Y given $X = x$ the limit of

$$P[Y \le y \mid X \in [x, x + \Delta x]]$$

as $\Delta x \longrightarrow 0$. This is a well-defined conditional probability, which is conventionally written as $F(y \mid x) = P[Y \le y \mid X = x]$. Find this limit in terms of the joint and marginal densities of X and Y, and show that the derivative of $F(y \mid x)$ is the same as the conditional density function that we defined in the section $q(y \mid x) = f(x, y) / p(x)$.

22. (*Mathematica*) For the uniform distribution on [0, 4] mentioned in the section, *Mathematica* has an interesting way of representing the piecewise defined p.d.f. f, using its Sign function. Enter the definition of f as in the section, then ask *Mathematica* for f[x] to see this representation. Verify that the function is what it should be. What does *Mathematica* think is the value of the density at 0 and 4? Does it matter?

3.3 Continuous Expectation

To conclude this transitional chapter from discrete to continuous probability, we would like to discuss the idea of expectation. Recall that for discrete random variables, we defined

$$E[X] = \sum_x x \cdot f(x) = \sum_x x \cdot P[X = x] \tag{1}$$

The idea was to express the average of states x that could be assumed by the random variable X, using the probability mass function $f(x)$ as a system of weights to give states of higher likelihood more influence on the average value. Now if we have a continuous random variable with a state space that is a bounded interval for instance, we could think of approximating the average value by discretizing the state space into a finite number of states x_i, $i = 1, ..., n$ separated by a common spacing Δx. Because of the meaning of the density function f, the product $f(x_i) \Delta x$ is the approximate probability weight for state x_i, and therefore the approximate expected value in the discretized state space would be

$$\sum_{x_i} x_i \cdot f(x_i) \Delta x$$

In the limit as the number of states $n \longrightarrow \infty$ and the spacing $\Delta x \longrightarrow 0$, this sum converges to the integral in the definition below.

Definition 1. Let X be a continuous random variable with probability density function f. Then the *expected value* of X is the following integral, if it exists.

$$E[X] = \int x f(x) \, dx \tag{2}$$

(The integral is taken over the entire state space of X.) We also use the term *mean of X* and the notation μ for E[X]. The *expected value of a function g* of X is

$$E[g(X)] = \int g(x) f(x) \, dx \tag{3}$$

In formula (3), we will again be most interested in the *moments* of the distribution, i.e., expected powers of X and expected powers of $X - \mu$. Among the moments of chief interest are the mean itself, which measures the center of the probability distribution, and the *variance*:

$$\sigma^2 = \text{Var}(X) = E[(X - \mu)^2] = \int (x - \mu)^2 f(x)\, dx \qquad (4)$$

which measures the spread of the distribution around μ. The square root σ of the variance σ^2 is the *standard deviation* of the distribution.

In this section we will discuss the meaning and computation of certain expectations, paralleling the discussion that began in Section 2.1 and continued in Section 2.6 for discrete random variables. We will feature the same basic results on means, variances, covariance and correlation. Most of the proofs of the results are entirely similar to the discrete case with sums replaced by integrals, and therefore they will usually either be omitted or left to the exercises.

Example 1 As a first example, let us find the mean of the uniform(a, b) distribution.

The p.d.f. of the distribution is $f(x) = 1/(b-a)$ for $x \in [a, b]$ and zero otherwise. Therefore the mean is

$$\mu = E[X] = \int_a^b x\, \frac{1}{b-a}\, dx = \frac{1}{b-a}\, \frac{x^2}{2}\, \Big|_a^b = \frac{b^2 - a^2}{2\,(b-a)} = \frac{b+a}{2} \qquad (5)$$

It is very intuitive that the mean of the uniform(a, b) distribution should come out to be the midpoint of the interval $[a, b]$, since all states in the interval contribute equally to the continuous average that is the mean. You are asked to compute the variance of X in Exercise 1. This computation will be easier if you use the formula in the next activity, which you should derive now. ∎

Activity 1 Show using basic properties of the integral that $\text{Var}(X) = E[X^2] - \mu^2$.

Example 2 Exercise 10 in Section 2.6 asked for a simulation of sample means for larger and larger sample sizes, to show the convergence of the sample mean to the theoretical mean μ. This is a good time to reconsider that investigation in the context of continuous expectation. We now write a command (essentially the same one that would be the solution of that exercise) to successively simulate sample values from a given distribution, pause every m sample values to recompute the sample mean, and update the list of sample means to date.

```
Needs["Statistics`ContinuousDistributions`"]
```

```
SimMeanSequence[
  distribution_, nummeans_, m_] :=
    Module[{nextsample, meanlist,
  runningsum, currnumobs},
    currnumobs = 0;
    runningsum = 0;
    meanlist = {};
    While[currnumobs < nummeans,
        nextsample =
    RandomArray[distribution, m];
        currnumobs = currnumobs + m;
        runningsum =
    runningsum + Apply[Plus, nextsample];
        AppendTo[meanlist,
    runningsum / currnumobs]];
    ListPlot[meanlist, PlotStyle ->
    PointSize[.02], PlotJoined -> True,
    DefaultFont → {"Times-Roman", 8}]]
```

Figure 17 shows the results of executing the command on the uniform(2, 6) distribution, whose mean is the midpoint 4. We simulate 1000 sample values, and update the sample mean every 10 values. The sample mean seems to be converging to the distributional mean of 4. You should try repeating the command several times, without resetting the seed, to see this phenomenon recurring, and also see the amount of variability in the sample means early on in the sequence. This and other similar experiments provide empirical evidence for the claim that the mean of a probability distribution is that value to which the mean of a random sample converges as more and more observations are taken. This result is a pivotal theorem in probability and statistics, called the Strong Law of Large Numbers. It will be discussed more fully in Chapter 5. ∎

```
SeedRandom[439873];
SimMeanSequence[
  UniformDistribution[2, 6], 1000, 10];
```

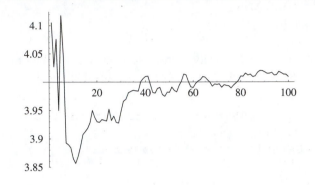

Figure 17 –Simulated sample mean as a function of sample size/10

Example 3 Recall from Section 3.2 the Weibull distribution with parameters α and β. Its p.d.f. is the following:

$$f(x) = \alpha \, \beta^{-\alpha} \, x^{\alpha-1} \, e^{-(x/\beta)^{\alpha}}, \quad x \geq 0$$

Its mean is

$$\mathrm{E}[X] = \int_0^{\infty} x f(x)\, dx = \int_0^{\infty} \alpha \, \beta^{-\alpha} \, x^{\alpha} \, e^{-(x/\beta)^{\alpha}} \, dx$$

It turns out that a substitution $u = (x/\beta)^{\alpha}$ expresses the integral in terms of a standard mathematical function denoted $\Gamma(r)$, which is defined by

$$\Gamma(r) = \int_0^{\infty} u^{r-1} \, e^{-u} \, du \tag{6}$$

(See Exercise 4.) The gamma function will also play an important role in later development. Among its many interesting properties is that if n is an integer, $\Gamma(n) = (n-1)!$. More generally, for all $\alpha > 1$, $\Gamma(\alpha) = (\alpha-1)\,\Gamma(\alpha-1)$, and also it can be shown that $\Gamma(1/2) = \sqrt{\pi}$. In Exercise 4 you will be asked to finish the derivation that the mean of the Weibull(α, β) distribution is

$$\mu = \beta \, \Gamma(1 + \tfrac{1}{\alpha}) \tag{7}$$

If, as in Example 3 of Section 3.2 on shock absorber lifetimes, $\alpha = 2$ and $\beta = 30$, then we can use *Mathematica* to compute the mean lifetime as

```
μ = 30 * Gamma[3 / 2]
N[μ]
```

$15 \sqrt{\pi}$

26.5868

The sketch of the Weibull density with these parameters in Figure 18 shows that 26.59 is a reasonable estimate of the center of the distribution. ∎

```
f[x_] := PDF[WeibullDistribution[2, 30], x];
Plot[f[x], {x, 0, 70},
    DefaultFont → {"Times-Roman", 8}];
```

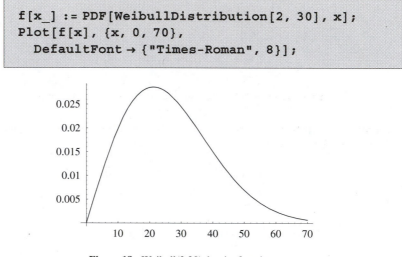

Figure 18 –Weibull(2,30) density function

It is worth pointing out that the Statistics`ContinuousDistributions` package has commands

Mean[distribution]   and   Variance[distribution]

that return the mean and variance of any distribution in the package. For example, for the uniform distribution on [a, b],

> ```
> {Mean[UniformDistribution[a, b]],
>  Variance[UniformDistribution[a, b]]}
> ```

$$\left\{ \frac{a+b}{2}, \ \frac{1}{12} (-a+b)^2 \right\}$$

Example 4   Compute the variance of the random variable X whose probability density function is $f(x) = c/x^4$ if $x > 1$, and 0 otherwise.

By the computational formula in Activity 1, we must compute the first moment $\mu = E[X]$ and the second moment $E[X^2]$. But first we should determine the constant c. In order for the density to integrate to 1, we must have

$$1 = c \int_1^\infty 1/x^4 \, dx = c \left(\frac{x^{-3}}{-3} \right)_1^\infty = c \cdot \frac{1}{3} \Longrightarrow c = 3$$

Then,

$$\mu = \int_1^\infty x \cdot \frac{3}{x^4} \, dx = \int_1^\infty \frac{3}{x^3} \, dx = 3 \left(\frac{x^{-2}}{-2} \right)_1^\infty = \frac{3}{2}$$

$$E[X^2] = \int_1^\infty x^2 \cdot \frac{3}{x^4} \, dx = \int_1^\infty \frac{3}{x^2} \, dx = 3 \left(\frac{x^{-1}}{-1} \right)_1^\infty = 3$$

Therefore by the computational formula for variance in Activity 1,

$$\sigma^2 = E[X^2] - \mu^2 = 3 - \frac{9}{4} = \frac{3}{4}$$

Notice that if we tried to do a similar computation of the variance of a random variable with density $f(x) = c/x^2$, $x > 1$, and 0 otherwise we would run into a problem. When computing $E[X^2]$ for instance, we would try to integrate $x^2 \cdot \frac{c}{x^2} = c$ from 1 to infinity, which would be infinite. Therefore the second moment would not exist, nor would the variance exist. ■

Activity 2   For densities of the form $f(x) = c/x^p$, $x > 1$, and 0 otherwise, for what powers p does the variance exist and for what powers does it not exist? How about the mean?

We can take our cue from the single-variable case in order to formulate a version of expectation in the context of jointly distributed random variables.

Definition 2. Let $X_1, X_2, \ldots, X_n$ be continuous random variables with joint density function $f(x_1, x_2, \ldots, x_n)$. Then the *expected value of a real-valued function g* of the random variables is

$$E[g(X_1, X_2, \ldots, X_n)] = \int \cdots \int g(x_1, x_2, \ldots, x_n) \cdot f(x_1, x_2, \ldots, x_n) \, dx_1 \, dx_2 \cdots dx_n \qquad (8)$$

where the integral is taken over all possible joint states $(x_1, x_2, \ldots, x_n)$.

Most of the time, except when the random variables are independent and the multiple integral simplifies to a product of single integrals, we will consider the bivariate case, in which formula (8) reduces to

$$E[g(X, Y)] = \int \int g(x, y) f(x, y) \, dx \, dy \qquad (9)$$

where again f is the joint density of X and Y. Perhaps the most interesting such expectation is the *covariance* of X and Y:

$$\text{Cov}(X, Y) = E[(X - \mu_x)(Y - \mu_y)] = \int \int (x - \mu_x)(y - \mu_y) f(x, y) \, dx \, dy \qquad (10)$$

which is as before a measure of the degree of association between X and Y. When X and Y tend to be large together and small together, the integrand will be large with high probability, and so the double integral that defines the covariance will be large. The definition of the *correlation* ρ between X and Y is also the same as before:

$$\rho = \text{Corr}(X, Y) = \frac{\text{Cov}(X,Y)}{\sigma_x \sigma_y} \qquad (11)$$

where the σ's are the respective standard deviations of X and Y.

Let's move to a couple of example computations, and then catalogue the general results about continuous expectation.

Example 5 Recall the random variables U and V from Example 4 of Section 3.2, which were the smaller and larger of two independent uniform(0,1) random variables. We found that the joint density of U and V was

$$f(u, v) = 2, \quad \text{if } 0 \le u \le v \le 1, \text{ and } 0 \text{ otherwise.}$$

Then the expected distance between the smaller and the larger is

$$E[V - U] = \int_0^1 \int_u^1 2(v - u) \, dv \, du = 2 \int_0^1 \left(\frac{v^2}{2} - vu \right) \Big|_u^1 \, du$$

$$= 2 \int_0^1 \frac{1}{2} - u + \frac{u^2}{2} \, du = \frac{2}{6}$$

Alternatively in the second expression on the line above we could have noted that the double integral breaks into a difference of two double integrals:

$$\int_0^1 \int_u^1 2\,v\,dv\,du - \int_0^1 \int_u^1 2\,u\,dv\,du$$

which is the same as $E[V] - E[U]$, and we could have computed the two easier double integrals to finish the problem. ∎

Activity 3 Finish the computation of $E[V] - E[U]$ in the manner suggested at the end of Example 5.

Example 6 Find the covariance and correlation of the random variables X and Y from Example 5 of Section 3.2, whose joint density is

$$f(x, y) = \tfrac{3}{2}\,(x^2 + y^2), \quad x, y \in [0, 1]$$

There are quite a few ingredients in the computation, but we will let *Mathematica* do the tedious integrations for us. It is helpful that X and Y have the same marginal distribution, hence the same mean and standard deviation. First let us find $\mu_x = \mu_y$. We can do this by integrating the function $g(x, y) = x$ times the density, which we define below, to get $E[X]$.

```
f[x_, y_] := 3/2 * (x² + y²)
```

```
mux = ∫₀¹ ∫₀¹ x * f[x, y] dy dx;   muy = mux
```

$\dfrac{5}{8}$

To get the variance of X, which equals the variance of Y, we can integrate to find $E[(X - \mu_x)^2]$:

$$\texttt{varx} = \int_0^1 \int_0^1 (\texttt{x} - \texttt{mux})^2 * \texttt{f[x, y] dy dx; vary = varx}$$

$$\frac{73}{960}$$

The covariance and correlation can be done by using formulas (10) and (11):

$$\texttt{covarxy} = \int_0^1 \int_0^1 (\texttt{x} - \texttt{mux}) * (\texttt{y} - \texttt{muy}) * \texttt{f[x, y] dy dx}$$

$$\rho = \frac{\texttt{covarxy}}{\sqrt{\texttt{varx}} * \sqrt{\texttt{vary}}}$$

$$-\frac{1}{64}$$

$$-\frac{15}{73}$$

Notice that X and Y have a slight negative correlation, and that, as usual, the correlation is less than or equal to 1 in magnitude. ∎

Here are the main theorems about expectation, listed in the continuous context. You will observe that they are the same as the ones we proved for discrete random variables. First is the linearity of expectation, which extends (see Exercise 13) to many random variables.

Theorem 1. If X and Y are continuous random variables with finite means, and c and d are constants, then

$$E[c \cdot X + d \cdot Y] = c \cdot E[X] + d \cdot E[Y] \tag{12}$$

If continuous random variables are independent, then the variance of a linear combination is easily expressed:

Theorem 2. If X and Y are independent continuous random variables with finite variances, and c and d are constants, then

$$\mathrm{Var}(c \cdot X + d \cdot Y) = c^2 \cdot \mathrm{Var}(X) + d^2 \cdot \mathrm{Var}(Y) \tag{13}$$

If you check back to the proof of the analogue of Theorem 2 for discrete random variables, you will find that it still works word-for-word here, because it relies only on the linearity of expectation. Theorem 2 also extends readily to many independent random variables; constant coefficients, if present, come out as squares and the variance of the sum breaks into the sum of the variances.

When continuous random variables are independent, expected products factor into products of expectations. We will include the proof of this theorem. It can be extended to many random variables as well (see Exercise 14).

Theorem 3. If X and Y are independent continuous random variables, then $E[g(X) h(Y)] = E[g(X)] E[h(Y)]$ provided these expectations exist.

Proof. If f is the joint density of X and Y, then by the independence assumption, $f(x, y) = p(x) q(y)$, where p and q are the marginal densities of X and Y; respectively. Therefore,

$$
\begin{aligned}
E[g(X) h(Y)] &= \int \int g(x) h(y) f(x, y) \, dy \, dx \\
&= \int \int g(x) h(y) p(x) q(y) \, dy \, dx \\
&= \int g(x) p(x) \, dx \int h(y) q(y) \, dy \\
&= E[g(X)] E[h(Y)]
\end{aligned}
$$

In the case where the random variables are not independent, the formula for the variance of a linear combination includes covariance terms as well, as indicated in the next theorem.

Theorem 4. If X_1, X_2, ..., X_n are jointly distributed random variables, then

$$
\mathrm{Var}(\textstyle\sum_{i=1}^{n} c_i X_i) = \sum_{i=1}^{n} c_i^2 \mathrm{Var}(X_i) + 2 \sum_{i=1}^{n} \sum_{j=1}^{i-1} c_i c_j \mathrm{Cov}(X_i, X_j) \qquad (14)
$$

A couple of familiar miscellaneous results about covariance are in the next theorem. You should read about another interesting property in Exercise 15. Formula (15) is an often useful computational formula for the covariance, and formulas (16) and (17) show that whereas the covariance is dependent on the units used to measure the random variables, the correlation is not. Therefore the correlation is a standardized index of association.

Theorem 5. Let X and Y be jointly distributed random variables with means μ_x and μ_y, and let a, b, c, and d be positive constants. Then

$$\text{Cov}(X, Y) \;=\; E[X \cdot Y] - \mu_x\,\mu_y \tag{15}$$

$$\text{Cov}(a\,X + b,\, c\,Y + d) = a \cdot c \cdot \text{Cov}(X, Y) \tag{16}$$

$$\text{Corr}(a\,X + b,\, c\,Y + d) = \text{Corr}(X, Y) \tag{17}$$

Lastly we have the result that gives meaning to the correlation ρ.

Theorem 6. If ρ is the correlation between random variables X and Y, then $|\rho| \le 1$. Moreover, ρ equals positive 1 or negative 1 if and only if, with certainty, Y is a linear function of X. If X and Y are independent, then $\rho = 0$.

Activity 4   If X and Y are independent random variables, what is $\text{Var}(X - Y)$? (Be careful.)

We would like to conclude this section by mentioning the version of conditional expectation for continuous, jointly distributed random variables. It follows along the same lines as in the discrete case.

Definition 3.   Let X and Y be continuous random variables with joint density $f(x, y)$, and let $q(y\,|\,x)$ be the conditional density of Y given $X = x$. Then the *conditional expectation* of a function g of Y given $X = x$ is

$$E[g(Y)\,|\,X = x] = \int g(y)\,q(y\,|\,x)\,dy \tag{18}$$

So, to find the average value of $g(Y)$ given an observed value of X, we take the weighted average of all possible values $g(y)$, weighted by the conditional density of y for the observed x. Again the law of total probability for expectation holds:

$$E[E[g(Y)\,|\,X]] = E[g(Y)] \tag{19}$$

where we think of $E[g(Y)\,|\,X]$ as the function $h(x) = E[g(Y)\,|\,X = x]$ composed with X, which then becomes a random variable $h(X)$ whose expectation can be taken relative to the marginal density of X. Loosely speaking, formula (19) states that we can find the expectation of $g(Y)$ by first conditioning on X, then unconditioning by averaging out over all X values.

Example 7 Returning to the random variables U and V again from Example 5, i.e., the smaller and larger of two independent, identically distributed uniform(0,1) random variables, we found in Section 3.2 that the marginal densities of U and V are:

$$p(u) = 2(1 - u), \ \ 0 \le u \le 1 \ \ \text{and} \ \ q(v) = 2v, \ \ 0 \le v \le 1$$

Since the joint density was $f(u, v) = 2$, if $0 \le u \le v \le 1$, the two conditional densities are

$$q(v \mid u) = \frac{f(u,v)}{p(u)} = \frac{2}{2(1-u)} = \frac{1}{1-u} \text{ for } v \in [u, 1]$$

$$p(u \mid v) = \frac{f(u,v)}{q(v)} = \frac{2}{2v} = \frac{1}{v} \text{ for } u \in [0, v]$$

This would mean for instance that

$$E[V \mid U = u] = \int_u^1 v \frac{1}{1-u} \, dv = \frac{1}{1-u} \frac{1-u^2}{2} = \frac{1+u}{2}$$

which is unsurprising, since V is uniform on $[u, 1]$ conditioned on $U = u$, and for the uniform distribution the mean is the midpoint of the state space. For similar reasons, the conditional mean of U given $V = v$ would be $v/2$. This says that if the larger of two uniform random numbers is known, then the smaller is expected to be half of the larger. The conditional second moment of U given $V = v$ is

$$E[U^2 \mid V = v] = \int_0^v u^2 \frac{1}{v} \, du = \frac{1}{v} \frac{v^3}{3} = \frac{v^2}{3}$$

Hence the conditional variance of U given $V = v$ is

$$E[U^2 \mid V = v] - (E[U \mid V = v])^2 = \frac{v^2}{3} - \left(\frac{v}{2}\right)^2 = \frac{v^2}{12} \ \blacksquare$$

Mathematica for Section 3.3

Command	Location
PDF[dist, x]	Statistics` ContinuousDistributions`
UniformDistribution[a, b]	Statistics` ContinuousDistributions`
WeibullDistribution[α, β]	Statistics` ContinuousDistributions`
RandomArray[dist, n]	Statistics` ContinuousDistributions`
Mean[dist]	Statistics` ContinuousDistributions`
Variance[dist]	Statistics` ContinuousDistributions`
Gamma[r]	kernel
SeedRandom[seed]	kernel
SimMeanSequence[dist, means, n]	Section 3.3

Exercises 3.3

1. Derive an expression for the variance of the uniform(a, b) distribution.

2. If X is a random variable with probability density function

$$f(x) = \frac{1}{\pi(1+x^2)}, \ x \in \mathbb{R}$$

find, if it exists, the mean of X. Does the variance of X exist?

3. Find the mean and variance of the random variable X with the triangular density function:

$$f(x) = \begin{cases} 1 + x & \text{if } -1 \leq x \leq 0 \\ 1 - x & \text{if } 0 < x \leq 1 \\ 0 & \text{otherwise} \end{cases}$$

4. By carrying out the integral by hand, verify the expression in (7) for the mean of the Weibull(α, β) distribution.

5. (*Mathematica*) Use *Mathematica* to find the first three moments $E[T]$, $E[T^2]$, and $E[T^3]$ for the random variable T of Exercise 4 of Section 3.2, whose density is

$$f(t) = \begin{cases} \frac{1}{100} \, t \, e^{-t/10} & \text{if } t \geq 0 \\ 0 & \text{otherwise} \end{cases}$$

6. Let μ be the mean of the uniform(0, 4) distribution, and let σ be the standard deviation. If X is a random variable with this distribution, find: (a) $P[\,|X - \mu| < \sigma]$; (b) $P[\,|X - \mu| < 2\sigma]$; (c) $P[\,|X - \mu| < 3\sigma]$

7. (*Mathematica*) Use simulation to approximate the mean of the Weibull(4, 6) distribution. Then use *Mathematica* and formula (7) to calculate the mean analytically. Graph the density function, noting the position of the mean on the x-axis.

8. (*Mathematica*) Recall the insurance policy problem of Exercise 3.2-9, in which there is a policy that pays \$40,000 to the holder's heirs if he dies within the next 25 years. The death time (in years beginning from the present) has the Weibull(1.5, 20) distribution. Now let us take into account the premiums paid by the policy holder, which for simplicity we suppose are collected at a constant rate of \$1800 per year. Find the expected profit for the insurance company.

9. Suppose that X and Y have joint density

$$f(x, y) = c(x^3 + y^3), \quad x, y \in [0, 2]$$

Find the covariance and correlation between X and Y.

10. (*Mathematica*) Build a simulator of differences $V - U$ for the random variables of Example 5. Look at several histograms of samples of such differences, and comment on the relationship of the histograms to the expected value of $V - U$ derived in that example.

11. Find $\text{Cov}(U, V)$ for the random variables of Example 5.

12. (*Mathematica*) Compute $E[X + 6Y]$ using *Mathematica* for the random variables X and Y of Exercise 18 of Section 3.2, whose joint density is

$$f(x, y) = \frac{c}{x+y}, \quad x, y \in [1, 2]$$

13. Prove that if $X_1, X_2, \ldots, X_n$ are continuous random variables with finite means and $c_1, c_2, \ldots, c_n$ are constants then

$$E[c_1 X_1 + c_2 X_2 + \cdots + c_n X_n]$$
$$= c_1 E[X_1] + c_2 E[X_2] + \cdots + c_n E[X_n]$$

14. Prove that if $X_1, X_2, \ldots, X_n$ are continuous random variables with finite means and if the X's are independent of each other, then

$$E[X_1 X_2 \cdots X_n] = E[X_1] E[X_2] \cdots E[X_n]$$

15. Show that if X and Y are random variables and $a, b, c,$ and d are constants then

$$\text{Cov}(a\,X + b\,Y,\, c\,X + d\,Y) =$$
$$a\,c\,\text{Var}(X) + (b\,c + a\,d)\,\text{Cov}(X,\,Y) + b\,d\,\text{Var}(Y)$$

16.   Let X and Y be random variables with joint density $f(x,\,y) = 2\,x \cdot 4\,e^{-4y}$, $x \in [0,\,1]$, $y \in [0,\,\infty)$. Find E$[2\,X\,Y]$.

17. If random variables X and Y are uncorrelated (i.e., $\rho = 0$), are the random variables $2\,X - 1$ and $3\,Y + 1$ also uncorrelated? Explain.

18. Show that if X and Y are continuous random variables with joint density $f(x,\,y)$, then

$$\text{E}[g(X) \mid X = x] = g(x)$$

19. Find E$[Y \mid X = x]$ for the random variables X and Y of Example 6.

CHAPTER 4
CONTINUOUS DISTRIBUTIONS

4.1 The Normal Distribution

With most of the general ideas of continuous distributions in hand, we can now move to a survey of some of the important continuous probability distributions. The continuous uniform distribution and the Weibull distribution that we met in the last chapter are certainly two of them. But in this section we study what is perhaps the most important: the normal distribution.

Recall that continuous random variables take values in subintervals of the real line (ideally at least). In many cases, those values tend to distribute across the line in a roughly symmetrical hill shape, with most of the observations near the center, and successively fewer as you move away from the center. Consider for example Figure 1(a), which is a scaled histogram of a large set of 241 National League batting averages on August 25, 1999. You may open up the first of the two closed cells above the figure to see the data that produced the plot. The center of the distribution seems to be around .275, and the great majority of the observations are between about .210 and .340, that is, about .065 on either side of the center.

```
Needs["KnoxProb`Utilities`"]
```

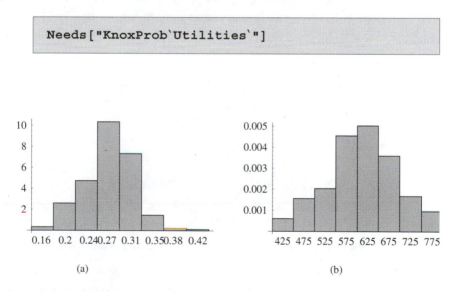

(a)   (b)

Figure 1 - (a) Empirical distribution of National League batting averages;
(b) Distribution of SAT scores

Observed distributions like this are pervasive in the social and physical sciences. Another example is the data set charted in Figure 1(b). (The first closed cell above the figure contains the list of raw data.) This time 167 SAT math scores for entering students at my college show the same sort of distribution, with a center around 640, fewer observations near the tails, and data distributed more or less symmetrically about the center (though there is a small left skew, created mainly by the preponderance of observations in the interval whose midpoint is 650).

Random sampling gives other instances of data distributed in this hill shape. For instance, below we take random samples of size 40 from the uniform(0, 2) distribution, compute the sample means for 100 such random samples, and plot a histogram of the resulting sample means. Almost all of the sample means are within about .25 of the mean $\mu = 1$ of the uniform distribution that we sampled from. This histogram shape is no coincidence, because a theorem called the *Central Limit Theorem*, discussed in Chapter 5, guarantees that the probability distribution of the sample mean is approximately normal.

```
SeedRandom[98996];
sampmeans = Table[
   Mean[RandomArray[UniformDistribution[0, 2],
      40]], {i, 1, 100}];
g1 = Histogram[sampmeans, 8, Type -> Scaled];
```

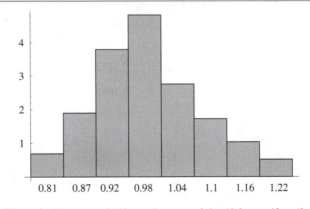

Figure 2 - Histogram of 100 sample means of size 40 from uniform(0, 2)

Activity 1   What is the standard deviation of the uniform(0, 2) distribution? What is the standard deviation of the sample mean $\overline{X}$ of a random sample of size 40 from this distribution? How many standard deviations of $\overline{X}$ does .25 represent?

The probability density function that provides a good fit to distributions such as the three above is the following.

Definition 1. The *normal distribution* with mean parameter μ and variance parameter σ^2 is the continuous distribution whose density function is

$$f(x) = \frac{1}{\sqrt{2\pi\sigma^2}}\, e^{\frac{-(x-\mu)^2}{2\sigma^2}}\,, \quad x \in \mathbb{R} \tag{1}$$

Therefore, if X is a normally distributed random variable (which we abbreviate as $X \sim N(\mu, \sigma^2)$) then the probability that X takes a value in the interval $[a, b]$ is

$$P[a \leq X \leq b] = \int_a^b \frac{1}{\sqrt{2\pi\sigma^2}}\, e^{\frac{-(x-\mu)^2}{2\sigma^2}}\, dx \tag{2}$$

This area is shown in Figure 3.

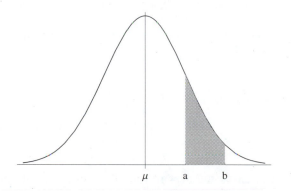

Figure 3 - Normal probability $P[a \leq X \leq b]$ is the shaded area under normal density

In *Mathematica*'s Statistics`ContinuousDistributions` package is the distribution object

NormalDistribution[μ, σ]

which represents the $N(\mu, \sigma^2)$ distribution. (We already loaded the package above when KnoxProb`Utilities` was loaded.) Note that the second argument is the standard deviation σ, which is the square root of the variance σ^2. The functions PDF, CDF, Random, and RandomArray can therefore be applied to this distribution object in the same way as before. To illustrate, suppose that a random variable X

has the normal distribution with mean $\mu = 0$ and variance $\sigma^2 = 4$, hence standard deviation $\sigma = 2$. The commands below define the density function in general and the particular c.d.f., integrate the density function to find $P[0 \le X \le 2]$, then find the same probability by using the c.d.f.

```
f[x_ , μ_ , σ_] :=
    PDF[NormalDistribution[μ, σ], x];
F[x_] := CDF[NormalDistribution[0, 2], x];
NIntegrate[f[x, 0, 2], {x, 0, 2}]
N[F[2] - F[0]]
```

0.341345

0.341345

(If you remove the numerical evaluation function N from the last command and reexecute, you will probably get a closed form that involves a function called Erf, which is defined in terms of a normal integral.)

By using a clever technique to change from a single to a double integral, followed by a substitution, it is possible (see Exercise 3) to show that the normal density in formula (1) integrates to 1 over the real line, as it should. Here we check this fact in *Mathematica* for the $N(0, 4)$ density we used above. You should try changing the mean and standard deviation parameters several times to make sure that this property still holds.

$$\int_{-\infty}^{\infty} f[x, 0, 2]\, dx$$

1

Figure 3, as well as the density formula itself, suggests that the normal density function is symmetric about its mean μ. Figure 4 gives you another look. We plot in three dimensions the normal density as a function of x and the mean μ for constant standard deviation $\sigma = 2$. In the foreground we see the symmetry of $f[x, -2, 2]$ about $\mu = -2$, and as μ increases to 2 we see the density curve cross section receding into the background in such a way that its center of symmetry is always at μ.

```
Plot3D[f[x, μ, 2], {x, -8, 8},
  {μ, -2, 2}, PlotPoints -> 30,
  ViewPoint -> {-0.012, -3.293, 0.779},
  AxesLabel -> {"x", "μ", None},
  DefaultFont → {"Times-Roman", 8},
  ColorOutput → GrayLevel];
```

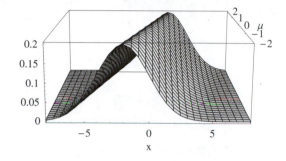

Figure 4 - Standard normal density as function of x and μ

Activity 2 Show that the points $\mu \pm \sigma$ are inflection points of the normal density with mean μ and standard deviation σ. (Use either hand computation or *Mathematica*.)

If you think about what Activity 2 says, you will reach the conclusion that the normal density curve spreads out as the standard deviation σ increases. Try a few *Mathematica* graphs to confirm this visually. Indeed, the larger is σ, the more probability weight is distributed to the tails of the distribution.

Example 1 In Activity 1 we asked some questions about the scenario of taking repeated random samples of size 40 from the uniform(0, 2) distribution and finding their sample means. Figure 2 was a histogram of 100 such sample means. Let us fill in the details now and see how an appropriate normal distribution fits the data.

The mean of the uniform(0, 2) distribution is 1, and the variance is $(2 - 0)^2 / 12 = 1/3$. Recall that if $\overline{X}$ is the mean of a random sample of size n from a distribution whose mean is μ and whose variance is σ^2, then the expected value of $\overline{X}$ is also μ and the variance of $\overline{X}$ is σ^2 / n. So in our situation, $\overline{X}$ has mean 1 and variance $(1/3)/40 = 1/120$, hence its standard deviation is $\sqrt{1/120} \approx .09$. The observed maximum distance of data points from $\mu = 1$ of about .25 that we noted in Figure 3 translates into a distance of around 3 standard deviations of $\overline{X}$. Now we superimpose the graph of the $N(1, 1/120)$ density function on the histogram of Figure 3. Figure 5 shows a fairly good fit, even with our small sample size of 40, and only 100 repetitions of the experiment. ∎

```
g2 = Plot[f[x, 1, √1 / 120],
    {x, .75, 1.25}, DisplayFunction -> Identity,
    DefaultFont → {"Times-Roman", 8}];
Show[g1, g2, DisplayFunction ->
    $DisplayFunction,
    Ticks -> {{0, .75, 1, 1.25}, Automatic}];
```

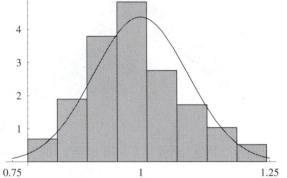

Figure 5 - $N(1, 1/120)$ density and histogram of 100 sample means of size 40 from uniform(0, 2)

Activity 3 Use the RandomArray command to simulate 1000 observations from the $N(0, 1)$ distribution, and plot a scaled histogram of this sample together with the $N(0, 1)$ density.

Example 2 A basic chemistry experiment is designed to measure the concentration of a sulfuric acid solution. Ten replications of this experiment by lab teams yield concentrations (in gram-equivalents/liter) of:

$$.154, .161, .157, .150, .143, .152, .158, .154, .153, .149$$

So, the measurements have a component which is the true concentration plus another component which is measurement error. A statistical model for the situation might be $X_i = c + \epsilon_i$, where X_i is the ith random measurement, c is the unknown actual concentration, and ϵ_i is the ith measurement error for $i = 1, 2, \ldots ,$ 10. We suppose for the rest of this example that the errors ϵ_i are independent, normally distributed random variables with mean 0 and some common variance σ^2. Let us use the observed random sample to estimate the actual concentration of the acid, to say something about the reliability of that estimate, and to predict the probability that a new lab team will observe a measurement that differs by at least .003 from the actual one.

Intuition tells us that the average concentration in the sample should estimate the true concentration. Let us compute that average.

```
concentrations = {.154, .161, .157, .150,
    .143, .152, .158, .154, .153, .149};
Mean[concentrations]
```

0.1531

Since the measurements have only three place accuracy, we should keep only three place accuracy in the mean. Why is the observed sample mean concentration $\bar{x}$ = .153 a good estimate? Observe that the expectation of $\bar{X}$ is

$$E[\bar{X}] =$$
$$E[\tfrac{1}{10} \textstyle\sum_{i=1}^{10} X_i] = \tfrac{1}{10} \textstyle\sum_{i=1}^{10} E[X_i] = \tfrac{1}{10} \textstyle\sum_{i=1}^{10} E[c + \epsilon_i] = \tfrac{1}{10} \textstyle\sum_{i=1}^{10} c = c$$

since $E[\epsilon_i] = 0$. So on average, $\bar{X}$ is expected to be the true c, which gives good support for using the sample mean to estimate c. How precise an estimate is $\bar{X}$? By the properties of expectation, $\text{Var}(X_i) = \text{Var}(\epsilon_i) = \sigma^2$. Hence, $\text{Var}(\bar{X}) = \sigma^2 / 10$. (Make sure you can justify this.) We do not know the error variance σ^2, but we can estimate it by the *sample variance*, which is an average squared deviation of the data values from the sample mean. The exact formula for the sample variance is

$$S^2 = \tfrac{1}{n-1} \sum_{i=1}^{n} (X_i - \bar{X})^2 \tag{3}$$

The so-called *sample standard deviation* S is the square root $S = \sqrt{S^2}$ of the sample variance, and serves as an estimate of σ. *Mathematica* can compute these statistics by the functions

Variance[datalist]   and   StandardDeviation[datalist]

which are contained in the Statistics`DescriptiveStatistics` package (which is also loaded with KnoxProb`Utilities`). Here are the sample variance and standard deviation for our list of measurements.

```
{Variance[concentrations],
 StandardDeviation[concentrations]}
```

```
{0.0000258778, 0.00508702}
```

So far we have that the estimated concentration of acid is .153 gram-equivalents/liter, and that a random measurement has a standard deviation of about .005. But then the sample mean of ten measurements has standard deviation $.005/\sqrt{10}$, which is about .0016. As we have seen before, a large portion of the time a random variable will take on a value within 2 standard deviations of its mean, and so we expect the sample mean to be within 2(.0016) = .0032 of the true concentration. Because the sample mean was .153, we can say with good confidence that the true concentration is between about .150 and .156.

To finish the example, we must compute the probability that a new lab result from a single team will differ from the true concentration c by .003 or more. This new measurement is a random variable X which decomposes as $c + \epsilon$, where ϵ is a normally distributed measurement error. Thus, the absolute difference between X and c is $|X - c| = |\epsilon|$. Using the assumed error mean of 0 and the estimated error standard deviation of .005, and also the symmetry of the normal density about its mean,

$$P[\,|\epsilon| \geq .003] = 2\,P[\epsilon \geq .003] = 2\,(1 - F(.003))$$

where F is the normal c.d.f. of the random error ϵ. This probability is

```
2 (1 - CDF[NormalDistribution[0, .005], .003])
```

```
0.548506
```

It is interesting to note that while it is very likely for the sample mean of 10 observations to be within .003 of the true mean c, it is not very likely $(1 - .548506 \approx .45)$ for a single observation to be this close to c. ∎

We have been calling the two parameters of the normal distribution by the names "mean" and "variance", but we have not yet justified that this is what they are. Here are *Mathematica* computations of the integrals that would define the mean and variance of the $N(0, 1)$ distribution. The second integral is actually $E[X^2]$, but this is the same as the variance in light of the fact that the first integral shows that $E[X] = 0$.

$$\mu = \int_{-\infty}^{\infty} x * \frac{1}{\sqrt{2\,\pi}} * E^{-x^2/2}\, dx$$

$$\text{sigsq} = \int_{-\infty}^{\infty} x^2 * \frac{1}{\sqrt{2\,\pi}} * E^{-x^2/2}\, dx$$

0

1

As expected, the mean is the same as the parameter μ, that is 0, and the variance is $\sigma^2 = 1$. You should try to integrate by hand to check these results; it will help to note that the first integrand is an odd function and the second one is an even function. We now state and prove the general result.

Theorem 1. If X is a random variable with the $N(\mu, \sigma^2)$ distribution, then

$$E[X] = \mu \text{ and } \operatorname{Var}(X) = \sigma^2 \tag{4}$$

Proof. To show that $E[X] = \mu$, it is enough to show that $E[X - \mu] = 0$. This expectation is

$$\int_{-\infty}^{\infty} (x - \mu)\, \frac{1}{\sqrt{2\,\pi\sigma^2}}\, e^{-(x-\mu)^2/2\sigma^2}\, dx$$

The substitution $z = (x - \mu)/\sigma$ changes the integral to

$$\int_{-\infty}^{\infty} z\, \frac{1}{\sqrt{2\,\pi}}\, e^{-z^2/2}\, dz$$

which equals zero from above, because it is the expected value of a $N(0, 1)$ random variable. For the second part, to show that $\operatorname{Var}(X) = E[(X - \mu)^2] = \sigma^2$ it is enough to show that $E[(X - \mu)^2 / \sigma^2] = 1$. The expectation on the left is

$$\int_{-\infty}^{\infty} \frac{(x - \mu)^2}{\sigma^2}\, \frac{1}{\sqrt{2\,\pi\sigma^2}}\, e^{-(x-\mu)^2/2\sigma^2}\, dx$$

Again we make the substitution $z = (x - \mu)/\sigma$ to get

$$\int\limits_{-\infty}^{\infty} z^2 \, \frac{1}{\sqrt{2\pi}} \, e^{-z^2/2} \, dz$$

As mentioned above, it can be verified that this integral equals 1, which completes the proof.

___

The next theorem, usually called the *standardization theorem* for normal random variables, shows that by subtracting the mean from a general $N(\mu, \sigma^2)$ random variable and dividing by the standard deviation, we manage to center and rescale the random variable in such a way that it is still normally distributed, but the new mean is 0 and the new variance is 1. The $N(0, 1)$ distribution therefore occupies an important role in probability theory and applications, and it is given the special name of the *standard normal distribution*.

___

Theorem 2.   If the random variable X has the $N(\mu, \sigma^2)$ distribution, then the random variable defined by

$$Z = \frac{X-\mu}{\sigma} \tag{5}$$

has the $N(0, 1)$ distribution.

Proof. We will use an important technique that will be discussed systematically in Section 4.3, called the *cumulative distribution function* method. In it we write an expression for the c.d.f. of the transformed random variable, take advantage of the known distribution of the original random variable to express that c.d.f. conveniently, and then differentiate it to get the density of the transformed variable.
    The c.d.f. of Z can be written as

$$F_Z(z) = P[Z \le z] = P[\tfrac{X-\mu}{\sigma} \le z] = P[X \le \mu + \sigma z] \tag{6}$$

But we know that X has the $N(\mu, \sigma^2)$ distribution. If $G(x)$ is the c.d.f. of that distribution, then formula (6) shows that $F_Z(z) = G(\mu + \sigma z)$. Differentiating, the density of Z is

$$f_Z(z) = F_Z{}'(z) = G'(\mu + \sigma z)\,\sigma = g(\mu + \sigma z)\,\sigma$$

where g is the $N(\mu, \sigma^2)$ density function. Substituting into formula (1) we obtain

$$f_Z(z) = g(\mu + \sigma z)\,\sigma = \frac{1}{\sqrt{2\pi\sigma^2}} \, e^{\frac{-(\mu+\sigma z - \mu)^2}{2\sigma^2}} \, \sigma = \frac{1}{\sqrt{2\pi}} \, e^{-z^2/2}$$

which is the $N(0, 1)$ density. Thus, Z is standard normal.

___

Activity 4 Use the c.d.f. technique of the proof of Theorem 2 to try to show a converse result: If $Z \sim N(0, 1)$ then the random variable $X = \sigma Z + \mu$ has the $N(\mu, \sigma^2)$ distribution.

Example 3 It has been a tradition in probability and statistics to go on at some length about standardizing normal random variables, and then to use tables of the standard normal c.d.f. to find numerical answers to probability questions. The availability of technology that can quickly give such numerical answers without recourse to standardization has reduced the necessity for doing this, but I cannot resist the temptation to show a couple of quick example computations.

Suppose that the National League batting averages in the introductory discussion are indeed normally distributed. Let us estimate their mean and variance by the sample mean and sample variance of the list of averages.

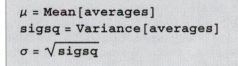

```
μ = Mean[averages]
sigsq = Variance[averages]
σ = √sigsq
```

0.271855

0.00176076

0.0419614

Rounding to three significant digits, we are assuming that a generic randomly selected batting average X has the $N(.272, .00176)$ distribution, so the values of μ and σ will be taken to be .272 and .0420, respectively. What is the probability that this random batting average exceeds .300?

Standardizing by subtracting the mean and dividing by the standard deviation yields:

$$P[X > .300] = P\left[\frac{X-\mu}{\sigma} > \frac{.300-\mu}{\sigma}\right] = P\left[Z > \frac{.300-.272}{.042}\right] = P[Z > .67]$$

Thus we have converted a non-standard normal probability involving X to a standard normal probability involving Z. Now many statistical tables of the standard normal c.d.f. are set up to give, for various values of z, the area under the standard normal density that is shaded in Figure 6(a), which is the probability $P[0 \leq Z \leq z]$.

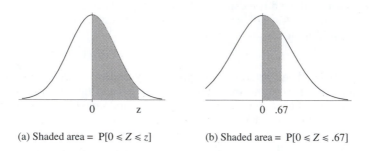

(a) Shaded area = P[0 ≤ Z ≤ z]   (b) Shaded area = P[0 ≤ Z ≤ .67]

Figure 6 - Standard normal probabilities

Such a table would tell us that the area in Figure 6(b), which is P[0 ≤ Z ≤ .67], is .2486. How do we find P[Z > .67] from this? Because of the symmetry of the standard normal density about 0, the area entirely to the right of 0 must be exactly .5. This breaks up into the shaded area in Figure 6(b) plus the unshaded area to the right of .67. Thus, .5 = .2486 + P[Z > .67]; hence P[Z > .67] = .5 − .2486 = .2514. To obtain this result from *Mathematica* is easier. We can either call for the complement of the standard normal c.d.f. at .67:

```
1 - CDF[NormalDistribution[0, 1], .67]
```

0.251429

or before standardizing we can ask for the complement of the normal c.d.f. with μ = .272 and σ = .0420 at the point .300:

```
1 - CDF[NormalDistribution[.272, .0420], .300]
```

0.252493

The slight difference in the numerical answers is due to the fact that the argument .67 in the first call to the standard normal c.d.f. was rounded from its exact value of 2/3.

As a second example computation, what is the probability that a randomly selected batting average is between .244 and .272? Standardizing, we obtain

$$P[.244 \leq X \leq .272] = P[\tfrac{.244-.272}{.042} \leq Z \leq \tfrac{.272-.272}{.042}] = P[-.67 \leq Z \leq 0]$$

Most standard normal tables do not give probabilities for intervals that lie to the left of 0, so you would usually not be able to read this off directly. But again by symmetry (refer to Figure 6(b)) the area between −.67 and 0 equals the area between 0 and .67. It happens that this is a number that has already been given above as .2486. Again, with *Mathematica* we have no need for tables, or for standardizing in this kind of computation, because we can compute the original probability by subtracting c.d.f. values directly:

```
CDF[NormalDistribution[.272, .0420], .272] -
    CDF[NormalDistribution[.272, .0420], .244]
```

0.247507

Again the small difference between the two answers is due to the rounding of .67. ∎

Activity 5 Suppose that a standard normal table gives the probabilities $P[0 \leq Z \leq 1.20] = .3849$ and $P[0 \leq Z \leq .42] = .1628$. Use the symmetry of the normal density to find both $P[-1.20 \leq Z \leq .42]$ and $P[Z > -.42]$.

Example 4 By no means is it irrelevant to know about standardization in the era of widely available technology. Standardization gives us the ability to make very general statements about normal densities based on properties of the standard normal density. For instance there is a rule of thumb called the "68-95 rule" which says that roughly 68% of the area under a normal curve lies within one standard deviation of the mean, and 95% of the area lies within two standard deviations of the mean. Let us see where this rule comes from.

Let X be a general $N(\mu, \sigma^2)$ random variable. Now the area under the normal curve with these parameters that is within one standard deviation of the mean is exactly the probability

$$P[\mu - \sigma \leq X \leq \mu + \sigma] = P[\tfrac{\mu-\sigma-\mu}{\sigma} \leq \tfrac{X-\mu}{\sigma} \leq \tfrac{\mu+\sigma-\mu}{\sigma}] = P[-1 \leq Z \leq 1] \qquad (7)$$

and the standard normal probability on the far right does not depend on the particular values of μ and σ. *Mathematica* computes this probability as:

```
N[CDF[NormalDistribution[0, 1], 1] -
  CDF[NormalDistribution[0, 1], -1]]
```

0.682689

By a standardization entirely similar to the one in formula (7),

$$P[\mu - 2\sigma \leqslant X \leqslant \mu + 2\sigma] = P[-2 \leqslant Z \leqslant 2] \tag{8}$$

and the probability on the right is

```
N[CDF[NormalDistribution[0, 1], 2] -
  CDF[NormalDistribution[0, 1], -2]]
```

0.9545

Thus, the "68-95" rule. ∎

Mathematica for Section 4.1

Command	Location
PDF[dist, x]	Statistics` ContinuousDistributions`
CDF[dist, x]	Statistics` ContinuousDistributions`
NormalDistribution[μ, σ]	Statistics` ContinuousDistributions`
UniformDistribution[a, b]	Statistics` ContinuousDistributions`
RandomArray[dist, n]	Statistics` ContinuousDistributions`
Mean[datalist]	Statistics` DescriptiveStatistics`
Variance[datalist]	Statistics` DescriptiveStatistics`
StandardDeviation[datalist]	Statistics` DescriptiveStatistics`
SeedRandom[]	kernel
Histogram[datalist, numrecs]	KnoxProb` Utilities`

Exercises 4.1

1. (*Mathematica*) The list below consists of 79 Nielsen television ratings of primetime shows selected from the time period of Aug. 16-Aug. 19, 1999. Does the distribution of ratings seem approximately normal? Regardless of the apparent

normality or lack of it, estimate the normal parameters, and superimpose the corresponding density function on a scaled histogram of the ratings.

5.9, 4.6, 4.6, 4.1, 3.1, 1.9, 7.2, 4.9, 4.9, 4.1, 2.0, 8.2, 7.3, 5.8, 5.5, 2.0, 1.9, 7.7, 2.0, 1.5, 7.7, 5.9, 6.1, 6.0, 5.2, 4.8, 2.0, 2.0, 7.7, 5.2, 2.0, 7.7, 6.3, 6.0, 4.7, 1.9, 1.6, 7.8, 4.2, 2.0, 9.6, 5.6, 4.3, 7.2, 6.3, 5.8, 4.9, 2.0, 1.5, 8.0, 7.5, 6.9, 6.3, 2.3, 1.9, 6.4, 8.0, 7.1, 7.7, 5.2, 5.2, 4.3, 2.3, 1.4, 8.6, 6.1, 2.3, 8.0, 6.8, 2.9, 2.4, 1.2, 6.8, 8.8, 3.1, 2.2, 7.4, 7.4, 4.9

2. (*Mathematica*) The data below are crime rates (#crimes/100 people) in 26 Illinois counties in 1998. Does the distribution of crime rates seem approximately normal? Estimate the normal parameters, and superimpose the corresponding density function on a scaled histogram of the crime rates.

7.08, 7.04, 6.27, 5.03, 4.75, 4.44, 4.43, 4.33, 4.28, 4.09, 3.87, 3.76, 3.67, 3.66, 3.37, 3.22, 2.88, 2.86, 2.73, 2.72, 2.65, 2.59, 2.55, 2.54, 2.42, 1.68

3. Show that the $N(0, 1)$ density integrates to 1 over all of $\mathbb{R}$; therefore it is a good density. (Hint: To show that $\int_{-\infty}^{\infty} f(x)\,dx = 1$ it is enough to show that the square of the integral is one. Then express the square of the integral as an iterated two-fold integral, and change to polar coordinates.)

4. Show that the $N(\mu, \sigma^2)$ density function is symmetric about μ.

5. Use calculus to show that the $N(\mu, \sigma^2)$ density function reaches its maximum value at μ, and compute that maximum value.

6. (*Mathematica*) Recall that if $X_1, X_2, \ldots, X_n$ is a random sample of size n, then the sample variance is $S^2 = \frac{1}{n-1} \sum_{i=1}^{n} (X_i - \overline{X})^2$. Simulate 100 samples of size 20 from the $N(0, 1)$ distribution, the $N(2, 1)$ distribution, and the $N(4, 1)$ distribution and in each case, sketch a histogram of the 100 observed values of S^2. Discuss the dependence of the empirical distribution of S^2 on the value of μ. Conduct a similar experiment and discuss the dependence of the empirical distribution of S^2 on the value of σ^2.

7. (*Mathematica* or tables) If X is a $N(\mu, \sigma^2)$ random variable, compute (a) $P[X - \mu < \sigma]$; (b) $P[X - \mu < 2\sigma]$; (c) $P[X - \mu < 3\sigma]$.

8. (*Mathematica* or tables) Suppose that compression strengths of human tibia bones are normally distributed with mean $\mu = 15.3$ (units of kg/mm^2 × 1000) and standard deviation $\sigma = 2.0$. Among four independently sampled tibia bones, what is the probability that at least three have strengths of 17.0 or better?

9. (*Mathematica* or tables) Assume that in a certain area the distribution of household incomes can be approximated by a normal distribution with mean $30,000 and standard deviation $6000. If five households are sampled at random, what is the probability that all will have incomes of $24,000 or less?

10. (*Mathematica*) Simulate 500 values of $Z = (X - \mu)/\sigma$ where $X \sim N(\mu, \sigma^2)$ for several cases of the parameters μ and σ. For each simulation, superimpose a scaled histogram of observed z values and a graph of the standard normal density curve. What do you see, and why should you have expected to see it?

11.   Suppose that a table of the standard normal distribution tells you that $P[0 \le Z \le 1.15] = .3749$, and $P[0 \le Z \le 2.46] = .4931$. Find
(a) $P[1.15 \le Z \le 2.46]$; (b) $P[-1.15 \le Z \le 2.46]$; (c) $P[Z > 2.46]$;
(d) $P[Z < -1.15]$.

12. Assume that X has the $N(1.8, 0.7^2)$ distribution. Without using *Mathematica* or tables, find $P[1.65 \le X \le 1.85]$ if a table of the standard normal distribution tells you that $P[0 \le Z \le .07] = .0279$ and $P[0 \le Z \le .21] = .0832$.

13. (*Mathematica*) For each of 200 simulated random samples of size 50 from the uniform(4, 6) distribution, compute the sample mean. Then plot a scaled histogram of the sample mean together with an appropriate normal density that you expect to be the best fitting one to the distribution of $\overline{X}$. Repeat the simulation for the Weibull(2, 30) distribution that we encountered in Example 3 of Section 3.3.

14. Suppose that the times people spend being served at a post office window are approximately normal with mean 1.5 minutes and variance .16 minutes. What are the probabilities that:
(a) a service requires either more than 1.9 minutes or less than 1.1 minutes?
(b) a service requires between .7 minutes and 2.3 minutes?

15. (*Mathematica*) The *deciles* of the probability distribution of a random variable X are the points $x_{.1}, x_{.2}, x_{.3}, \ldots, x_{.9}$ such that

$$P[X \le x_{.1}] = .1, \ P[X \le x_{.2}] = .2, \ P[X \le x_{.3}] = .3,$$

etc. Find the deciles of the $N(0, 1)$ distribution.

16. The pth *percentile* of the probability distribution of a random variable X is the point x_p such that $P[X \le x_p] = p$. Find a general relationship between the pth percentile x_p of the $N(\mu, \sigma^2)$ distribution and the pth percentile z_p of the standard normal distribution.

4.2 Bivariate Normal Distribution

There is a two-dimensional version of the normal distribution that is of importance in the study of how continuous random variables depend on one another. This so-called *bivariate normal distribution* is the topic of this section.

```
homedata = {{2050, 2650 , 1639},
   {2080, 2600, 1088},
   {2150, 2664 , 1193}, {2150, 2921, 1635},
   {1999, 2580, 1732}, ..., {739, 970, 541}};
```

```
{prices, area, taxes} = Transpose[homedata];
```

To begin, consider the set of triples above, which are sale prices of homes (in hundreds of dollars), square footage of the home, and yearly real estate taxes sampled randomly in 1993 by the Albuquerque, New Mexico Board of Realtors. The first cell is actually a dummy cell to give you an idea of the structure of the data set; the full data set is in the closed cell beneath it, and it consists of 107 such triples. The *Mathematica* command following the data list defines variables called "prices", "area", and "taxes" as the lists of first, second, and third components of "homedata". Preliminary investigation (try it) suggests that the variables themselves are somewhat skewed to the right. In an effort to bring down the large observations and achieve a more symmetrical distribution, we define new variables as the logs of the original variables.

```
logprices = N[Log[prices]];
logarea = N[Log[area]];
logtaxes = N[Log[taxes]];
```

```
Needs["KnoxProb`Utilities`"]
```

Here we use the command

DotPlot[datalist]

in the KnoxProb`Utilities` package to display the distribution of the two logged variables. This command simply plots all points on a horizontal axis, stacking points vertically when necessary.

```
Show[GraphicsArray[{DotPlot[
    logprices, DisplayFunction → Identity,
    DefaultFont → {"Times-Roman", 8}],
  DotPlot[logarea, DisplayFunction → Identity,
    DefaultFont → {"Times-Roman", 8}]}],
  DisplayFunction → $DisplayFunction];
```

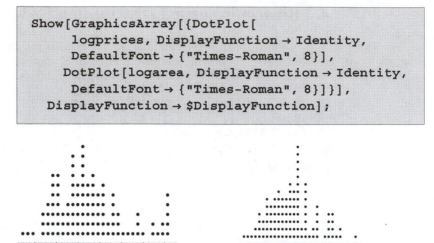

Figure 7 - Dot plots of logged home prices (left) and areas (right)

Consider the logged price and area variables. The dot plots of the individual components in Figure 7 seem to indicate that it is reasonable to suppose that each has a normal distribution, although the home price distribution has a few very large values. But if we plot the pairs in the plane, we see an interesting phenomenon (see Figure 8). The pairs distribute themselves near a line, in an elliptically shaped cloud. You can think of the dot plots in Figure 7 as projections of the point cloud in Figure 8 onto the two coordinate axes. If you try the same thing with the logged price and tax data you will see a similar result. We observe that the two random variables do not appear to act independently; when log(area) is high then log(price) is also high and when one is low the other is low. Thus we have one of many examples in which two random variables *X* and *Y* are normally distributed when considered alone, however they are correlated. What joint distribution has this property?

```
ListPlot[Transpose[{logarea, logprices}],
  DefaultFont → {"Times-Roman", 8}];
```

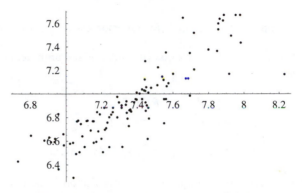

Figure 8 - Scatterplot of logged prices vs. logged areas

Bivariate Normal Density

One way of forming and using a probabilistic model for this situation is to postulate a particular two-dimensional distribution called the *bivariate normal distribution* for the pair (X, Y), which gives rise to the distribution of samples of pairs as in Figure 8. It will be proved that both X and Y have marginal normal distributions. We will also have the capability of modeling the linear relationship between X and Y, and using it for predictive purposes.

Definition 1. The pair of random variables (X, Y) is said to have the *bivariate normal distribution* with parameters $\mu_x, \mu_y, \sigma_x^2, \sigma_y^2$, and ρ if its joint density function is

$$f(x, y) = \frac{1}{2\pi\sigma_x\sigma_y\sqrt{1-\rho^2}} \exp\left[\frac{-1}{2(1-\rho^2)}\left(\frac{(x-\mu_x)^2}{\sigma_x^2} - 2\rho\frac{(x-\mu_x)(y-\mu_y)}{\sigma_x\sigma_y} + \frac{(y-\mu_y)^2}{\sigma_y^2}\right)\right] \quad (1)$$

Activity 1 Let the parameter $\rho = 0$ in formula (1). What do you notice about the joint density $f(x, y)$?

As usual with joint densities, the probability that the pair (X, Y) falls into a set B is

$$P[(X, Y) \in B] = \iint_B f(x, y)\, dx\, dy \quad (2)$$

which is the volume bounded below by the set B in the x-y plane and above by the surface $z = f(x, y)$.

Mathematica knows about this distribution. In the standard package Statistics`MultinormalDistribution` is the object

MultinormalDistribution[meanvector, covariancematrix]

where the argument *meanvector* is the list $\{\mu_x, \mu_y\}$ and the argument *covariancematrix* is the 2×2 matrix

$$\begin{pmatrix} \sigma_x{}^2 & \rho\,\sigma_x\,\sigma_y \\ \rho\,\sigma_x\,\sigma_y & \sigma_y{}^2 \end{pmatrix}$$

(written as $\{\,\{\sigma_x{}^2,\,\rho\,\sigma_x\,\sigma_y\},\,\{\rho\,\sigma_x\,\sigma_y,\,\sigma_y{}^2\}\,\}$ in *Mathematica*)

We will see shortly that these parameters are aptly named. The parameters μ_x and μ_y are the means of X and Y, σ_x^2 and σ_y^2 are the variances of X and Y, and ρ is the correlation between X and Y, which means the matrix entry $\rho\,\sigma_x\,\sigma_y$ is the covariance. (You will show that Corr(X, Y) = ρ in Exercise 9.)

Here for example is a definition of the bivariate normal density with $\mu_x = 0$, $\mu_y = 2$, $\sigma_x^2 = 1$, $\sigma_y^2 = 4$, and $\rho = .6$. Then $\rho\sigma_x\,\sigma_y = (.6)\,(1)\,(2) = 1.2$. Note the use of the list $\{x, y\}$ as the argument for the two variable form of the PDF function.

```
Needs["Statistics`MultinormalDistribution`"]
```

```
meanvector = {0, 2};
covariancematrix = {{1, 1.2}, {1.2, 4}};
f[x_, y_] := PDF[MultinormalDistribution[
    meanvector, covariancematrix], {x, y}];
```

We can understand the graph of f in two ways: by just plotting the density surface or by looking at its contour graph, i.e., the curves of equal probability density. (The closed cell shows the code for generating Figure 9.)

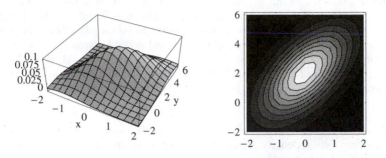

Figure 9 - Bivariate normal density surface and contour plot

The probability mass is concentrated most densely in elliptical regions around the point (0, 2), which is the mean. The major and minor axes are tilted, and there is more spread in the *y*-direction than in the *x*-direction. (Look at the tick marks carefully.) Simulated data from this distribution show the same tendency. Here are 200 points randomly sampled from the same distribution and plotted.

```
simlist = RandomArray[MultinormalDistribution[
    meanvector, covariancematrix], {200}];
ListPlot[simlist, PlotStyle → {PointSize[.015]},
    DefaultFont → {"Times-Roman", 8}];
```

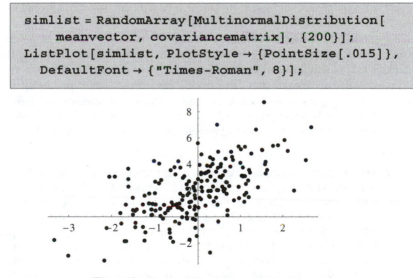

Figure 10 - Simulated bivariate normal data

Note the similarity between the simulated data in Figure 10 and the real data in Figure 8. It appears that the bivariate normal distribution could be a good model for such real data. The role of the mean vector (μ_x, μ_y) in determining the center of the distribution and the point cloud is clear. Try the next activity to investigate the exact role of the variance parameters and the correlation parameter.

Activity 2 Sketch contour graphs of the bivariate normal density, holding μ_x and μ_y at 0, σ_x and σ_y at 1, and varying ρ between $-.9$ and $.9$. Try it again with $\sigma_x = \sigma_y = 2$ and $\sigma_x = \sigma_y = 3$. What seems to be the effect of changing ρ when the standard deviations are held fixed and equal? Next fix $\rho = .7$, and plot contours for the cases: $\sigma_x = 1, \sigma_y = 2$ and $\sigma_x = 1, \sigma_y = 3$ and $\sigma_x = 1, \sigma_y = 4$. For fixed ρ, what is the effect of changing the ratio of standard deviations?

Exercise 8 asks you to show from the density formula (1) that the contours of the bivariate normal density are indeed ellipses centered at (μ_x, μ_y), rotated at an angle α with the x-axis satisfying

$$\cot(2\,\alpha) = \frac{\sigma_x^2 - \sigma_y^2}{2\,\rho\sigma_x\,\sigma_y} \tag{3}$$

Therefore when the variances are equal, the cotangent is 0, hence $2\alpha = \pi/2$, and so the angle of rotation is $\alpha = \pi/4$, as you probably noticed in Activity 2. You may have also noticed that the variance parameters affected the lengths of the two axes of symmetry of the ellipses: as σ_x^2 increases, the length of the axis parallel to the rotated x-axis grows, and similarly for σ_y^2. As in the one-dimensional case, these parameters control the spread of the distribution.

The role of the parameters σ_x^2 and σ_y^2 is easiest to see in a very special case. When you did Activity 1 you should have deduced that when $\rho = 0$ the joint density $f(x, y)$ simplifies to the product $p(x)\,q(y)$ of a $N(\mu_x, \sigma_x^2)$ density for X and a $N(\mu_y, \sigma_y^2)$ density for Y. It is clear in this case that X and Y are independent, and the μ and σ^2 parameters determine the centers and spreads of their marginal distributions. It is not a general rule that lack of correlation implies independence, but it is true when the two random variables have the bivariate normal distribution.

Activity 3 In the closed cell below is a data list of pairs of atmospheric ozone levels collected by the Economic Research Service of the United States Department of Agriculture for a number of counties in the East Coast in the year 1987. The first member of each pair is an average spring 1987 ozone level, and the second is the summer ozone for the same county. Plot the pairs as points in the plane to observe the relationship between the two variables: spring and summer ozone level. Also, do a dot plot or histogram of the individual season levels. Do the underlying assumptions of the model for random bivariate normal pairs seem to hold for this data?

Example 1 Let us estimate for the logged area/homeprice data the probability that the logged area X is between 6 and 7, and simultaneously the logged price Y is between 6 and 7.

Have faith for a moment in the earlier claims that $X \sim N(\mu_x, \sigma_x^2)$ and $Y \sim N(\mu_y, \sigma_y^2)$ and $\rho = \text{Corr}(X, Y)$. We can estimate the means of X and Y by the component sample means below.

```
μ₁ = Mean[logarea];
μ₂ = Mean[logprices];
meanvector = {μ₁, μ₂}
```

`{7.37361, 6.92895}`

Here are the sample variances and standard deviations, which estimate the variances and standard deviations of X and Y.

```
varx = Variance[logarea];
vary = Variance[logprices];
σ₁ = StandardDeviation[logarea];
σ₂ = StandardDeviation[logprices];
{varx, vary}
{σ₁, σ₂}
```

`{0.0883847, 0.100527}`

`{0.297296, 0.31706}`

The correlation ρ must also be estimated. Recall that its defining formula is

$$\rho = \frac{E[(X-\mu_x)(Y-\mu_y)]}{\sigma_x \sigma_y}$$

The expectation in the numerator is an average product of differences between X and its mean with differences between Y and its mean. Given n randomly sampled pairs (X_i, Y_i) a sensible estimate of the correlation is the *sample correlation coefficient R*, defined by

$$R = \frac{\sum_{i=1}^{n}(X_i-\overline{X})(Y_i-\overline{Y})/(n-1)}{S_x S_y} = \frac{\sum_{i=1}^{n}(X_i-\overline{X})(Y_i-\overline{Y})}{\sqrt{\sum_{i=1}^{n}(X_i-\overline{X})^2}\sqrt{\sum_{i=1}^{n}(Y_i-\overline{Y})^2}} \tag{4}$$

So the probabilistic weighted average of products of differences from means is replaced by the average among the sample pairs of products of differences from sample means, and the standard deviations are estimated by sample standard deviations.

Mathematica has a Correlation function in the package called "Statistics`MultiDescriptiveStatistics`". Its syntax is

Correlation[variable1,variable2]

where the arguments are each lists of observations to be correlated. Here we estimate the covariance matrix for the home data set.

```
Needs["Statistics`MultiDescriptiveStatistics`"]
```

```
R = Correlation[logarea, logprices]
covariancematrix =
   {{varx, R * σ₁ * σ₂}, {R * σ₁ * σ₂, vary}}
```

0.867155

{{0.0883847, 0.0817384}, {0.0817384, 0.100527}}

Actually, we need not have gone to this trouble of writing the estimated covariance matrix out longhand, because in the Statistics`MultiDescriptiveStatistics` package is another command

CovarianceMatrix[datalist]

which does the whole job of computing the estimated covariance matrix, as you can see below. Its argument is the full data list of pairs, which we create using Transpose.

```
CovarianceMatrix[
  Transpose[{logarea, logprices}]]
```

{{0.0883847, 0.0817384}, {0.0817384, 0.100527}}

Now we can define the bivariate normal density function using the parameters we have estimated.

```
f[x_, y_] := PDF[MultinormalDistribution[
    meanvector, covariancematrix], {x, y}]
```

The question asks for P[6 < X < 7, 6 < Y < 7], which would be the following integral.

```
NIntegrate[f[x, y], {x, 6, 7}, {y, 6, 7}]
```

```
0.102633
```

For your reference, the graph of the bivariate normal density that approximates the joint distribution of the logged area and price variables appears in Figure 11. ∎

```
Plot3D[f[x, y], {x, 6, 8}, {y, 6, 8},
    PlotRange → All, PlotPoints → 25,
    DefaultFont → {"Times-Roman", 8},
    AxesLabel → {"logarea", "logprice", None},
    ColorOutput → GrayLevel];
```

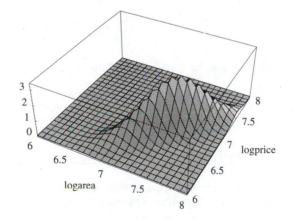

Figure 11 - Approximate joint density of (log(area), log(price))

Activity 4 Referring to Example 1, find the probability that both X and Y are between 7 and 8. Check the reasonableness of your answer by referring to Figure 11.

Marginal and Conditional Distributions

We will now show that if the pair of random variables (X, Y) has the bivariate normal distribution with the usual parameters, then $X \sim N(\mu_x, \sigma_x{}^2)$ and $Y \sim N(\mu_y, \sigma_y{}^2)$. As an incidental but important by-product, we will obtain the conditional distributions of each random variable given the other.

The argument rests on the following algebraic identity, which is tedious but straightforward to check (it comes from completing the square on y in the exponent of the bivariate normal density formula (1))

$$\frac{(x-\mu_x)^2}{2\sigma_x{}^2} + \frac{\left(y-\left[\mu_y+\frac{\rho\sigma_y}{\sigma_x}(x-\mu_x)\right]\right)^2}{2\sigma_y{}^2(1-\rho^2)}$$

$$= \frac{1}{2(1-\rho^2)}\left(\frac{(x-\mu_x)^2}{\sigma_x{}^2} - 2\rho(x-\mu_x)\frac{(y-\mu_y)}{\sigma_x\sigma_y} + \frac{(y-\mu_y)^2}{\sigma_y{}^2}\right)$$

Now the negative of the second expression is just the exponent in the density formula. To shorten notation write

$$\mu_{y|x} = \mu_y + \frac{\rho\sigma_y}{\sigma_x}(x-\mu_x) \quad \text{and} \quad \sigma_{y|x}{}^2 = \sigma_y{}^2(1-\rho^2) \qquad (5)$$

Substituting the left side of the identity into (1), we see that the bivariate normal joint density can be written in factored form as

$$f(x, y) = \frac{1}{\sqrt{2\pi}\,\sigma_x}\exp\left(-\frac{(x-\mu_x)^2}{2\sigma_x{}^2}\right) \cdot \frac{1}{\sqrt{2\pi}\,\sigma_{y|x}}\exp\left(-\frac{(y-\mu_{y|x})^2}{2\sigma_{y|x}{}^2}\right) \qquad (6)$$

The marginal density of X can be found by computing the integral $\int_{-\infty}^{\infty} f(x, y)\, dy$. But the two leading factors in (6) come out of this integral, and the integral of the product of the two rightmost factors is 1, because they constitute a normal density with mean $\mu_{y|x}$ and variance $\sigma_{y|x}{}^2$. Hence the marginal density of X is

$$p(x) = \frac{1}{\sqrt{2\pi}\,\sigma_x}\exp\left(-\frac{(x-\mu_x)^2}{2\sigma_x{}^2}\right)$$

which is the normal density with mean μ_x and variance σ_x^2.

But in fact more is true. Equation (6) states that

$$f(x, y) = p(x) \cdot \frac{1}{\sqrt{2\pi}\,\sigma_{y|x}}\exp\left(-\frac{(y-\mu_{y|x})^2}{2\sigma_{y|x}{}^2}\right)$$

where $p(x)$ is the marginal density of X. Therefore by the definition of conditional density,

$$q(y \mid x) = \frac{f(x,y)}{p(x)} = \frac{1}{\sqrt{2\pi}\,\sigma_{y|x}}\exp\left(-\frac{(y-\mu_{y|x})^2}{2\sigma_{y|x}{}^2}\right) \qquad (7)$$

Hence the conditional density of Y given $X = x$ is also normal, with parameters $\mu_{y|x}$ and $\sigma_{y|x}{}^2$ given by formula (5).

The formula for $\mu_{y|x}$ in (5) is particularly interesting, because it gives a predictive linear relationship between the variables; if $X = x$ is known to have occurred, then the conditional expected value of Y is $\mu_y + \frac{\rho \sigma_y}{\sigma_x}(x - \mu_x)$. It is rather surprising however that the conditional variance $\sigma_y^2(1 - \rho^2)$ of Y does not depend on the particular value that X takes on. But it is intuitively reasonable that the largest value of the conditional variance is σ_y^2 in the case $\rho = 0$ where the variables are uncorrelated, and the conditional variance decreases to 0 as ρ increases to 1. The more strongly that X and Y are correlated, the less is the variability of Y when the value of X is known.

Symmetric arguments yield the marginal distribution of Y and the conditional p.d.f. of X given $Y = y$, in the theorem below, which summarizes all of our results.

Theorem 1. Let (X, Y) have the bivariate normal density in (1) with parameters $\mu_x, \mu_y, \sigma_x^2, \sigma_y^2$ and ρ. Then,

(a) the marginal distribution of X is $N(\mu_x, \sigma_x^2)$;

(b) the marginal distribution of Y is $N(\mu_y, \sigma_y^2)$;

(c) ρ is the correlation between X and Y;

(d) the curves of constant probability density are ellipses centered at (μ_x, μ_y) and rotated at an angle α with the x-axis satisfying $\cot(2\alpha) = \frac{\sigma_x^2 - \sigma_y^2}{2\rho\sigma_x\sigma_y}$;

(e) if $\rho = 0$, then X and Y are independent;

(f) the conditional density of Y given $X = x$ is normal with conditional mean and variance

$$\mu_{y|x} = \mu_y + \frac{\rho\sigma_y}{\sigma_x}(x - \mu_x) \quad \text{and} \quad \sigma_{y|x}{}^2 = \sigma_y^2(1 - \rho^2)$$

(g) the conditional density of X given $Y = y$ is normal with conditional mean and variance

$$\mu_{x|y} = \mu_x + \frac{\rho\sigma_x}{\sigma_y}(y - \mu_y) \quad \text{and} \quad \sigma_{x|y}{}^2 = \sigma_x^2(1 - \rho^2)$$

Activity 5 Finish the details of the proof that Y has a normal distribution, and that the conditional distribution of X given $Y = y$ is normal.

Example 2 You may have heard of the statistical problem of *linear regression*, which involves fitting the best linear model to a data set of pairs (x_i, y_i). The presumption is that the data are instances of a model

$$Y = a + bX + \epsilon \tag{8}$$

where X is a random variable that has the $N(\mu_x, \sigma_x^2)$ distribution, and ϵ is another normally distributed random variable with mean 0 and some variance σ^2, whose correlation with X is zero. The game is to estimate the coefficients a and b and to use them to predict new y-values for given x values. Let us see that the bivariate normal model implies a linear regression model, and also gives information about the coefficients.

Suppose that X and Y are bivariate normal. Define a random variable ϵ by:

$$\epsilon = (Y - \mu_y) - \frac{\rho \sigma_y}{\sigma_x} (X - \mu_x)$$

Since X and Y are normally distributed, and ϵ is a linear combination of them, it can be shown that ϵ is normally distributed. (We will show this fact in the independent case in the next section, though we omit the proof in the correlated case.) It is also easy to see that $E[\epsilon] = 0$ (why?). Finally, ϵ is uncorrelated with X because

$$
\begin{aligned}
\text{Cov}(\epsilon, X) &= E[\epsilon(X - \mu_x)] \\
&= E\left[(Y - \mu_y)(X - \mu_x) - \frac{\rho \sigma_y}{\sigma_x}(X - \mu_x)\right] \\
&= \text{Cov}(X, Y) - \frac{\rho \sigma_y}{\sigma_x}\sigma_x^2
\end{aligned}
$$

The last line follows because $\text{Cov}(X, Y) = \rho \sigma_x \sigma_y$. Therefore Y and X are related linearly by

$$(Y - \mu_y) = \frac{\rho \sigma_y}{\sigma_x}(X - \mu_x) + \epsilon \implies Y = \mu_y + \frac{\rho \sigma_y}{\sigma_x}(X - \mu_x) + \epsilon \qquad (9)$$

and X and ϵ satisfy the distributional assumptions of linear regression. In particular, the slope coefficient $b = \rho \sigma_y / \sigma_x$, and the predicted value of Y given $X = x$ is the conditional mean $\mu_{y|x} = \mu_y + \frac{\rho \sigma_y}{\sigma_x}(x - \mu_x)$. Notice that the farther x is from μ_x, the more the predicted y differs from μ_y. ∎

Activity 6   Simulate 100 pairs (X, Y) using model (8) with $a = 1$, $b = 2$, $\mu_x = 0$, $\sigma_x^2 = 4$, $\sigma^2 = 1$. You should do this by simulating independent X and ϵ observations, using each pair (X, ϵ) to compute the corresponding Y, and repeating 100 times. Does a graph of the resulting pairs (X_i, Y_i) provide convincing evidence that indeed X and Y are bivariate normal?

Let us close this section with an example of a different set of data on which Theorem 1 sheds light.

Example 3 The data below have to do with highway death rates in fatalities per million vehicle miles over a forty year span beginning in 1945. The first component is an index to the year, the second component is the fatality rate in the state of New Mexico in that year, and the third is the death rate in the entire U.S. in that year. We wonder how highway death rates in New Mexico relate to those in the

U.S. as a whole. Is a bivariate normal model appropriate? Can we make predictions about highway fatalities in New Mexico (or in the U.S.)?

```
hwydata = {{1, 14.5, 11.3}, {2 , 14.1, 9.8},
   {3, 12.6 , 8.8}, {4 , 12.7, 8.1},
   {5 , 12.9, 7.5}, {6, 12.6 , 7.6},
   {7, 13.0 , 7.5}, {8 , 11.1 , 7.4},
   {9 , 11.7 , 7.0}, {10 , 10.3, 6.3},
   {11 , 9.4 , 6.4}, {12, 9.6 , 6.4},
   {13, 9.7 , 6.0}, {14 , 8.9 , 5.6},
   {15 , 9.3, 5.4}, {16, 8.0 , 5.3},
   {17, 6.9 , 5.2}, {18 , 8.2 , 5.3},
   {19, 6.8 , 5.5}, {20, 6.9 , 5.7},
   {21, 8.3 , 5.5}, {22, 7.2 , 5.7},
   {23 , 7.3 , 5.5}, {24, 7.4 , 5.4},
   {25, 8.0 , 5.3}, {26, 7.6, 4.9},
   {27, 6.7, 4.7}, {28, 6.6, 4.5},
   {29, 6.7, 4.3}, {30, 5.7, 3.6},
   {31, 5.8, 3.4}, {32, 5.3, 3.3},
   {33, 6.1, 3.4}, {34, 5.6, 3.4},
   {35, 5.7, 3.5}, {36, 5.5, 3.5},
   {37, 4.7, 3.3}, {38, 4.9, 2.9},
   {39, 4.4, 2.7}, {40, 3.8, 2.7}};
```

```
{year, newmexico, us} = Transpose[hwydata];
```

The scatter plot in Figure 12 shows a rather strong linear dependence between the two variables, which supports a bivariate normal model. But the histograms following in Figure 13 show the same kind of right skewed marginal distributions that we observed in Example 1.

```
ListPlot[Transpose[{newmexico, us}],
   AxesLabel → {"New Mex", "U.S."},
   PlotStyle → {PointSize[.02]},
   DefaultFont → {"Times-Roman", 8}];
```

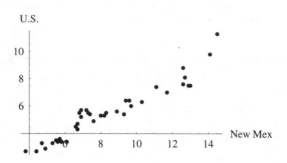

Figure 12 - U.S. vs. New Mexico highway fatality rates, 1945–1984

```
Show[GraphicsArray[{Histogram[newmexico, 5,
    DisplayFunction → Identity], Histogram[
    us, 5, DisplayFunction → Identity]}],
  DisplayFunction → $DisplayFunction];
```

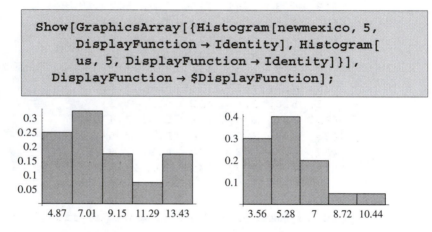

Figure 13 - Marginal distributions of New Mexico (left) and U.S. (right) fatality rates

So we will work from here onward with the logged variables. You can check that the histograms show far more symmetry than those for the original variables. The new scatterplot is in Figure 14, and it confirms that we can safely pursue a bivariate normal model.

```
loggeddata =
  Transpose[{Log[newmexico], Log[us]}];
{lognewmex, logus} = Transpose[loggeddata];
ListPlot[loggeddata,
  PlotStyle → {PointSize[.02]},
  DefaultFont → {"Times-Roman", 8}];
```

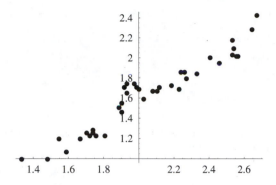

Figure 14 - Scatter plot of logs of variables in Figure 12

Now let us estimate the bivariate normal parameters. We use the Mean, StandardDeviation, and Correlation commands as before.

```
{μ₁, μ₂, σ₁, σ₂, ρ} = {Mean[lognewmex],
   Mean[logus], StandardDeviation[lognewmex],
   StandardDeviation[logus],
   Correlation[lognewmex, logus]}
```

{2.05947, 1.64157, 0.346194, 0.356417, 0.960863}

Notice the very high value $\rho = .96$ which indicates the strength of the relationship. By Theorem 1(a), since the logged New Mexico death rate X has approximately the $N(2.05947, .346194^2)$ distribution, we can find the probability $P[1.7 \le X \le 2.3]$ for example as

```
CDF[NormalDistribution[μ₁, σ₁], 2.3]
 - CDF[NormalDistribution[μ₁, σ₁], 1.7]
```

0.606852

Since X is the log of the actual death rate, say $X = \log(R)$, .606851 is also the probability that $R \in [e^{1.7}, e^{2.3}] \approx [5.47, 9.97]$.

We can also estimate the linear regression relationship between the logged variables. Let Y be the log of the U.S. death rate and write using Theorem 1(f) the conditional mean of Y given $X = x$ as:

```
Y[x_] := μ₂ + (ρ σ₂)/σ₁ (x - μ₁);

Y[x]
```

$$1.64157 + 0.989239 \, (-2.05947 + x)$$

Then the predicted value of the logged U.S. death rate in a year in which the logged New Mexico death rate was 2.2 is

```
Y[2.2]
```

$$1.78059$$

Notice how the occurrence of an *x* value somewhat above the mean of *X* led to a predicted *Y* value above the mean of *Y*. This predicted *Y* is actually the mean of the conditional distribution of *Y* given *X* = 2.2. The variance of that distribution is

```
σ₂² (1 - ρ²)
```

$$0.00974876$$

Therefore, given *X* = 2.2 we have for instance that the conditional probability that *Y* is at least as large as μ_y is

```
1 - CDF[

    NormalDistribution[1.78059, √.00974876], μ₂]
```

$$0.92043$$

Knowledge of this *X* value has drastically altered the distribution of *Y*. ∎

Activity 7 How would you predict the actual (not logged) U.S. death rate in the last example given the actual New Mexico death rate?

Mathematica for Section 4.2

Command	Location
PDF[dist, x]	Statistics` ContinuousDistributions`
(also in)	Statistics` MultinormalDistribution`
CDF[dist, x]	Statistics` ContinuousDistributions`
NormalDistribution[μ, σ]	Statistics` ContinuousDistributions`
MultinormalDistribution[μ, cov]	Statistics` MultinormalDistribution`
RandomArray[dist, {sampsize}]	Statistics` MultinormalDistribution`
Mean[datalist]	Statistics` DescriptiveStatistics`
Variance[datalist]	Statistics` DescriptiveStatistics`
StandardDeviation[datalist]	Statistics` DescriptiveStatistics`
Correlation[var1, var2]	Statistics` MultiDescriptiveStatistics`
CovarianceMatrix[datalist]	Statistics` MultiDescriptiveStatistics`
Histogram[datalist, numrecs]	KnoxProb` Utilities`
DotPlot[list]	KnoxProb` Utilities`

Exercises 4.2

1. (*Mathematica*) Consider the Albuquerque home price data from the start of the section. Produce a scatter plot of the logged taxes vs. the logged prices. Do you see evidence that the bivariate normal distribution could be a reasonable model for the joint distribution of these two variables? Estimate the parameters of that model. Predict the log tax and the tax value when the price is 900.

2. (*Mathematica*) Two company stocks have weekly rates of return X and Y which are bivariate normal with means .001 and .003, variances .00002 each, and correlation $-.2$. Simulate 500 observations of the pair (X, Y) and make a scatterplot of your simulated data. Compute the probability that simultaneously the rate of return X exceeds .0015 and the rate of return Y is less than .002.

3. (*Mathematica*) The November 15, 1999 issue of *The Sporting News* listed statistics for 32 NFL quarterbacks on several variables, including the pairs below, which are percentage of passes completed and average yards gained per pass. Conventional wisdom would indicate that safer, short passes would have a higher completion percentage. Produce a scatterplot of the data, noting any such tendency. Estimate the parameters of a bivariate normal model, and compute
(a) the probability that a randomly selected quarterback completes at least 60% of his passes;
(b) the probability that a randomly selected quarterback averages at least 7.2 yards per pass play;

(c) the predicted completion percentage of a quarterback who averages 7.0 yards per pass.

```
quarterbacks = {{68.8, 8.66}, {61.0, 8.08},
   {60.5, 8.54}, {59.2, 7.42}, {60.1, 8.26},
   {60.5, 7.03}, {59.4, 7.29}, {58.3, 6.78},
   {62.0, 5.78}, {58.2, 7.49}, {58.7, 7.61},
   {58.4, 7.67}, {61.3, 6.64}, {54.6, 7.32},
   {57.6, 6.41}, {60.5, 6.86}, {58.7, 7.51},
   {57.4, 6.94}, {62.0, 7.38}, {62.3, 6.61},
   {60.2, 5.54}, {54.5, 6.23}, {54.5, 6.44},
   {53.7, 7.16}, {52.1, 5.61}, {51.9, 6.22},
   {53.9, 5.70}, {54.0, 6.03}, {51.6, 5.43},
   {52.8, 5.38}, {52.3, 5.26}, {50.6, 6.36}};
{pct, yds} = Transpose[quarterbacks];
```

4. (*Mathematica*) U.S. Department of Commerce data on exports and imports of coal (in thousand tons) in quarters from 1993 to 1998 are below. Do the data suggest a linear relationship between the variables? Estimate that relationship, and use it to predict coal imports in a quarter when exports are 18,000. Compare your answer to the mean imports, and comment on your findings.

```
coal =
   {{18870, 1213}, {19946, 1093}, {18522, 2142},
   {17181, 2861}, {14877, 1850}, {17940, 1577},
   {19704, 2304}, {18838, 1853}, {18988, 1795},
   {23184, 1609}, {22175, 1725}, {24201, 2071},
   {20516, 1713}, {23039, 1552}, {23504, 2071},
   {23414, 1790}, {20011, 1331}, {20603, 1708},
   {22354, 2222}, {20576, 2226}, {18621, 1839},
   {20749, 2193}, {19898, 2145}, {18780, 2547}};
```

5. (*Mathematica*) The pairs below are, respectively, the age-adjusted mortality rate and sulfur dioxide pollution potential in 60 cities (data from the U.S. Department of Labor Statistics). Do you find evidence that these variables can be modeled by a bivariate normal distribution? Try a log transformation and check again. Estimate the parameters for the log transformed data. Does it seem as if mortality is related to the amount of sulfur dioxide present? What is the conditional variance and standard deviation of log mortality given the log sulfur dioxide level?

```
pollution = {{922, 59}, {998, 39}, {962, 33},
    {982, 24}, {1071, 206}, {1030, 72},
    {935, 62}, {900, 4}, {1002, 37}, {912, 20},
    {1018, 27}, {1025, 278}, {970, 146},
    {986, 64}, {959, 15}, {860, 1}, {936, 16},
    {872, 28}, {959, 124}, {942, 11}, {892, 1},
    {871, 10}, {971, 5}, {887, 10}, {953, 1},
    {969, 33}, {920, 4}, {844, 32}, {861, 130},
    {989, 193}, {1006, 34}, {861, 1}, {929, 125},
    {858, 26}, {961, 78}, {923, 8}, {1113, 1},
    {995, 108}, {1015, 161}, {991, 263},
    {894, 44}, {939, 18}, {946, 89}, {1026, 48},
    {874, 18}, {954, 68}, {840, 20}, {912, 86},
    {791, 3}, {899, 20}, {904, 20}, {951, 25},
    {972, 25}, {912, 11}, {968, 102}, {824, 1},
    {1004, 42}, {896, 8}, {912, 49}, {954, 39}};
```

6. (*Mathematica*) Simulate 200 data pairs from the bivariate normal distribution with $\mu_x = \mu_y = 0$, $\sigma_x^2 = \sigma_y^2 = 1$, and each of $\rho = 0, .1, .3, .5, .7, .9$. For each value of ρ, produce individual histograms of the two variables, and a scatterplot of the pairs. Report on the key features of your graphs and how they compare for changing ρ.

7. (*Mathematica*) The data below are scores on the mathematics ACT test, and scores on a mathematics placement examination (out of 17 points) for a group of 70 students. The data are of course integer valued; however do a scatterplot and individual dot plots show evidence that the bivariate normal distribution could be a reasonable approximate model for the two variables? Find the approximate probability that a randomly selected student scores at least 25 on the ACT, find the probability that a randomly selected student scores at least 10 on the placement exam, and find the conditional probability that a student with a 25 on the ACT scores at least 9 on the placement exam.

```
testscores =
    {{27, 10}, {28, 15}, {30, 13}, {16, 5}, {26, 5},
     {23, 11}, {20, 3}, {21, 4}, {21, 5}, {31, 12},
     {28, 10}, {29, 10}, {23, 8}, {20, 5}, {17, 4},
     {19, 4}, {28, 13}, {25, 3}, {32, 14}, {23, 9},
     {26, 10}, {27, 13}, {17, 5}, {25, 10}, {27, 6},
     {21, 4}, {18, 5}, {32, 13}, {30, 10}, {21, 5},
     {22, 6}, {28, 13}, {30, 10}, {27, 11},
     {21, 4}, {26, 8}, {29, 16}, {26, 10},
     {31, 12}, {27, 6}, {29, 12}, {31, 12},
     {22, 8}, {22, 7}, {23, 10}, {24, 8}, {26, 6},
     {22, 4}, {18, 7}, {25, 10}, {26, 10}, {20, 3},
     {28, 13}, {29, 11}, {28, 12}, {27, 13},
     {28, 7}, {28, 11}, {20, 8}, {22, 3}, {25, 4},
     {26, 8}, {23, 6}, {26, 9}, {32, 10}, {19, 4},
     {16, 4}, {28, 6}, {32, 14}, {25, 7}};
```

8. Verify that the contours of the bivariate normal density are ellipses centered at (μ_x, μ_y), rotated at an angle α with the x-axis satisfying formula (3).

9. Show that the parameter ρ in the bivariate normal density is indeed the correlation between X and Y. (Hint: simplify the computation by noting that

$$\rho = E\left[\frac{X-\mu_x}{\sigma_x} \frac{Y-\mu_y}{\sigma_y}\right]$$

and making the two-variable substitution $z_1 = (x - \mu_x)/\sigma_x$, $z_2 = (y - \mu_y)/\sigma_y$. Then complete the square in the exponent.)

10. What do you expect to see if you take cross-sections of the bivariate normal density surface by planes perpendicular to the x-axis? Are the resulting curves probability density functions? If not, what transformation can you apply to make them into density functions?

11. Suppose that Y and X satisfy a linear regression model given by $Y = \mu_y + \frac{\rho\sigma_y}{\sigma_x}(X - \mu_x) + \epsilon$. Compute the correlation between them.

12. *Mathematica* is capable of using its Random command to simulate from the bivariate normal distribution with given μ_x, μ_y, σ_x, σ_y, and ρ. Explain how you could simulate such data yourself using the linear regression model in equation (9).

13. If X and Y are bivariate normal, write the integral for $P[X \leq x, Y \leq y]$, then change variables in the integral as in the hint for Exercise 9. What density function is now in the integrand? Use this to generalize an important result from Section 4.1.

14. A generalization of the bivariate normal distribution to three or more jointly distributed normal random variables is possible. Consider three random variables X_1, X_2, and X_3, and let

$$\mu = \begin{pmatrix} \mu_1 \\ \mu_2 \\ \mu_3 \end{pmatrix} \quad \text{and} \quad \Sigma = \begin{pmatrix} \sigma_1{}^2 & \rho_{12}\,\sigma_1\,\sigma_2 & \rho_{13}\,\sigma_1\,\sigma_3 \\ \rho_{12}\,\sigma_1\,\sigma_2 & \sigma_2{}^2 & \rho_{23}\,\sigma_2\,\sigma_3 \\ \rho_{13}\,\sigma_1\,\sigma_3 & \rho_{23}\,\sigma_2\,\sigma_3 & \sigma_3{}^2 \end{pmatrix}$$

be, respectively, the column vector of means of the X's and the matrix of variances and paired covariances of the X's. Here, ρ_{ij} is the correlation between X_i and X_j. Consider the *trivariate normal density in matrix form*:

$$f(x) = \frac{1}{(2\pi)^{3/2}\,\sqrt{\det(\Sigma)}}\, e^{-1/2\,(x-\mu)^t\,\Sigma^{-1}\,(x-\mu)}$$

in which x is a column vector of variables x_1, x_2, and x_3, the notation "det" stands for determinant, Σ^{-1} is the inverse of the covariance matrix, and the notation y^t means the transpose of the vector y. Show that if all paired correlations are 0 then the three random variables X_1, X_2, and X_3 are mutually independent.

15. Referring to the formula in Exercise 14 for the multivariate normal density function $f(x)$ in matrix form, show that in the two-variable case $f(x)$ agrees with the formula for the bivariate normal density function in (1). (In place of $(2\pi)^{3/2}$ in the denominator, put $(2\pi)^{2/2}$ for the two-variable case. In general, it would be $(2\pi)^{n/2}$ for n variables.)

4.3 New Random Variables from Old

From the very beginning of this book we have set the stage for the subject of this section: new random variables created from others by a transformation. Recall the following examples, among others:

■ In Section 1.1 we looked at the empirical distribution of the minimum among 5 randomly sampled values from $\{1, 2, \ldots, 50\}$. The transformation is $Y = \min\{X_1, X_2, X_3, X_4, X_5\}$.

■ In Example 1 of Section 1.4 we simulated the distribution of the range of a sample of five integers from $\{1, 2, \ldots, 50\}$. The transformation is $Y = \max\{X_1, X_2, X_3, X_4, X_5\} - \min\{X_1, X_2, X_3, X_4, X_5\}$.

■ In Example 3 of Section 2.1 we simulated random numerical grades that had a target distribution. The grade G was a piecewise-defined function of a uniform random variable U.

■ In Example 4 of Section 3.2 we used simulation to support an intuitive argument that random variables U and V, which were the smaller and larger of two uniform random variables, had a uniform density over a triangular region. In this case we have a joint transformation taking the two uniform random variables onto two others, U and V.

■ And in the proof of Theorem 2 of Section 4.1 we used cumulative distribution functions to show that if $X \sim N(\mu, \sigma^2)$, then the standardized $Z = (X - \mu)/\sigma$ has the $N(0, 1)$ distribution.

In this section we will build on your earlier exposure to develop some of the basic ideas and techniques of *transformation theory* in probability. The problem is to find the probability distribution of a random variable that is a function of other random variables. In the first subsection we look at the *cumulative distribution function technique*, which was first introduced in the proof of the normal standardization theorem. We will highlight its implications to simulating continuous random variables. Then in the second subsection we define *moment-generating functions* and explore their usefulness in finding distributions of sums of independent random variables.

Activity 1 Look up and reread the bulleted references above. Try to find other examples from earlier in the book of transformed random variables.

C.D.F. Technique and Simulation

Consider a random variable X whose distribution is known, and a transformed random variable

$$Y = g(X) \tag{1}$$

The idea of the cumulative distribution function technique is to write down the c.d.f. $F_Y(y) = P[Y \leq y]$ of Y, then substitute $Y = g(X)$ in, and express the probability $P[g(X) \leq y]$ as a function of y by using the known distribution of X. We can do this in either the discrete or continuous case, but we will focus mostly on continuous random variables in this section. In the case that X and Y are continuous, once we have the c.d.f. of Y, differentiation produces the density function. We illustrate in the next example.

Example 1   Let X be a continuous random variable with the p.d.f. $f(x) = 2x$, $x \in [0, 1]$, $f(x) = 0$ otherwise. Find the probability density function of $Y = X^2$.

Since X takes on only non-negative values here, we have

$$F_Y(y) = P[Y \le y] = P[X^2 \le y] = P\left[0 \le X \le \sqrt{y}\,\right] \qquad (2)$$

The probability on the far right is

$$P[0 \le X \le \sqrt{y}\,] = \int_0^{\sqrt{y}} 2x\,dx = x^2 \Big|_0^{\sqrt{y}} = y \qquad (3)$$

and this computation is valid for all $y \in [0, 1]$. So, the c.d.f. of Y is $F_Y(y) = y$ for these y's, and clearly F_Y vanishes for $y < 0$ and $F_Y = 1$ for $y > 1$. (Why?) Differentiation of the c.d.f. of Y gives the density of Y as $f_Y(y) = 1$ for $y \in [0, 1]$. Our conclusion is that if X has the distribution given in the problem, then $Y = X^2$ has the uniform(0,1) distribution. ■

Numbered equations (2) and (3) in the last example illustrate beautifully the heart of the c.d.f. method; make sure you understand how they are obtained and why we are looking at them. The next activity should serve as a good check.

Activity 2 This is a converse to the result in Example 1. Let $X \sim$ uniform(0,1) and consider the random variable $Y = \sqrt{X}$.   Show that Y has the p.d.f. $f_Y(y) = 2y$, $y \in [0, 1]$, $f_Y(y) = 0$ otherwise.

Example 2 Let X and Y be independent uniform(0,1) random variables, and let U be the smaller and V the larger of X and Y. We will show that the pair (U, V) has the constant joint density function $f_{U,V}(u, v) = 2$, $0 \le u \le v \le 1$.

Because U is the smaller and V is the larger of X and Y, the state space is clearly the set of pairs (u, v) indicated in the statement: $0 \le u \le v \le 1$. Now just as it is true in the single variable case that a density $f(x)$ is the derivative $F'(x)$ of the c.d.f., it is also true in the two variable case that the joint density is the second order mixed partial of the joint c.d.f., that is:

$$f(x, y) = \frac{\partial^2 F}{\partial x\, \partial y} \qquad (4)$$

You may see this by applying the Fundamental Theorem of Calculus twice to the double integral that defines the joint c.d.f. In the case at hand we have that the joint c.d.f. of U and V is

$$F_{U,V}(u, v) =$$
$$P[U \le u, V \le v] = P[X \le u, Y \le v, \ X \le Y] + P[X \le v, Y \le u, X > Y],$$

by separately considering the two cases where X is the smaller and where Y is the smaller. Now to find the total probability, we would integrate the joint density $f(x, y) = 1$ over the shaded region in Figure 15, but that is just the area of the shaded region. This area, by adding two rectangle areas, is

$$F_{U,V}(u, v) = u\,v + u(v - u) = 2\,u\,v - u^2$$

The partial derivative of this with respect to v is $2\,u$, and the second order mixed partial is

$$f_{U,V}(u, v) = \frac{\partial^2 F_{U,V}}{\partial u\, \partial v} = \frac{\partial}{\partial u}(2\,u) = 2$$

as desired. ∎

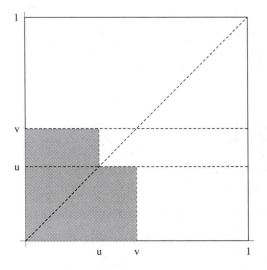

Figure 15 - Region of integration for $U \le u$, $V \le v$, when $u \le v$

Activity 3 In Example 2, though there is no probability density when $u > v$, the c.d.f. $P[U \le u, V \le v]$ is actually defined and non-zero even when $u > v$. Draw a graph similar to Figure 15 for this case, and find an expression for $P[U \le u, V \le v]$. Then take the second order mixed partial and see what happens.

The c.d.f. technique is particularly useful in the proof of the next important theorem on simulation.

Theorem 1. Let U~uniform(0,1), and let F be a continuous, strictly increasing c.d.f. Then $X = F^{-1}(U)$ is a random variable with the distribution associated with F.

Proof. Since F is strictly increasing and continuous, F^{-1} exists and is strictly increasing and continuous. The c.d.f. of $X = F^{-1}(U)$ is

$$P[X \le x] = P[F^{-1}(U) \le x] = P[F(F^{-1}(U) \le F(x)] = P[U \le F(x)] = F(x)$$

since the c.d.f. of U is $F_U(u) = u$, $u \in [0, 1]$. Thus, X has the distribution function F.

Theorem 1 completes the story of simulation of continuous random variables. By the random number generating methods discussed in Chapter 1, one can simulate a pseudo-random observation U from uniform(0,1). To obtain a simulated X that has a given distribution characterized by F, just apply the transformation $X = F^{-1}(U)$.

Example 3 An important continuous distribution that we will look at later in this chapter is the *exponential(λ) distribution*, whose p.d.f. is

$$f(t) = \lambda e^{-\lambda t}, \ t \ge 0 \tag{5}$$

This turns out to be the distribution of the amount of time $T_n - T_{n-1}$ that elapses between successive arrivals of a Poisson process. The Poisson process itself can be simulated if we can simulate an exponential random variable S. By Theorem 1, $S = F^{-1}(U)$ has this distribution, where U~unif(0,1) and F is the c.d.f. associated with f. It remains only to find F^{-1}. First, the c.d.f. is

$$F(t) = \int_0^t \lambda e^{-\lambda u} \, du = 1 - e^{-\lambda t}, t \ge 0 \tag{6}$$

Hence,

$$t = F(F^{-1}(t)) \Longrightarrow t = 1 - e^{-\lambda F^{-1}(t)} \Longrightarrow F^{-1}(t) = \tfrac{-1}{\lambda} \log(1 - t)$$

So, we can simulate an exponential random variable by

$$S = \tfrac{-1}{\lambda} \log(1 - U) \tag{7}$$

Below is a *Mathematica* command that uses what we have done to simulate a list of *n* exponential(λ) observations. Observe that the data histogram of the simulated values has the exponentially decreasing shape that one would expect, given the form of the density function. ∎

```
SimulateExp[n_, λ_] :=
  Table[ -1
         ── * Log[1 - Random[]], {n}]
          λ
```

```
Needs["KnoxProb`Utilities`"]
```

```
SeedRandom[13645]
```

```
datalist = SimulateExp[200, .5];
g1 = Histogram[datalist, 8,
     Type → Scaled, Endpoints → {.5, 11},
     DisplayFunction → Identity];
g2 = Plot[.5 E^-.5 t, {t, 0, 10},
     DefaultFont → {"Times-Roman", 8},
     DisplayFunction → Identity];
Show[g1, g2, DisplayFunction →
     $DisplayFunction];
```

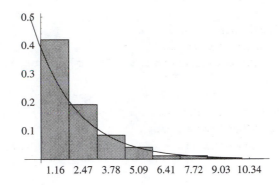

Figure 16 - Histogram of 200 simulated exp(1/2) observations with exp(1/2) density function

Example 4 We have used *Mathematica*'s Random[dist] command and NormalDistribution[μ,σ] object before to simulate $N(\mu, \sigma^2)$ observations. We now know how to simulate normal observations ourselves using uniform(0,1) pseudo-random numbers. But we must compute the inverse c.d.f. numerically rather than in closed form. This is because the normal c.d.f. is

$$F(x) = \int_{-\infty}^{x} \frac{1}{\sqrt{2\pi\sigma^2}} \, e^{-(t-\mu)^2/2\sigma^2} \, dt$$

which is not expressible in closed form, nor is its inverse. Fortunately, *Mathematica* has the function

Quantile[distribution, x]

in the Statistics`ContinuousDistributions` package, which returns the inverse c.d.f. evaluated at *x* for the given distribution. For example,

```
CDF[NormalDistribution[0, 2], 1.5]
Quantile[NormalDistribution[0, 2], .773373]
```

```
0.773373
```

```
1.5
```

These computations are telling us that if F is the $N(0, 2^2)$ c.d.f., then $F(1.5) = .773373$, and correspondingly $F^{-1}(.773373) = 1.5$. The following command simulates a list of n normal observations, using Theorem 1, i.e., by computing $F^{-1}(U)$ for the desired number of pseudo-random uniform random variables U. The dotplot of the observations in Figure 17 illustrates that the shape of the empirical distribution is appropriate. ∎

```
SimNormal[n_, μ_, σ_] :=
  Table[Quantile[
    NormalDistribution[μ, σ], Random[]], {n}]
```

```
datalist = SimNormal[200, 0, 2];
DotPlot[datalist,
  DefaultFont → {"Times-Roman", 8}];
```

Figure 17 - 200 simulated N(0, 2^2) observations

Moment-Generating Functions

In many interesting applications of probability we need to find the distribution of the sum of independent random variables:

$$Y = X_1 + X_2 + \cdots + X_n$$

or the sample mean of independent, identically distributed random variables:

$$\overline{X} = \frac{X_1 + X_2 + \cdots + X_n}{n}$$

In Example 2 of Section 2.6 we simulated such sample means, and also observed by properties of expectation that $E[\overline{X}] = \mu$ and $Var(\overline{X}) = \sigma^2/n$. But can we draw conclusions analytically about the entire distribution of $\overline{X}$? Remember that at least for large sample sizes, it appeared from the simulation experiment that this distribution was approximately normal.

A new probabilistic device, defined below, is very useful for problems involving the distributions of sums of random variables.

Definition 1. The *moment-generating function* (m.g.f.) of a probability distribution is the function

$$M(t) = E[e^{tX}] \tag{8}$$

valid for all real numbers t such that the expectation is finite.

Notice that M is just a real-valued function of a real variable t. It would be calculated as

$$M(t) = \sum_{k \in E} e^{tk} q(k) \tag{9}$$

for a discrete distribution with p.m.f. q, and

$$M(t) = \int_E e^{tx} f(x)\,dx \tag{10}$$

for a continuous distribution with density function f.

Let us compute m.g.f.'s for two important distributions from earlier in the book: binomial and normal.

Example 5 Find the moment-generating function of the $b(n, p)$ distribution.

By formula (9) and the binomial theorem,

$$
\begin{aligned}
M(t) = E[e^{tX}] &= \sum_{k=0}^{n} e^{tk} \binom{n}{k} p^k (1-p)^{n-k} \\
&= \sum_{k=0}^{n} \binom{n}{k} (p\,e^t)^k (1-p)^{n-k} \\
&= (p\,e^t + (1-p))^n
\end{aligned}
\tag{11}
$$

The formula in (11) is valid for all $t \in \mathbb{R}$. ∎

Example 6 Find the moment-generating function of the $N(\mu, \sigma^2)$ distribution.

It suffices to find the moment-generating function $M_Z(u)$ of the standard normal distribution, because the general normal m.g.f. may be written as

$$M(t) = E[e^{tX}] =$$
$$E[e^{\sigma t(X-\mu)/\sigma} e^{\mu t}] = e^{\mu t} E[e^{\sigma t(X-\mu)/\sigma}] = e^{\mu t} E[e^{\sigma t Z}] = e^{\mu t} M_Z(\sigma t)$$

The standard normal m.g.f. is

$$
\begin{aligned}
M_Z(u) = \mathrm{E}[e^{uZ}] &= \int_{-\infty}^{\infty} \frac{e^{uz}}{\sqrt{2\pi}}\, e^{-z^2/2}\, dz \\
&= \int_{-\infty}^{\infty} \frac{1}{\sqrt{2\pi}}\, e^{-(z^2 - 2uz + u^2)/2} \cdot e^{u^2/2}\, dz \\
&= e^{u^2/2} \int_{-\infty}^{\infty} \frac{1}{\sqrt{2\pi}}\, e^{-(z-u)^2/2}\, dz \\
&= e^{u^2/2}, \quad u \in \mathbb{R}
\end{aligned}
$$

since the last integral on the right is that of a $N(u, 1)$ density, which equals 1. Therefore the general $N(\mu, \sigma^2)$ m.g.f. is

$$
M(t) = e^{\mu t}\, M_Z(\sigma t) = e^{\mu t}\, e^{(\sigma t)^2/2} = e^{\mu t + \sigma^2 t^2/2}, \quad t \in \mathbb{R} \quad \blacksquare \tag{12}
$$

Activity 4 Derive the m.g.f. of the Bernoulli(p) distribution.

The phrase "moment-generating function" comes from the fact that

$$
M'(0) = \mathrm{E}[X], \quad M''(0) = \mathrm{E}[X^2], \tag{13}
$$

etc.; in general the nth derivative of the m.g.f. evaluated at 0 is $M^{(n)}(0) = \mathrm{E}[X^n]$. So, the function $M(t)$ generates all of the moments of the distribution of X. You will prove this in Exercise 9.

As interesting as the previous fact might be, it takes second place to the following theorem, whose proof is beyond the level of this book: the m.g.f. is *unique to the distribution*, meaning that no two distributions share the same m.g.f. This implies that if we can recognize the m.g.f. of a transformed random variable Y by knowing the m.g.f.'s of those random variables that Y depends on, then we have the distribution of Y. This line of reasoning is known as the *moment-generating function technique*. We use it in two examples below to prove important results about the normal distribution.

Example 7 Let $X \sim N(\mu, \sigma^2)$. Use the moment-generating function technique to show that $Y = aX + b$ is normally distributed.

From formula (12), the m.g.f. of X is $M(t) = \exp(\mu t + \frac{1}{2}\sigma^2 t^2)$. Therefore the m.g.f. of Y is

$$
\begin{aligned}
M_Y(t) = E[e^{tY}] \ &= \ E[e^{t(aX+b)}] \\
&= \ e^{tb} \, E[e^{(ta)X}] \\
&= \ e^{tb} \, M(ta) \\
&= \ e^{tb} \, e^{\mu ta + 1/2 \sigma^2 (ta)^2} \\
&= \ e^{(a\mu+b)t + 1/2(\sigma^2 a^2)t^2}
\end{aligned}
$$

The last expression is the m.g.f. of the $N(a\mu + b, a^2\sigma^2)$ distribution; hence by the uniqueness of m.g.f's, Y must have this distribution. ∎

Activity 5 Use the moment-generating function technique to show the normal standardization theorem: if $X \sim N(\mu, \sigma^2)$, then $Z = \frac{X-\mu}{\sigma} \sim N(0, 1)$.

So by Example 7, a linear transformation of a normal random variable results in another normal random variable. In fact, linear combinations of many independent normal random variables are also normal, as we show in the next example. (The hypothesis of independence can be dropped, but we will not prove that in this book.)

Example 8 Let $X_i \sim N(\mu_i, \sigma_i^2)$, $i = 1, 2, \ldots, n$, and suppose that the X_i's are mutually independent. Prove that the random variable

$$
Y = a_1 X_1 + a_2 X_2 + \cdots + a_n X_n
$$

is normal with mean and variance

$$
\mu_Y = \sum_{i=1}^{n} a_i \mu_i, \quad \sigma_Y^2 = \sum_{i=1}^{n} a_i^2 \sigma_i^2
$$

First recall that if $X_1, X_2, \ldots, X_n$ are mutually independent, then

$$
E[g_1(X_1) g_2(X_2) \cdots g_n(X_n)] = E[g_1(X_1)] E[g_2(X_2)] \cdots E[g_n(X_n)] \tag{14}
$$

Therefore the m.g.f. of Y is

$$
\begin{aligned}
M_Y(t) = E[e^{tY}] \ &= \ E[e^{t(a_1 X_1 + a_2 X_2 + \cdots + a_n X_n)}] \\
&= \ E[\textstyle\prod_{i=1}^{n} e^{t a_i X_i}] \\
&= \ \textstyle\prod_{i=1}^{n} E[e^{(t a_i X_i)}] \\
&= \ \prod_{i=1}^{n} e^{t a_i \mu_i + \frac{1}{2}\sigma_i^2 (a_i t)^2} \\
&= \ e^{t \sum a_i \mu_i + \frac{1}{2}(\sum \sigma_i^2 a_i^2)t^2}
\end{aligned}
\tag{15}
$$

The latter is the m.g.f. of $N(\Sigma\, a_i\, \mu_i,\, \Sigma\, \sigma_i^2\, a_i^2)$, which completes the proof. ∎

A special case of the last example deserves our attention.   Let $X_1, X_2, \ldots, X_n$ be a random sample from the $N(\mu, \sigma^2)$ distribution. In particular, the X_i's are independent. Consider the sample mean $\overline{X}$. It can be rewritten as

$$\overline{X} = \frac{\Sigma X_i}{n} = \frac{1}{n}\, X_1 + \frac{1}{n}\, X_2 + \cdots + \frac{1}{n}\, X_n$$

In Example 8, let $a_i = 1/n$, $\mu_i = \mu$, and $\sigma_i^2 = \sigma^2$ for all $i = 1, 2, \ldots, n$. Then the conclusion of Example 8 implies that

$$\overline{X} \sim N\!\left(\sum_{i=1}^{n} \frac{1}{n}\, \mu,\; \sum_{i=1}^{n} \frac{1}{n^2}\, \sigma^2\right) = N\!\left(\mu,\, \frac{\sigma^2}{n}\right) \tag{16}$$

This is one of the most important results in probability and statistics. The sample mean of a normal random sample is a random variable with a normal distribution also, whose mean is the same as the mean μ of the distribution being sampled from, and whose variance σ^2/n is $1/n$ times the underlying variance. We have seen these results about the mean and variance before, but we have just now proved the normality of $\overline{X}$. Let us close this section with an illustration of the power of this observation.

Example 9 A balance is suspected of being in need of calibration. Several weight measurements of a known 1 gram standard yield the data below. Give a probability estimate of the likelihood that the balance needs recalibration. (Assume that the standard deviation of measurements is $\sigma = .0009$ grams.)

```
datalist = {1.0010, .9989, 1.0024,
    1.0008, .9992, 1.0015, 1.0020, 1.0004,
    1.0018, 1.0005, 1.0013, 1.0002};
```

Apparently a correctly calibrated balance should yield observations with a mean of 1 gram on the standard weight. There are 12 observations, which we will assume constitute a random sample from a normal distribution with $\sigma = .0009$. It is helpful to draw a legal analogy in order to respond to the question. Assume that the balance is innocent until proven guilty beyond a reasonable doubt. If the 12 observations together are unlikely to have occurred if the balance was correctly calibrated, then we convict the balance of the crime of miscalibration. We can summarize the information contained in the raw data by the sample mean $\overline{X}$; then if the particular $\overline{x}$, a possible value of the random variable $\overline{X}$, that we observe is a very unlikely one when $\mu = 1$, we would conclude that μ is most probably not 1.

First, we let *Mathematica* give us the observed value $\overline{x}$ of the sample mean.

```
Mean[datalist]
```

1.00083

So the observed sample mean is a bit bigger than 1. But is it big enough to constitute "beyond reasonable doubt"? As a random variable, under the "assumed
innocence" of the balance, $\overline{X}$ would have the $N(1, (.0009)^2/12)$ distribution by
formula (16). Now the likelihood that $\overline{X}$ comes out to be identically equal to
1.00083 is 0 of course, since $\overline{X}$ is continuously distributed. But that is not a fair use
of the evidence. It is more sensible to ask how large is the probability that $\overline{X} \geq$
1.00083 if the true $\mu = 1$. If that is very small then we have a lot of probabilistic
evidence against $\mu = 1$. We can compute:

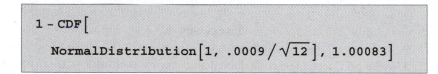

```
1 - CDF[
    NormalDistribution[1, .0009/√12], 1.00083]
```

0.00069995

We now see that if $\mu = 1$ were true, it would only be about .0007 likely for us to
have observed a sample mean as large or larger than we did. This is very strong
evidence that the true μ is greater than 1, and the balance should be recalibrated.
We might say that in deciding this, our risk of an erroneous decision is only about
.0007.

It would be a good idea for you to reread this example a couple of times,
paying particular attention to the use of statistical evidence to draw a conclusion. It
illustrates the kind of thinking that you will see frequently if you take a course in
statistics. ∎

Activity 6 How sensitive is the decision we made in Example 9 to the largest
observation 1.0024? Delete that and rerun the problem. Do we still prove the
balance guilty beyond a reasonable doubt? What is the new error probability?

Mathematica for Section 4.3

Command	Location
CDF[dist, x]	Statistics` ContinuousDistributions`
PDF[dist, x]	Statistics` ContinuousDistributions`
NormalDistribution[μ, σ]	Statistics` ContinuousDistributions`
Quantile[distribution, q]	Statistics` ContinuousDistributions`
Mean[datalist]	Statistics` DescriptiveStatistics`
Random[]	kernel
SimulateExp[n, λ]	Section 4.3
SimNormal[n, μ, σ]	Section 4.3

Exercises 4.3

1. Use the c.d.f. technique to find the density function of (a) $Y = X^2$; (b) $Y = \sqrt[3]{X}$ if X is a continuous random variable with density function $f(x) = 3\,x^2$, $x \in [0, 1]$.

2. Use the c.d.f. technique to find the density function of $U = Z^2$ if $Z \sim N(0, 1)$. This density is called the *chi-square density*, and will be of interest to us in a later section.

3. Explain, with justification, how to simulate a uniform(a, b) observation using a uniform(0, 1) observation.

4. Prove this converse of the simulation theorem. If F is the strictly increasing, continuous c.d.f of the random variable X, then the random variable $U = F(X)$ has the uniform(0,1) distribution.

5. (*Mathematica*) Write a *Mathematica* command to simulate a list of n observations from the distribution whose density function is $f(x) = 3\,x^2$, $x \in [0, 1]$. Run the command several times, and compare histograms of data lists to the density function.

6. If a random temperature observation in Fahrenheit is normally distributed with mean 70 and standard deviation 5, find the probability that the temperature recorded in degrees Celsius exceeds 20.

7. Derive the moment-generating function of the uniform(a, b) distribution.

8. Derive the moment-generating function of the geometric(p) distribution.

9. Prove that if X is a random variable with moment-generating function $M(t)$, then the nth derivative of the m.g.f. at 0 is the nth moment, i.e., $M^{(n)}(0) = E[X^n]$.

10. Derive the moment-generating function of the Bernoulli(p) distribution (see Activity 4), and use it to show that the sum of n i.i.d. Bernoulli(p) random variables has the binomial(n, p) distribution.

11. Derive the moment-generating function of the Poisson(μ) distribution, and use it to show that if $X_1 \sim \text{Poisson}(\mu_1)$, $X_2 \sim \text{Poisson}(\mu_2)$, ..., $X_n \sim \text{Poisson}(\mu_n)$ are independent random variables, then $X_1 + X_2 + \cdots + X_n$ has a Poisson distribution with parameter $\mu_1 + \mu_2 + \cdots + \mu_n$.

12. Recall from Section 4.1 the example data set on National League batting averages. Accepting as correct the estimates of the mean batting average (.271855) and the standard deviation (.0419614), what is the probability that a sample of 20 such batting averages will have a sample mean of less than .250? Greater than .280?

13. As in Exercise 4.1.14, suppose that the times people spend being served at a post office window are normal with mean 1.5 minutes and variance .16 minutes. What are the probabilities that
(a) the average service time among ten people requires more than 1.7 minutes;
(b) the total service time among ten people requires less than 16 minutes?

4.4 Gamma Distributions

Main Properties

The family of *gamma distributions* is one of the most important of all in probability and statistics. In this section we will learn about the distributional properties of the general gamma distribution, then we will discuss some of the special features and applications of an important instance: the *exponential distribution*. Another instance of the gamma family, the *chi-square distribution*, will be introduced in the next section.

Definition 1. A continuous random variable X has the *gamma distribution* if its probability density function is:

$$f(x; \alpha, \beta) = \frac{1}{\beta^\alpha \, \Gamma(\alpha)} \, x^{\alpha-1} \, e^{-x/\beta}, \ x > 0 \tag{1}$$

where $\alpha, \beta > 0$ are constant parameters. We abbreviate $X \sim \Gamma(\alpha, \beta)$.

As we mentioned earlier in the book, the gamma function in the denominator of (1) is defined by the integral

$$\Gamma(\alpha) = \int_0^\infty t^{\alpha-1} \, e^{-t} \, dt \tag{2}$$

which can be obtained by integration by parts for many values of α. We will consider its properties in more detail in a moment.

Activity 1 Use the substitution $u = x/\beta$ to show that the integral of the gamma density is 1.

The $\Gamma(n, 1/\lambda)$ distribution, where n is a positive integer, is called by some authors the *Erlang distribution* with parameters n and λ, but we will not use that terminology here.

Let's take a look at the graph of the gamma density function. In the ContinuousDistributions package there is a GammaDistribution[α, β] object, to which you can apply the PDF function as below. In Figure 18 is the gamma p.d.f. for parameters $\alpha = 2$ and $\beta = 3$. Note the asymmetrical, right-skewed shape of the density graph.

```
Needs["KnoxProb`Utilities`"];
f[x_] := PDF[GammaDistribution[2, 3], x];
PlotContsProb[f[x], {x, 0, 18}, {1, 5},
   Ticks → {{1, 3, 5, 7, 9, 11, 13, 15}, Automatic},
   DefaultFont → {"Times-Roman", 8}];
```

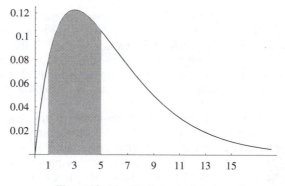

Figure 18 - The $\Gamma(2,3)$ density function

The area shaded is the probability $P[1 \leq X \leq 5]$ which can be found by integrating the density function over the interval $[1, 5]$.

$$N\left[\int_1^5 f[x] \, dx\right]$$

0.451707

Activity 2 Try to approximate the median m of the $\Gamma(2,3)$ distribution by using the CDF function. (Remember that the median of a distribution is such that exactly half of the total probability lies to its right.) Then check your guess by using the Quantile command.

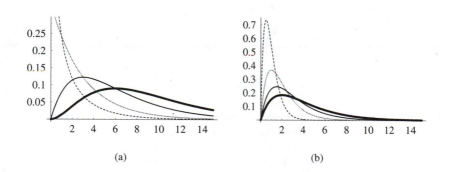

(a)   (b)

Figure 19 - (a) Gamma densities, $\alpha = .5$ (dashed), 1 (gray), 2 (thin), 3(thick) and $\beta = 3$; (b) Gamma densities, $\beta = .5$ (dashed), 1 (gray), 1.5 (thin), 2 (thick) and $\alpha = 2$

The graphs in Figure 19(a) illustrate the dependence of the gamma density on the α parameter for $\alpha = .5, 1, 2,$ and 3. In all cases we fix β at 3. The two cases where $\alpha \leq 1$ show a markedly different behavior: for $\alpha = .5$ the density is asymptotic to the y-axis, and for $\alpha = 1$ there is a non-zero y-intercept (What is it?). You can check that in fact $\alpha = 1$ is a break point, beyond which the density graph takes on the humpbacked shape of the other two curves. Note that the probability weight shifts to the right as α increases.

In Figure 19(b) we examine the dependence of the shape of the gamma density on the β parameter for $\beta = .5, 1, 2,$ and 3. In each case α is set at 2. As β increases, notice the spreading out of the density accompanied by a shift to the right.

The commands in the closed cell below this paragraph will produce 10 graphs, which you can select and animate to see the shift of the density to the right as α increases from .5 to 2.75. Feel free to use different α values, or use more graphs.

Activity 3 By editing the previous animation commands, animate graphs of the $\Gamma(3, \beta)$ density for β's between 1 and 5.5 with step size .5.

The gamma function upon which the gamma density depends has some interesting properties that permit the density to be written more explicitly in some cases. First,

$$\Gamma(1) = \int_0^\infty e^{-t}\, dt = 1. \tag{3}$$

Second, by using integration by parts, we can derive the following recursive relationship:

$$
\begin{aligned}
\Gamma(n) &= \int_0^\infty t^{n-1}\, e^{-t}\, dt \\
&= -t^{n-1}\, e^{-t} \big|_0^\infty + \int_0^\infty (n-1)\, t^{n-2}\, e^{-t}\, d\,t \\
&= 0 + (n-1) \int_0^\infty t^{n-2}\, e^{-t}\, d\,t \\
&= (n-1)\, \Gamma(n-1)
\end{aligned}
\tag{4}
$$

This relationship is true regardless of whether n is an integer; however, if n is an integer then we can also iterate backward to obtain:

$$\Gamma(n) = (n-1)\,(n-2)\cdots 2\cdot 1\cdot \Gamma(1) = (n-1)! \tag{5}$$

Defining 0! to be 1 as is the custom, this formula gives values to the gamma function for all positive integers n. It also turns out that $\Gamma(1/2) = \sqrt{\pi}$, as *Mathematica* shows below.

$$\int_0^\infty t^{-1/2} \, E^{-t} \, dt$$

$\sqrt{\pi}$

The first two moments of the gamma distribution depend on the parameters α and β in an easy way.

Theorem 1. If $X \sim \Gamma(\alpha, \beta)$ then
 (a) $E[X] = \alpha \beta$;
 (b) $\text{Var}(X) = \alpha \beta^2$.

Proof: (a) We have

$$E[X] = \int_0^\infty t \cdot \frac{1}{\beta^\alpha \, \Gamma(\alpha)} \, t^{\alpha-1} \, e^{-t/\beta} \, dt.$$

But this integral can be rearranged into

$$E[X] = \beta \cdot \frac{\Gamma(\alpha+1)}{\Gamma(\alpha)} \cdot \int_0^\infty \frac{1}{\beta^{\alpha+1} \, \Gamma(\alpha+1)} \, t^\alpha \, e^{-x/\beta} \, dt.$$

The integral on the right is that of a $\Gamma(\alpha+1, \beta)$ density; hence it reduces to 1. Therefore,

$$E[X] = \beta \cdot \frac{\Gamma(\alpha+1)}{\Gamma(\alpha)} = \beta \cdot \frac{\alpha \Gamma(\alpha)}{\Gamma(\alpha)} = \alpha\beta.$$

(b) We leave this as Exercise 3.

The next result on sums of independent gamma random variables is crucial to much of statistical inference and applied probability. It says that if the random variables share the same β parameter, then the sum also has the gamma distribution. This will yield a very fundamental property of the Poisson process in our third subsection, and it will loom large in the study of inference on the normal variance parameter in Section 4.5.

Theorem 2. (a) The moment-generating function of the $\Gamma(\alpha,\beta)$ distribution is

$$M(t) = \mathrm{E}[e^{tX}] = \frac{1}{(1-\beta t)^\alpha} \tag{6}$$

(b) If $X_1, X_2, \ldots, X_n$ are independent random variables with, respectively, $\Gamma(\alpha_1, \beta)$, $\Gamma(\alpha_2, \beta)$, $\ldots$, $\Gamma(\alpha_n, \beta)$ distributions, then

$$Y = \sum_{i=1}^{n} X_i \sim \Gamma(\alpha_1 + \alpha_2 + \cdots + \alpha_n, \beta)$$

Proof. (a) The m.g.f. is

$$M(t) = \int_0^\infty e^{tx} \cdot \frac{1}{\beta^\alpha \, \Gamma(\alpha)} \, x^{\alpha-1} \, e^{-x/\beta} \, dx = \int_0^\infty \frac{1}{\beta^\alpha \, \Gamma(\alpha)} \, x^{\alpha-1} \, e^{-x(1/\beta - t)} \, dx$$

In the integral on the right, for fixed t we make the change in parameter $\delta = \beta/(1-\beta t)$. Notice that $1/\delta = 1/\beta - t$. Then,

$$
\begin{aligned}
M(t) &= \frac{1}{\beta^\alpha} \cdot \int_0^\infty \frac{1}{\Gamma(\alpha)} \, x^{\alpha-1} \, e^{-x/\delta} \, dx \\
&= \frac{\delta^\alpha}{\beta^\alpha} \cdot \int_0^\infty \frac{1}{\delta^\alpha \, \Gamma(\alpha)} \, x^{\alpha-1} \, e^{-x/\delta} \, dx \\
&= \left(\frac{\delta}{\beta}\right)^\alpha = \frac{1}{(1-\beta t)^\alpha}.
\end{aligned}
$$

(b) The m.g.f. of the sum is

$$M_Y(t) = \mathrm{E}[e^{t(X_1 + \cdots + X_n)}] = \mathrm{E}[e^{tX_1} \cdot e^{tX_2} \cdots e^{tX_n}]$$

By independence of the X_i's and part (a), this becomes

$$M_Y(t) = \mathrm{E}[e^{tX_1}] \cdot \mathrm{E}[e^{tX_2}] \cdots \mathrm{E}[e^{tX_n}] = \frac{1}{(1-\beta t)^{\Sigma \alpha_i}}$$

which is the desired m.g.f.

The Exponential Distribution

The simplest member of the gamma family is also one of the most useful and interesting. Taking $\alpha = 1$ and $\lambda = 1/\beta$ in the gamma density formula, we get the following.

Definition 2. A random variable X has the *exponential distribution* if its probability density function is

$$f(x; \lambda) = \lambda\, e^{-\lambda x},\ x > 0 \tag{7}$$

where $\lambda > 0$ is a constant parameter. We abbreviate $X \sim \exp(\lambda)$.

The related *Mathematica* object in the Statistics`ContinuousDistributions` package is called ExponentialDistribution[λ]. For example, the following defines the $\exp(\lambda)$ p.d.f. in *Mathematica*.

```
f[x_, λ_] := PDF[ExponentialDistribution[λ], x]
```

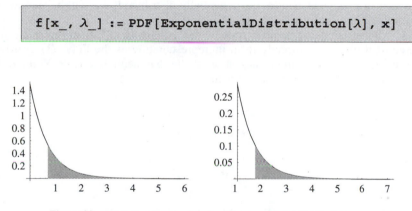

Figure 20 - Proportional area property of the exponential distribution

The exponential distribution has a property that is a bit peculiar. Consider the picture in Figure 20 of the $\exp(1.5)$ density, and two typical times $t = 1.1$ and $t + s = 1.8$, hence $s = .7$. In the first graph, the axis origin is set at $(0,0)$ as usual, and we shade the area to the right of $s = .7$, and on the second graph we move the axis origin to $(t, 0) = (1.1, 0)$ and shade the area to the right of $t + s = 1.8$. Though the vertical scales are different, the graphs look identical. A more careful way of saying this is that the share that the first shaded region has of the total area of 1 appears to be the same as the share that the second shaded region has of the total area to the right of $t = 1.1$. That is,

$$\frac{P[X > s]}{1} = \frac{P[X > t+s]}{P[X > t]}.$$

The right side is also a conditional probability, which implies that

$$P[X > s] = P[X > t+s \,|\, X > t]. \tag{8}$$

Activity 4 Verify formula (8) algebraically for an arbitrary exp(λ) distribution.

The intuitive interpretation of equation (8) is as follows. Suppose that we are waiting for our friend Godot, who will arrive at an exponentially distributed time X. We wait for t units of time and Godot does not come. Someone else comes by, and asks how likely it is that it will be s more time units (total time $t + s$) until Godot gets here. Thus, we are asked about the conditional probability on the right. We must answer the unconditional probability $P[X > s]$, which is the same as if we had just begun waiting for Godot at time instant t. Our waiting has done no good, probabilistically. This characteristic of the exponential distribution is called *memorylessness*. It is a great help in analyzing customer service situations in which times between successive arrivals are exponentially distributed. We will have more to say about this in a moment, and also in Chapter 6.

Some of the basic properties of the exponential(λ) distribution are below. Three of them follow directly from its representation as the $\Gamma(1, 1/\lambda)$ distribution. The c.d.f. in (11) is a direct integration of the exponential density. You should be sure you know how to prove all of them.

If $X \sim \exp(\lambda)$, then

$$E[X] = 1/\lambda \tag{9}$$

$$\text{Var}(X) = 1/\lambda^2 \tag{10}$$

$$F(x) = P[X \leq x] = 1 - e^{-\lambda x}, \; x > 0 \tag{11}$$

$$M(t) = E[e^{tX}] = \frac{1}{(1 - t/\lambda)} \tag{12}$$

Example 1 If the waiting time for service at a restaurant is exponentially distributed with parameter $\lambda = .5$/min. then the expected waiting time is $1/.5 = 2$ minutes, and the variance of the waiting time is $1/(.5)^2 = 4$ so that the standard deviation is 2 minutes also. Using the *Mathematica* c.d.f. function, the probability that the waiting time will be between 1 and 4 minutes is computed as:

```
F[x_, λ_] := CDF[ExponentialDistribution[λ], x];
F[4, .5] - F[1, .5]
```

0.471195

Of course, we would get the same result by integrating the density between 1 and 4, as below. ∎

```
⌠4
⎮  f[x, .5] dx
⌡1
```

0.471195

More on the Poisson Process

The most important area of application for the exponential distribution is the *Poisson process*. Recall from Section 2.4 that this is essentially a counting process that gives the total number of arrivals of some entity that have come by each fixed time t. Here is a more precise definition than we gave earlier.

Definition 3. A *Poisson process* $N = N(t,\omega)$ with rate λ is for each fixed t a random variable and for each $\omega \in \Omega$ a function from $[0, \infty)$ to the non-negative integers, such that $N(0, \omega) = 0$ and

(a) the function $t \to N(t, \omega)$ increases by jumps of size 1 only;
(b) there is an independent, identically distributed sequence of $\exp(\lambda)$ random variables $S_1, S_2, S_3, \cdots$ such that the times of jumps are

$$T_n = S_1 + S_2 + \cdots + S_n, \quad n = 1, 2, \cdots$$

So, a Poisson arrival counting process begins at state 0, waits for an exponentially distributed length of time S_1, then at time $T_1 = S_1$ jumps to 1, where it waits another exponential length of time S_2 until at time $T_2 = S_1 + S_2$ it jumps to state 2, etc. Such processes are often used in the field of queueing theory to model arrivals of customers to a service facility. The memorylessness of the exponential distribution means that at any instant of time, regardless of how long we may have been waiting, we will still wait an exponential length of time for the next arrival to come.

Since the inter-jump times determine the process completely, one would suppose that it is now possible to derive the result stated earlier, namely that the probability distribution of N_t, the total number of arrivals by time t, is Poisson. In fact, the following interesting interplay between the exponential, gamma, and Poisson distributions comes about.

Theorem 3. (a) The distribution of the nth jump time T_n is $\Gamma(n, 1/\lambda)$.
(b) The distribution of N_t = number of arrivals in $[0, t]$ is Poisson(λt).

Proof. (a) Since the $\exp(\lambda)$ distribution is also the $\Gamma(1, 1/\lambda)$ distribution, Theorem 2(b) implies that

$$T_n = S_1 + S_2 + \cdots + S_n \sim \Gamma(n, 1/\lambda)$$

(b) It is easy to see that the event $N_t \geq n$ is the same as the event $T_n \leq t$, since both of these are true if and only if the nth arrival has come by time t. Then,

$$
\begin{aligned}
P[N_t = n] &= P[N_t \geq n] - P[N_t \geq n + 1] \\
&= P[T_n \leq t] - P[T_{n+1} \leq t].
\end{aligned}
$$

Because T_{n+1} has the $\Gamma(n + 1, \lambda)$ distribution, integration by parts yields

$$
\begin{aligned}
P[T_{n+1} \leq t] &= \int_0^t \frac{\lambda^{n+1}}{n!} x^n e^{-\lambda x} \, d x \\
&= -\frac{\lambda^{n+1}}{n!} \cdot \frac{1}{\lambda} \cdot e^{-\lambda x} \cdot x^n \Big|_0^t + \int_0^t \frac{\lambda^{n+1}}{n!} \frac{1}{\lambda} n x^{n-1} e^{-\lambda x} \, dx \\
&= -\frac{(\lambda t)^n}{n!} e^{-\lambda t} + \int_0^t \frac{\lambda^n}{(n-1)!} x^{n-1} e^{-\lambda x} \, d x \\
&= -\frac{(\lambda t)^n}{n!} e^{-\lambda t} + P[T_n \leq t]
\end{aligned}
$$

Substituting this into the previous equation yields

$$P[N_t = n] = \frac{(\lambda t)^n}{n!} e^{-\lambda t}$$

which is the nth term of the Poisson(λt) distribution, as desired.

Example 2 Suppose that users of a computer lab arrive according to a Poisson process with rate $\lambda = 15$/hr. Then the expected number of users in a 2-hour period is $\lambda t = 15(2) = 30$ (recall that the Poisson parameter is equal to the mean of the distribution). The probability that there are 10 or fewer customers in the first hour is $P[N_1 \leq 10]$, i.e.:

$$N\left[\sum_{n=0}^{10} \frac{15^n}{n!} E^{-15}\right]$$

0.118464

The probability that the second arrival comes before time .1 (i.e., in the first 6 minutes) is $P[T_2 < .1]$, which is

$$\mathbf{N}\left[\int_0^{.1} \frac{15^2}{1!}\, x^{2-1}\, \mathbf{E}^{-15\,x}\, d\mathbf{x}\right]$$

```
0.442175
```

■

Activity 5 For the computer lab example above, what is the standard deviation of the number of users in a 2-hour period? What is the probability that there are 5 or fewer customers in the first half hour? What is the probability that the third arrival comes somewhere between the first 10 and the first 20 minutes?

Example 3 Again in the context of the computer lab, let us find the smallest value of the arrival rate λ such that the probability of having at least 5 arrivals in the first hour is at least .95. Measuring time in hours, we want the smallest λ such that

$$P[N_1 \geq 5] \geq .95 \;\; \Rightarrow \;\; P[N_1 \leq 4] \leq .05$$

by complementation. Since N_1 has the Poisson($\lambda \cdot 1$) distribution, the probability that there are four or fewer arrivals is the sum of the following terms:

$$\mathbf{P4orfewer[\lambda\_]} := \sum_{k=0}^{4} \frac{\mathbf{E}^{-\lambda} * \lambda^k}{k!}$$

To get some idea of where the appropriate value of λ lies, we graph the function for values of λ between 5 and 15.

```
Plot[{P4orfewer[λ], .05}, {λ, 5, 15},
    DefaultFont → {"Times-Roman", 8}];
```

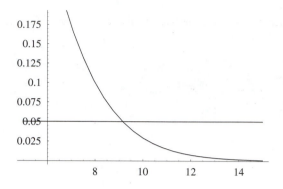

Figure 21 - Probability of 4 or fewer arrivals in first hour as a function of arrival rate

So the solution seems to be about $\lambda = 9$. We can find a root of the equation P4orfewer[λ] $- .05 = 0$ with an initial guess of 9 in order to get closer. We find that to the fifth decimal place $\lambda = 9.15351$. ∎

```
FindRoot[P4orfewer[λ] - .05, {λ, 9}]
```

$\{\lambda \rightarrow 9.15351\}$

Mathematica for Section 4.4

Command	Location
GammaDistribution[α, β]	Statistics`ContinuousDistributions`
ExponentialDistribution[λ]	Statistics`ContinuousDistributions`
PDF[dist, x]	Statistics`ContinuousDistributions`
CDF[dist, x]	Statistics`ContinuousDistributions`
PlotContsProb[f, domain, between]	KnoxProb`Utilities`
P4orfewer[λ]	Section 4.4

Exercises 4.4

1. (*Mathematica*) Suppose that X is a random variable with the $\Gamma(1.5, 1.8)$ distribution. Evaluate: (a) $P[X \leq 3]$; (b) $P[X > 5]$; (c) $P[3 \leq X \leq 4]$.

2. Use the moment-generating function of the gamma distribution in Theorem 2 to derive the formulas for the mean and variance of the distribution.

3. Derive the formula $\text{Var}(X) = \alpha\beta^2$ for the variance of a gamma random variable by direct integration.

4. A project requires two independent phases to be completed, and the phases must be performed back-to-back. Each phase completion time has the exp(4.5) distribution. Find the probability that the project will not be finished by time 0.5.

5. Redo Exercise 4 under the assumption that the two phases may be performed concurrently.

6. Use calculus to show that if $X \sim \Gamma(2, \beta)$ then $P[X \geq 3]$ is an increasing function of β.

7. If we are waiting for a bus which will arrive at an exponentially distributed time with mean 10 minutes, and we have already waited for 5 minutes, what is the probability that we will wait at least 10 more minutes? What is the expected value of the remaining time (beyond the 5 minutes we have already waited) that we will wait?

8. (*Mathematica*) If $X \sim \Gamma(4, 1/6)$, find: (a) $P[1 \leq X \leq 7]$; (b) $P[X > 2]$.

9. Find the c.d.f. of the $\Gamma(2, 1/\lambda)$ distribution analytically.

10. If $X_1 \sim \Gamma(2, 1/1.1)$ and $X_2 \sim \Gamma(3, 1/1.1)$ what are the distribution, mean, and variance of $X_1 + X_2$?

11. (*Mathematica*) For what value of β does the $\Gamma(3, \beta)$ distribution have median equal to 2?

12. (*Mathematica*) If $N = N(t, \omega)$ is a Poisson process with rate 6, find:
(a) $P[N_{1.2} = 3]$; (b) $P[T_4 \leq .25]$.

13. (*Mathematica*) Implement a *Mathematica* command to simulate and graph a Poisson process.

14. (*Mathematica*) Use the SimulatePoissonProcess command of Exercise 13 to simulate around 40 paths of a Poisson process with rate 2 between times 0 and 4. Count how many of your paths had exactly 0, 1, and 2 arrivals by time 1. How well do your simulations fit the theoretical probabilities of 0, 1, and 2 arrivals?

15. (*Mathematica*) Find the smallest rate parameter λ for a Poisson process such that the probability of 8 or fewer arrivals in the time interval [0, 2] is no more than .9.

16. Customer arrivals forming a Poisson process of rate 2/minute come to a single server. The server takes a uniformly distributed amount of time between 20 and 30 seconds for each customer to perform service. Compute the expected number of new arrivals during a service time. (Hint: condition and uncondition on the duration of the service time.)

17. Show that for a Poisson process, if you take the fact that $N_t \sim$ Poisson($\lambda\, t$) for all t as a primitive assumption, then it follows that the time of the first arrival must have the exponential distribution.

4.5 Chi-Square, Student's t, and F-distributions

In Example 9 of Section 4.3 we showed how to use the mean $\overline{X}$ of a random sample to do *statistical inference* on μ, the underlying mean parameter of the normal distribution from which the sample was taken. Specifically we wondered, if the default value of mean balance measurement was $\mu = 1$, how likely it would be that $\overline{X}$ could be as extreme as the data indicated. The smaller was that likelihood, the less faith we had in the assumed value $\mu = 1$. Statistical inference is the problem of using random samples of data to draw conclusions in this way about population parameters like the normal mean μ.

There are other important inferential situations in statistics that depend on the three distributions that we will cover in this section. We will only scratch the surface of their statistical application; the rest is best left for a full course in statistics. But you will complete this section with a sense of how the *chi-square distribution* gives information about σ^2 in $N(\mu, \sigma^2)$, how the *Student's t-distribution* enables inference on μ in $N(\mu, \sigma^2)$ without the necessity of knowing the true value of σ^2, and how the *F-distribution* applies to problems where two population variances σ_1^2 and σ_2^2 are to be compared. So our plan in this section will be to define each of these distributions in turn, examine their main properties briefly, and then illustrate their use in statistical inference problems.

Chi-Square Distribution

If you observe the sample variance $S^2 = \sum_{i=1}^{n}(X_i - \overline{X})^2 \Big/ (n-1)$ of a random sample $X_1, X_2, \ldots, X_n$ from a normal distribution many times, you will find the observed s^2 values distributing themselves in a consistent pattern. The command below simulates a desired number of sample variances in a straightforward way. The argument *numvars* refers to the number of sample variances we wish to simulate, *sampsize* is the common sample size n for each sample variance, and μ and σ are the normal distribution parameters. In the electronic version of the text you should try running it yourself several times.

```
Needs["KnoxProb`Utilities`"];
```

```
SimSampleVariances[
    numvars_, sampsize_, μ_, σ_] :=
  Table[Variance[RandomArray[
      NormalDistribution[μ, σ],
      sampsize]], {i, 1, numvars}];
```

```
SeedRandom[984562];
```

```
Histogram[SimSampleVariances[200, 10, 0, 1], 8];
```

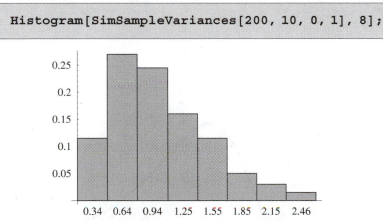

Figure 22 - Histogram of 200 sample variances of samples of size 40 from $N(0, 1)$

The right-skewed shape of the empirical distribution in Figure 22 should remind you of the gamma distribution. In fact, we argue in a moment that a rescaled version of S^2 has a probability distribution which is a special case of the gamma family, defined below.

Definition 1. The *chi-square distribution* with *degrees of freedom parameter r*, abbreviated $\chi^2(r)$, is the $\Gamma(r/2, 2)$ distribution. Hence the $\chi^2(r)$ density function is

$$f(x) = \frac{1}{2^{r/2}\,\Gamma(r/2)}\, x^{r/2-1}\, e^{-x/2}, \quad x > 0 \tag{1}$$

Activity 1 Why is the mean of the $\chi^2(r)$ distribution equal to r? What is the variance of the distribution?

Mathematica has an object in the Statistics'ContinuousDistributions' package for this distribution, whose syntax is

ChiSquareDistribution[r]

Then for example the following defines the density function in terms of both the variable x and the parameter r. Figure 23 graphs several densities for values of r from 5 to 8. Notice how, in line with the fact that you just observed in Activity 1, the probability weight shifts to the right as r increases (the mean is r) and also spreads out as r increases (the variance is $2r$). In the electronic version you may open up the closed cell that generated these graphs and try experimenting with other values of the r parameter.

```
f[x_, r_] := PDF[ChiSquareDistribution[r], x]
```

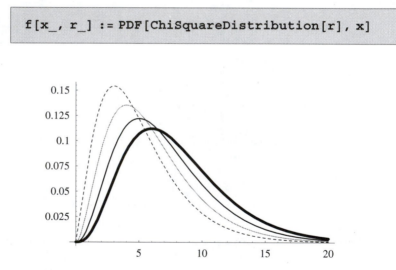

Figure 23 - $\chi^2(r)$ densities; $r = 5$ (dashed), $r = 6$ (gray), $r = 7$ (thin), $r = 8$ (thick)

Exercise 2 of Section 4.3 instructed you to show that if $Z \sim N(0, 1)$, then the random variable $U = Z^2$ has a density which turns out to be $\chi^2(1)$. If you did not do that exercise yet, you should do it now.   (Start by finding $P[U \le u] = P[-\sqrt{u} \le Z \le \sqrt{u}] = P[Z \le \sqrt{u}] - P[Z \le -\sqrt{u}]$ as an integral, then use the Fundamental Theorem of Calculus to differentiate with respect to u. Note that $\Gamma(1/2) = \sqrt{\pi}$.) This is the key to understanding what the χ^2-distribution has to do with the sample variance S^2.

Recall that by the normal standardization theorem,

$$Z_i = \frac{X_i - \mu}{\sigma} \sim N(0, 1) \tag{2}$$

for each sample member X_i; hence, by the result of Exercise 2,

$$U_i = Z_i^2 = \left(\frac{X_i - \mu}{\sigma}\right)^2 \sim \chi^2(1) \tag{3}$$

for each $i = 1, 2, ..., n$.   Moreover, in Exercise 2 you are asked to use moment-generating functions to show that the sum of independent chi-square random variables is also chi-square distributed, with degrees of freedom equal to the sum of the degrees of freedom of the terms. So, the random variable

$$U = \sum_{i=1}^{n} U_i = \sum_{i=1}^{n} \left(\frac{X_i - \mu}{\sigma}\right)^2 \sim \chi^2(n) \tag{4}$$

Now look at the random variable

$$V = \frac{(n-1)S^2}{\sigma^2} = \frac{n-1}{\sigma^2} \frac{\sum_{i=1}^{n}(X_i - \bar{X})^2}{n-1} = \sum_{i=1}^{n}\left(\frac{X_i - \bar{X}}{\sigma}\right)^2 \tag{5}$$

Observe that V differs from U only in the sense that in V, μ has been estimated by $\bar{X}$. Since U has a χ^2-distribution, we might hope that V does also, especially in view of the simulation evidence. The following theorem gives the important result.

Theorem 1. If $X_1, X_2, ..., X_n$ is a random sample from the $N(\mu, \sigma^2)$ distribution, and S^2 is the sample variance, then

$$\frac{(n-1)S^2}{\sigma^2} \sim \chi^2(n-1) \tag{6}$$

and moreover S^2 is independent of $\bar{X}$.

    The full proof of Theorem 1 requires more information about multidimensional normal distributions than we have now. The discussion prior to the statement of the theorem provided good motivation for the fact that the distribution of $(n-1)S^2/\sigma^2$ should be of χ^2 type, but why is the degrees of freedom parameter $r = n - 1$ (as opposed to n, for instance)?
    To shed some light on this, consider the case $n = 2$. Then the sample variance is

$$S^2 = \frac{1}{1}\sum_{i=1}^{2}(X_i - \overline{X})^2 \;=\; (X_1 - \overline{X})^2 + (X_2 - \overline{X})^2$$

$$= (X_1 - \tfrac{X_1 + X_2}{2})^2 + (X_2 - \tfrac{X_1 + X_2}{2})^2 \tag{7}$$

$$= (\tfrac{2X_1 - X_1 - X_2}{2})^2 + (\tfrac{2X_2 - X_1 - X_2}{2})^2$$

$$= \frac{(X_1 - X_2)^2}{2}$$

after a bit of simplification. Also, $X_1 \sim N(\mu, \sigma^2)$ and $X_2 \sim N(\mu, \sigma^2)$, which means that $X_1 - X_2 \sim N(0, 2\sigma^2)$. Consequently,

$$\frac{X_1 - X_2}{\sqrt{2\sigma^2}} \sim N(0, 1)$$

Recalling that the square of a standard normal random variable is $\chi^2(1)$, it follows that for the case $n = 2$,

$$\frac{(n-1)S^2}{\sigma^2} = \frac{(2-1)S^2}{\sigma^2} = \frac{(X_1 - X_2)^2}{2\sigma^2} \sim \chi^2(1)$$

The estimation of μ by $\overline{X}$ and subsequent reduction of terms in S^2 from 2 to 1 in formula (7) resulted in the loss of one "degree of freedom" from the sample size $n = 2$. This argument can be made more elegant and extended to the case of larger n.

Activity 2 Adapt the command SimSampleVariances from earlier to simulate observations of $(n-1)S^2/\sigma^2$. Choose your own n and σ^2 and superimpose a graph of a scaled histogram of the simulated observations on a graph of the χ^2 density to observe the fit.

When we talk about the *t*-distribution in the next subsection, the second part of Theorem 1 becomes important: S^2 is independent of $\overline{X}$. At first glance this almost seems ludicrous, because both S^2 and $\overline{X}$ depend functionally on the sample values $X_1, X_2, \ldots, X_n$. To see that it is possible for S^2 and $\overline{X}$ to be independent, consider again the case $n = 2$, where S^2 is a function of $X_1 - X_2$ by formula (7), while $\overline{X}$ is a function of $X_1 + X_2$. We will argue using moment-generating functions that $Y_1 = X_1 - X_2$ and $Y_2 = X_1 + X_2$ are independent. (The following paragraph may be omitted on first reading.)

For two jointly distributed random variables Y_1 and Y_2 there is a natural two-variable analogue of the m.g.f., namely the *joint moment-generating function*, defined by

$$M(t_1, t_2) = \mathrm{E}[e^{t_1 Y_1 + t_2 Y_2}] \tag{8}$$

It has the same basic properties as the single variable m.g.f.; most importantly, it is unique to the distribution. In particular, if we find that a joint m.g.f. happens to match the joint m.g.f. of two independent random variables (which is the product of their individual m.g.f.'s), then the original two random variables must have been independent. This observation plays out very nicely with $Y_1 = X_1 - X_2$ and $Y_2 = X_1 + X_2$, where X_1 and X_2 are independent $N(\mu, \sigma^2)$ random variables. Then by formula (8), the joint m.g.f. of Y_1 and Y_2 is

$$
\begin{aligned}
M(t_1, t_2) = \mathrm{E}[e^{t_1 \, Y_1 + t_2 \, Y_2}] \ &= \ \mathrm{E}[e^{t_1 (X_1 - X_2) + t_2 (X_1 + X_2)}] \\
&= \ \mathrm{E}[e^{(t_1 + t_2) X_1 + (t_2 - t_1) X_2}] \\
&= \ \mathrm{E}[e^{(t_1 + t_2) X_1}] \, \mathrm{E}[e^{(t_2 - t_1) X_2}] \\
&= \ e^{\mu(t_1 + t_2) + \frac{1}{2}\sigma^2 (t_1 + t_2)^2} \; e^{\mu(t_2 - t_1) + \frac{1}{2}\sigma^2 (t_2 - t_1)^2} \\
&= \ e^{0 \cdot t_1 + \frac{1}{2}(2\sigma^2) t_1^2} \; e^{2\mu \cdot t_2 + \frac{1}{2}(2\sigma^2) t_2^2}
\end{aligned}
$$

In the first line we use the defining formulas for Y_1 and Y_2, we gather like terms and then use independence of X_1 and X_2 to factor the expectation in line 3, we use the known formula for the normal m.g.f. in line 4, and the fifth line is a straightforward rearrangement of terms in the exponent of the fourth line, which you should verify. But the final result is that the last formula is the product of the $N(0, 2\sigma^2)$ m.g.f. and the $N(2\mu, 2\sigma^2)$ m.g.f. So, Y_1 and Y_2 must be independent with these marginal distributions. It follows that the sample mean $\overline{X}$ and the sample variance S^2 are independent when $n = 2$, and the result can be extended to higher n.

Example 1 The data below from the Illinois state police are 1998 crime rates per 100 people in a number of counties in the state of Illinois. We might choose to view these crime rates as a sample of crime rates from all counties in the U.S., which we suppose are approximately normally distributed. (Although these data certainly do not constitute a random sample, a truly random sample could be obtained, and we use this data set for illustrative purposes only.) Can we obtain a reasonable estimate of the variability of crime rates, as measured by σ? Is there a way of measuring our confidence in the estimate?

```
crimerates = {7.08, 7.04, 6.27, 5.03, 4.75,
    4.44, 4.43, 4.33, 4.28, 4.09, 3.87, 3.76,
    3.67, 3.66, 3.37, 3.22, 2.88, 2.86, 2.73,
    2.72, 2.65, 2.59, 2.55, 2.54, 2.42, 1.68};
DotPlot[crimerates, DefaultFont →
    {"Times-Roman", 8}];
```

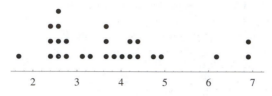

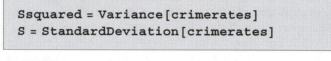

Figure 24 - Crime rates per 100 people in 26 Illinois counties, 1998

This rather small data set is displayed in the form of a dot plot in Figure 24. There are a couple of rather large observations indicating a right skewness in the empirical distribution, but this is possible in a small sample and there are at least no gross departures from normality that are visible here; so we will proceed under the assumption that county crime rates are normally distributed. The sample variance and standard deviation are:

```
Ssquared = Variance[crimerates]
S = StandardDeviation[crimerates]
```

```
1.92459
```

```
1.3873
```

Hence the observed $s = 1.3873$ is a point estimate of the population variability σ. But as you have probably experienced in such contexts as opinion polls, a more informative estimate would be equipped with a margin of error, so that we can say with high confidence that the true σ is in an interval. This gives us information about the precision of the point estimate s.

This is where Theorem 1 comes in. The transformed random variable $Y = (n - 1) S^2 / \sigma^2$ has the $\chi^2(25)$ distribution here, so that we can find two endpoints a and b such that the interval $[a, b]$ encloses Y with a desired high probability. Then we will be able to say something about σ^2 itself, and in turn about σ. Let us demand 90% confidence that Y lands in $[a, b]$. Even under this specification, a and b can be chosen in infinitely many ways, but a natural way to choose them is to set $P[Y > b] = .05$ and $P[Y < a] = .05$, i.e., to split the 10% = .10 error probability into two equal pieces. Then we can find a and b by computing

```
a = Quantile[ChiSquareDistribution[25], .05]
b = Quantile[ChiSquareDistribution[25], .95]
```

14.6114

37.6525

Now we have that

$$.90 = P[a \le Y \le b] = P\left[a \le \frac{(n-1)S^2}{\sigma^2} \le b\right] = P\left[\frac{(n-1)S^2}{b} \le \sigma^2 \le \frac{(n-1)S^2}{a}\right] \quad (9)$$

by rearranging the inequalities in the middle expression. The interval $[(n-1)S^2/b, (n-1)S^2/a]$ is called a *confidence interval* of level 90% for the parameter σ^2, and a confidence interval for σ itself follows by just taking the square roots of the endpoints. For our data the interval for σ is:

$$\left\{ \frac{\sqrt{25} * S}{\sqrt{b}}, \frac{\sqrt{25}\, S}{\sqrt{a}} \right\}$$

{1.13043, 1.81465}

The interpretation is that it is 90% likely for the random interval $[\sqrt{n-1}\, S/\sqrt{b}, \sqrt{n-1}\, S/a]$ to contain the parameter σ. For our observed value of S, we say loosely that we are 90% confident that σ lies between 1.13043 and 1.81465. When you take a course in statistics you will hear much more about confidence intervals. But the main idea is always as in derivation (9). Choose endpoints to achieve a target probability that a transformed random variable is between the endpoints, then rearrange inequalities to isolate the desired parameter in order to get an interval estimate of the parameter. ∎

Student's *t*-Distribution

In 1908, statistician W. S. Gossett, an employee of the Guinness brewery, studied the distribution of the sample mean. Recall that the expected value of $\overline{X}$ is the population mean μ, and the variance of $\overline{X}$ is σ^2/n, but Gossett considered standardizing $\overline{X}$ by an estimate $S/\sqrt{n}$ of its standard deviation, rather than the usually unknown $\sigma/\sqrt{n}$. He was led to discover a continuous probability distribution called the *t-distribution*, which we study now. But for proprietary reasons he was unable to publish his result openly. In the end he did publish under the pseudonym "Student", and for this reason the distribution is sometimes called the *Student's t-distribution*.

In Figure 25 is the result of a simulation experiment in which 200 observations of the random variable

$$T = \frac{\bar{X} - \mu}{S/\sqrt{n}} \tag{10}$$

are obtained and plotted, where in each of the 200 sampling experiments a random sample of size $n = 20$ is drawn from the normal distribution with mean 10 and standard deviation 2. The closed cell below this paragraph contains the code, but do not open it until you have tried Exercise 13, which asks you to write such a simulation program. This is an important exercise for your understanding of sampling for inference on a normal mean.

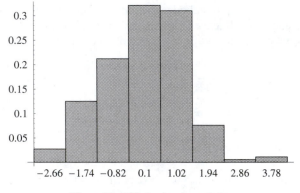

Figure 25 - 200 simulated t-statistics

Note in particular the approximate symmetry of the empirical distribution of T about 0, and the bell-shape similar to the shape of the normal density. The distribution defined below has a density with these very properties.

Definition 2. The *Student's t-distribution* is the probability distribution of the transformed random variable

$$T = \frac{Z}{\sqrt{U/r}} \tag{11}$$

where $Z \sim N(0, 1)$, $U \sim \chi^2(r)$, and Z and U are independent. We abbreviate the distribution as $t(r)$. The parameter r is called the *degrees of freedom* of the distribution.

Mathematica has an object called

StudentTDistribution[r]

in its Statistics`ContinuousDistributions` package, to which we can apply functions like PDF, CDF, Quantile, and Random in the usual way. Here for instance is a definition of the density function, and an evaluation of it for general x so that you can see the rather complicated form. (The standard mathematical function Beta[a,b] that you see in the output is defined as $(\Gamma[a]\,\Gamma[b])/\Gamma[a+b]$.)

```
f[x_, r_] := PDF[StudentTDistribution[r], x]
```

```
f[x, r]
```

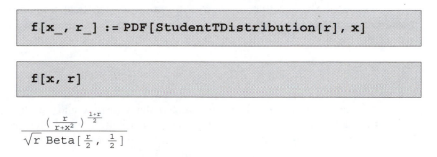

$$\frac{\left(\frac{r}{r+x^2}\right)^{\frac{1+r}{2}}}{\sqrt{r}\ \text{Beta}\left[\frac{r}{2},\ \frac{1}{2}\right]}$$

It can be shown that the *t*-density converges to the standard normal density as $r \longrightarrow \infty$. The closed cell above Figure 26 actually contains code for generating graphs of the *t*-density and the $N(0, 1)$ density for degrees of freedom $r = 4, 8, 12, 16, 20, 24,$ and 28, but we have only kept the first one. You can reexecute the command, select and animate these cells to observe the convergence of the *t*-density to the standard normal density as r becomes large. (The *t*-density is the light curve and the normal is the dark one.) Observe that the *t*-density is a little heavier in the tails, but the probability weight shifts toward the middle as r increases. And the two distributions are very hard to tell apart for even such moderate r as 20.

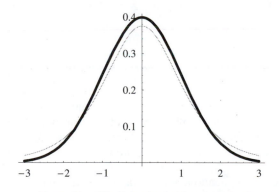

Figure 26 - $N(0, 1)$ density (bold) and $t(4)$-density.

Activity 3   Use *Mathematica* to find out how large r must be so that the probability that a $t(r)$ random variable lies between -1 and 1 is within .01 of the probability that a standard normal random variable lies between -1 and 1.

How is the formula for the $t(r)$ density obtained?   Using the c.d.f. technique of Section 4.3, you can write an integral for the c.d.f. $P[T \le t] = P\left[Z / \sqrt{U/r} \le t\right]$, then differentiate with respect to t.   You are asked to do this in Exercise 8.   But the density formula is of secondary importance to what we turn to now.

To see the connection of the t-distribution to statistical inference on the mean, consider the sample mean $\overline{X}$ and sample variance S^2 of a random sample of size n from the $N(\mu, \sigma^2)$ distribution.   We know from previous work that

$$\overline{X} \sim N(\mu, \sigma^2 / n) \quad \text{and} \quad \frac{(n-1)S^2}{\sigma^2} \sim \chi^2(n-1) \tag{12}$$

and also $\overline{X}$ and S^2 are independent.   So, we can standardize $\overline{X}$ and use the result as Z in the defining formula (11) for the t-distribution, and we can take $U = (n-1)S^2 / \sigma^2$ and $r = n-1$ in that formula.   The result is that the random variable in formula (13) has the $t(n-1)$ distribution:

$$T = \frac{\frac{\overline{X} - \mu}{\sigma / \sqrt{n}}}{\sqrt{\frac{(n-1)S^2}{\sigma^2 (n-1)}}} \sim t(n-1) \tag{13}$$

Simplifying the complex fraction, we obtain the following theorem.

Theorem 2.   If $\overline{X}$ and S^2 are the sample mean and sample variance of a random sample of size n from the $N(\mu, \sigma^2)$ distribution, then the random variable

$$T = \frac{\overline{X} - \mu}{S / \sqrt{n}} \tag{14}$$

has the $t(n-1)$ distribution.

The random variable T in (14) is striking in the sense that it does not depend on σ^2, nor does the t-distribution have σ^2 as a parameter.   So, if we want to infer whether an observed $\overline{X}$ is reasonably consistent with an hypothesized μ, we need not make any assumptions about the value of σ^2.   We illustrate statistical reasoning about μ using the t-distribution in the next example.

Example 2   Suppose that an auto maker has a new model that is claimed to average 28 miles per gallon of gasoline on the highway.   What do you think of the veracity of the claim, if 25 cars were driven under similar conditions, and the average mileage was 26.5 with a sample standard deviation of 2.7?

If the claim is true, then the random variable

$$T = \frac{\bar{X}-28}{S/\sqrt{n}}$$

would have the $t(24)$ distribution, since the sample size is 25. If we observe a T of unusually large magnitude, then doubt is cast on the claim that $\mu = 28$. And we can quantify how much doubt by computing the probability that a $t(24)$ distributed random variable is as large or larger in magnitude than the actual value we observe. The smaller is that probability, the less we believe the claim. Our particular data give an observed value

```
t = ────────────
      26.5 – 28
    2.7 / √25
```

-2.77778

The probability that $T \le -2.7778$ is

```
CDF[StudentTDistribution[24], -2.7778]
```

0.00522692

So if truly $\mu = 28$, we would only observe such an extreme T with probability about .005. It is therefore very likely that the true value of μ is less than 28. ∎

Activity 4 The sample average mileage of 26.5 in the last example was within one S of the hypothesized $\mu = 28$. In view of this, why was the evidence so strong against the hypothesis? Is there strong evidence against the hypotheses that $\mu = 27$?

F-Distribution

The last of the three special probability distributions for statistical inference is defined next.

Definition 3.   Let $U \sim \chi^2(m)$ and $V \sim \chi^2(n)$ be two independent chi-square distributed random variables. The *F-distribution* with *degree of freedom parameters* m and n (abbreviated $F(m, n)$) is the distribution of the transformed random variable

$$F = \frac{U/m}{V/n} \qquad\qquad (15)$$

Mathematica's version of the *F*-distribution goes by the name

FRatioDistribution[m, n]

and is contained in the Statistics`ContinuousDistributions` package. We define the density function below and give a typical graph in Figure 27 for degrees of freedom $m = n = 20$.   Its shape is similar to that of the gamma density. Note in particular that the state space is the set of all $x \geq 0$.

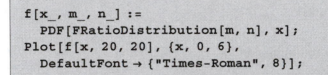

```
f[x_ , m_ , n_] :=
   PDF[FRatioDistribution[m, n], x];
Plot[f[x, 20, 20], {x, 0, 6},
   DefaultFont → {"Times-Roman", 8}];
```

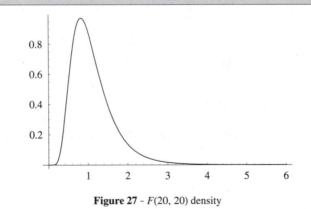

Figure 27 - $F(20, 20)$ density

Activity 5 Hold the m parameter fixed at 10, and plot the F-density for values of the n parameter equal to 5, 10, 15, 20, and 25 and observe the effect of changing the n parameter on the shape of the density. Similarly, hold n at 10 and let m take the values 5, 10, 15, 20, and 25 to see the effect on the density of changes to m.

We use *Mathematica* to show the exact form of the $F(m, n)$ density below, which can again be obtained by the c.d.f. technique, but that is less interesting than the use of the F-distribution in doing inference on two normal variances, which we discuss in the next paragraph.

```
f[x, m, n]
```

$$\frac{m^{m/2}\ n^{n/2}\ x^{-1+\frac{m}{2}}\ (n + m\,x)^{\frac{1}{2}\,(-m-n)}}{\text{Beta}[\frac{m}{2},\ \frac{n}{2}]}$$

There may be many reasons to be interested in comparing the variances σ_x^2 and σ_y^2 of two normal populations. If for example random variables X and Y are physical measurements on items manufactured using two different processes, then for quality purposes the process with smaller variability is the more desirable one. Or, perhaps the application of a treatment such as a blood sugar regulator has as its main goal the maintenance of more even sugar concentrations, so that a population of readings without the regulator should have greater variance than a population of readings with it.

Statistical comparisons of variances may be done in the following way. Let $X_1, X_2, \ldots, X_m$ and $Y_1, Y_2, \ldots, Y_n$ be independent random samples from $N(\mu_x, \sigma_x^2)$ and $N(\mu_y, \sigma_y^2)$ distributions, respectively. Let the sample variances be S_x^2 and S_y^2. From Theorem 1 we know that

$$U = \frac{(m-1)\,S_x^2}{\sigma_x^2} \sim \chi^2(m-1) \quad \text{and} \quad V = \frac{(n-1)\,S_y^2}{\sigma_y^2} \sim \chi^2(n-1)$$

and also, by the independence of the two samples, S_x^2 and S_y^2 are independent; hence U and V are independent. Using this U and V in the definition of the F-distribution, and dividing by their respective degrees of freedom yields the following result.

Theorem 3. If $X_1, X_2, \ldots, X_m$ is a random sample from $N(\mu_x, \sigma_x^2)$, and $Y_1, Y_2, \ldots, Y_n$ is an independent random sample from the $N(\mu_y, \sigma_y^2)$ distribution, and the sample variances are S_x^2 and S_y^2, then the random variable

$$F = \frac{S_x^2/\sigma_x^2}{S_y^2/\sigma_y^2} \tag{16}$$

has the $F(m-1, n-1)$ distribution.

Exercise 19 asks you to simulate a number of F-ratios as in (16) and compare the empirical distribution to the $F(m-1, n-1)$ density. You should see a good fit.

An important special case of the use of the F-random variable in (16) occurs when we want to make a judgment about whether it is feasible that two population variances σ_x^2 and σ_y^2 are equal to one another. Under the hypothesis that they are equal, they cancel one another out in (16), leaving $F = S_x^2 / S_y^2$. If F is either unreasonably large or unreasonably close to 0, i.e., if we observe an F too far out on either the left or right tail of the distribution in Figure 27, then we disbelieve the hypothesis of equality of the two population variances. The probability of observing such an F quantifies our level of disbelief. The next example illustrates the reasoning, and recalls the idea of confidence intervals from earlier in the section.

Example 3 I am the coordinator for a mathematics placement examination given every year to entering students at my college. Not only is the average level of mathematics preparation of students of concern in teaching, but also the variability of their preparation. Below are samples of placement test scores from two years. Assuming that the population of all scores is normally distributed for both years, let us see if there is significant statistical evidence that the variability of scores is different from one year to another based on this data.

```
year1 = {9, 14, 16, 8, 19, 7, 14, 17,
    16, 7, 13, 16, 11, 9, 30, 14, 9, 9, 9,
    11, 13, 16, 6, 17, 18, 11, 12, 18, 3,
    12, 20, 8, 13, 14, 21, 11, 27, 26, 5};
year2 = {16, 21, 12, 19, 13, 14, 9, 29, 21, 5, 7,
    11, 4, 8, 17, 5, 10, 8, 13, 19, 14, 16, 15, 8,
    6, 8, 15, 9, 6, 23, 19, 8, 19, 7, 16, 27, 13,
    13, 28, 9, 7, 18, 12, 9, 10, 7, 8, 20, 19};
Length[year1]
Length[year2]
VarX = N[Variance[year1]]
VarY = N[Variance[year2]]
```

39

49

35.4103

39.5323

The first sample has size $m = 39$ and the second has size $n = 49$, and so the F-ratio in (16) has the $F(38, 48)$ distribution. The two sample variances of 35.4103 and 39.5323 seem to be close, but are they sufficiently close for us to believe that σ_x^2 for year 1 is the same as σ_y^2 for year 2?

There are two bases upon which we can give an answer, which are somewhat different statistical approaches, but which turn out to be equivalent. The first approach is to give an interval estimate of the ratio σ_x^2/σ_y^2 in which we are highly confident that this ratio falls. Under the hypothesis of equal population variances, the ratio is 1, and thus 1 should fall into our confidence interval. If it doesn't, then we disbelieve that the population variances are the same, and the level of the confidence interval measures our probability of being correct. To find a confidence interval for σ_x^2/σ_y^2 of level 90% for example, we can use *Mathematica* to find, for the $F(38, 48)$ distribution, two numbers a and b such that $P[F < a] = .05$ and $P[F > b] = .05$; then $P[a \le F \le b] = .90$.

```
a = Quantile[FRatioDistribution[38, 48], .05]
b = Quantile[FRatioDistribution[38, 48], .95]
```

```
0.594734
```

```
1.65219
```

You might say that $a \approx .59$ and $b \approx 1.65$ are our extremes of reasonableness. It is not often that an F-random variable with these degrees of freedom takes a value beyond them. Then we can write

$$.90 = P[a \le F \le b] = P\left[a \le \tfrac{S_x^2/\sigma_x^2}{S_y^2/\sigma_y^2} \le b\right] = P\left[\tfrac{S_x^2}{b\,S_y^2} \le \tfrac{\sigma_x^2}{\sigma_y^2} \le \tfrac{S_x^2}{a\,S_y^2}\right] \qquad (17)$$

after rearranging the inequalities. So the two endpoints of the confidence interval are formed by dividing the ratio of sample variances by b and a, respectively, which we do below.

```
{0.542148, 1.5061}
```

Since 1 is safely inside this 90% confidence interval, we have no statistical evidence against the hypothesis $\sigma_x^2 = \sigma_y^2$.

The second approach is to specify a tolerable error probability for our decision as to whether the population variances are equal, such as 10%. Specifically, we want a rule for deciding, such that if the two variances are equal, we can only err by deciding that they are unequal with probability .10. If $\sigma_x^2 = \sigma_y^2$, then as noted earlier, $F = S_x^2/S_y^2$ has the $F(38, 48)$ distribution. We already know two numbers a and b, namely $a = .594734$ and $b = 1.65219$, such that the chance that F can be as extreme as these or more so is .10 in total. So if our decision rule is to accept the hypothesis of equal variance as long as F stays between a and b, then the chance of rejecting the hypothesis incorrectly is .10. We can compute

```
F = VarX / VarY
```

```
0.895729
```

and since .895729 lies between a and b, we do accept the hypothesis $\sigma_x^2 = \sigma_y^2$.
Notice that

$$a \le \frac{S_x^2}{S_y^2} \le b \iff \frac{S_x^2}{b\,S_y^2} \le 1 \le \frac{S_x^2}{a\,S_y^2}$$

Referring to (17), we see that our decision rule for accepting the hypothesis
$\sigma_x^2 = \sigma_y^2$ using this second approach is exactly the same as the first approach
using confidence intervals. ∎

Activity 6 In the last example check to see whether 1 is inside confidence
intervals of level .80 and .70. If it is, why is this information more helpful than
merely knowing that 1 is inside a 90% confidence interval?

Mathematica for Section 4.5

Command	Location
PDF[dist, x]	Statistics` ContinuousDistributions`
CDF[dist, x]	Statistics` ContinuousDistributions`
Quantile[distribution, q]	Statistics` ContinuousDistributions`
ChiSquareDistribution[r]	Statistics` ContinuousDistributions`
StudentTDistribution[r]	Statistics` ContinuousDistributions`
FRatioDistribution[m, n]	Statistics` ContinuousDistributions`
NormalDistribution[μ, σ]	Statistics` ContinuousDistributions`
RandomArray[dist, {sampsize}]	Statistics` ContinuousDistributions`
Variance[datalist]	Statistics` DescriptiveStatistics`
StandardDeviation[datalist]	Statistics` DescriptiveStatistics`
Histogram[datalist, numrecs]	KnoxProb` Utilities`
DotPlot[list]	KnoxProb` Utilities`
SimSampleVariances[n, size, μ, σ]	Section 4.5

Exercises 4.5

1. (*Mathematica*) If X has the $\chi^2(10)$ distribution, find:
   (a) $P[8 \le X \le 12]$;
   (b) $P[X \ge 15]$;
   (c) the 25th percentile of the distribution, i.e., the point q such that
   $P[X \le q] = .25$;
   (d) two points a and b such that $P[a \le X \le b] = .95$.

2. Find the moment-generating function of the $\chi^2(r)$ distribution. Use it to show that if $X_1 \sim \chi^2(r_1)$, $X_2 \sim \chi^2(r_2)$, ..., $X_n \sim \chi^2(r_n)$ are independent, then the random variable $Y = X_1 + X_2 + \cdots + X_n$ has the $\chi^2(r_1 + r_2 + \cdots + r_n)$ distribution.

3. (*Mathematica*) Derive a general formula for a 95% confidence interval for the variance σ^2 of a normal distribution based on a random sample of size n. Then write a *Mathematica* command to simulate such a confidence interval as a function of the normal distribution parameters and the sample size n. Run it 20 times for the case $\mu = 0$, $\sigma^2 = 1$, $n = 50$. How many of your confidence intervals contain the true $\sigma^2 = 1$? How many would you have expected to contain it?

4. (*Mathematica*) In Example 2 of Section 4.1 we analyzed a data set of sulfuric acid concentrations. Assuming that the measurements form a random sample from a normal distribution, find a 90% confidence interval estimate of the population standard deviation σ. The data are reproduced below for your convenience.

```
acids = {.154, .161, .157, .150,
         .143, .152, .158, .154, .153, .149}
```

5. (*Mathematica*) Recall the Albuquerque home price data set at the beginning of Section 4.2. The logged selling prices seemed to be roughly normal. Do you find significant statistical evidence that the standard deviation of the log price variable exceeds .2? Explain.

6. (*Mathematica*) If $\overline{X}$ and S^2 are the sample mean and sample variance of a random sample of size 20 from the $N(\mu, \sigma^2)$ distribution, find $P[\overline{X} \ge \mu, S^2 \ge \sigma^2]$.

7. (*Mathematica*) For the crime rate data in Example 1, do you find statistically significant evidence that the mean rate is more than 3? Explain.

8. Derive the formula for the t-density using the approach suggested in the section.

9. (*Mathematica*) If $T \sim t(23)$, find: (a) $P[T > 2.1]$; (b) $P[T < -1.6]$; (c) a point t such that $P[-t \le T \le t] = .80$.

10. (*Mathematica*) The data below are free-throw shooting percentages for Atlanta Hawks basketball players as of March 16, 1999. Viewing these as a sample from a universe in which a normally distributed continuum of percentages is possible, do you find statistically significant evidence that the distribution of percentages has a mean that is less than .75?

```
freethrows = {.826, .622, .659, .705, .673, .744,
    .629, .929, .788, .778, .800, .421, .750}
```

11. Derive the form of a 90% confidence interval for the mean μ of a normal distribution based on a sample of size n and the t-distribution.

12. Show that the t-distribution is symmetric about 0.

13. (*Mathematica*) Simulate a list of 200 T random variables as in formula (14), where samples of size 20 are taken from the $N(0, 1)$ distribution. Superimpose a scaled histogram of the data on a graph of the appropriate t-density to observe the fit.

14. In Exercise 5 of Section 4.2 were data on sulfur dioxide pollution potentials for a sample of 60 cities. They are reproduced below for your convenience. You should have found in that exercise that the variable $X = \log(SO_2)$ had an approximate normal distribution. Comment as a probabilist on the reasonableness of the claim that the mean of X is 3.

```
SO2 = {59, 39, 33, 24, 206, 72, 62, 4, 37,
    20, 27, 278, 146, 64, 15, 1, 16, 28, 124,
    11, 1, 10, 5, 10, 1, 33, 4, 32, 130, 193,
    34, 1, 125, 26, 78, 8, 1, 108, 161, 263,
    44, 18, 89, 48, 18, 68, 20, 86, 3, 20,
    20, 25, 25, 11, 102, 1, 42, 8, 49, 39};
```

15. (*Mathematica*) Find the smallest value of the degree of freedom parameter r such that the $t(r)$ density is within .01 of the $N(0, 1)$ density at every point x.

16. (*Mathematica*) If a random variable F has the $F(12, 14)$ distribution, find: (a) $P[F > 1]$; (b) $P[1 \leq F \leq 3]$; (c) numbers a and b such that $P[F < a] = .05$ and $P[F > b] = .05$.

17. (*Mathematica*) Below are two sets of hotel room rates for the Hyatt and Sheraton chains, advertised in major newspapers in 1999. Do you find significant statistical evidence that there is a difference in variability of these room rates? Comment on the appropriateness of the techniques of this section.

```
hyatt = {119, 99, 99, 189, 109, 139, 119,
     149, 155, 149, 139, 89, 79, 89, 109, 119,
     155, 99, 79, 109, 99, 135, 99, 89, 179,
     79, 99, 105, 89, 135, 99, 79, 119, 99, 99,
     69, 89, 129, 119, 79, 149, 79, 109, 89};
sheraton = {79, 69, 69, 89, 69, 89, 69, 99, 79,
     119, 69, 89, 115, 89, 89, 59, 69, 219, 119,
     89, 89, 75, 69, 99, 79, 109, 149, 75, 65, 69,
     79, 145, 79, 89, 99, 85, 159, 159, 159, 129,
     79, 69, 79, 79, 115, 79, 109, 74, 79, 89, 99,
     69, 69, 49, 55, 129, 79, 79, 65, 69, 79};
```

18. Discuss how you would go about testing statistically whether one normal variance σ_1^2 is two times another σ_2^2, or whether alternatively σ_1^2 is more than $2\sigma_2^2$.

19. (*Mathematica*) Write a *Mathematica* command to simulate a desired number of F-ratios as in formula (16). Then superimpose a histogram of 200 such F-ratios on the appropriate F-density, where the two random samples used to form S_x^2 and S_y^2 are both of size 20 and are taken from $N(0, 4)$ and $N(0, 9)$ distributions, respectively.

20. (*Mathematica*) Below are Nielsen television ratings from the week of August 16, 1999 for shows from the CBS network, and from the ABC network. Find a 90% confidence interval for the ratio of population variances, assuming that the data are independent random samples from normal populations. Conclude from your interval whether the data are consistent with the hypothesis of equality of the two variances.

```
CBS = {5.9, 4.9, 8.2, 7.7, 5.9, 6.0, 7.7,
       5.6, 4.9, 7.5, 5.2, 6.8, 7.4, 3.8, 4.0,
       4.1, 5.4, 3.6, 5.0, 5.7, 8.8, 8.3, 8.3};
ABC = {4.6, 7.2, 7.3, 6.1, 7.7, 6.0, 4.2,
       4.3, 6.3, 8.0, 6.9, 6.4, 8.0, 5.2,
       8.6, 6.8, 4.9, 6.9, 8.7, 5.7, 4.8,
       4.9, 4.2, 5.9, 3.9, 4.7, 9.4, 6.4};
```

21. If a random variable F has the $F(m, n)$ distribution, what can you say about the random variable $1/F$, and why?

CHAPTER 5
ASYMPTOTIC THEORY

5.1 Strong and Weak Laws of Large Numbers

The basic facts, concepts, and computational methods of probability have mostly been covered in Chapters 1 to 4. This short chapter will attempt to set a context for our subject, by using the two most important theorems of probability, the Law of Large Numbers and the Central Limit Theorem, as rallying points around which we study historical development. The final chapter will continue the process of context development in a different way, by examining several problem areas in applied mathematics that have arisen mostly in the latter part of this century, in which probability has taken a central place. Because our focus is on history and context rather than the development of new problem solving techniques, this chapter will be discursive and will have very few exercises. Yet it is an important one for you to read to complete your understanding.

The source material in this chapter comes from several places: some is just part of the oral tradition, and other information has come from some useful sources cited in the bibliography: Burton, Cooke, Hacking, Katz, and the fine probability book by Snell.

One might say that probability has always been with us, because human-kind has always been engaged in games of chance involving dice, cards, and other media. According to Burton (p. 415), archaeological digs of prehistoric sites frequently produce examples of forerunners of today's dice called *astraguli*, which are made from the tarsal bones of hooved animals. There is evidence in Egypt reaching as far back as 3500 B.C. that people played a board game called "Hounds and Jackals" using these astraguli. Dice in today's form seem to have been with us since at least the early Roman empire. There and elsewhere gambling became such a problem that the authorities moved to limit or outlaw it, to little avail. Card playing was introduced into Europe by contact with Eastern civilizations: Chinese, Indian, and also Egyptian, and by the 1300's it had become very popular. In fact, according to Burton again (p. 417), after Johann Gutenberg printed his famous Bible in 1440, he immediately went to work printing cards.

Financial problems involving insurance, investment, and mortality also arose early in Western history, as the merchant class developed. Among the earliest and most common such problems was the practice adopted by communities of raising public money by borrowing lump sums from individuals, then agreeing to pay a lifetime annuity to the lender. In light of the importance of the problem, and its dependence on knowing the likelihoods that the lender dies in a given amount of time, it is strange that a theory of probability did not come about earlier than it did. These communities tended not to play the insurance game very well.

The mathematician Gerolamo Cardano (1501-1576) was one of the earliest to study problems of probability. Cardano was a brilliant and eccentric character: an astrologer, physician, and mathematician, who became Rector of the University of Padua. He did early work on the solution of the general cubic equation, and was involved in a dispute over what part was actually his work. But for us it is most interesting that he wrote a work called *Liber de ludo*, or *Book on Gambling*, in which among other things he introduced the idea of characterizing the probability of an event as a number p between 0 and 1, showed understanding in the context of dice of the theoretical concept of equiprobable outcomes, and anticipated that if the probability of an event is p and you perform the experiment a large number of times n, then the event will occur about $n\,p$ times. Thus, he predicted important results about binomial expectation and limiting probabilities that were more fully developed by others much later. But *Book on Gambling* was only published posthumously in 1663, by which time another very important event had already happened.

It is widely acknowledged that there is a specific time and place where probability as a well-defined mathematical area was born. Blaise Pascal (1623-1662) was a prodigy who at the age of 12 mastered Euclid's *Elements*. He was never in great health and died at the age of 39, but before that he gained acclaim not only as a mathematician, but as an experimental physicist, religious philosopher, and activist. In 1654 he began a correspondence with the great mathematician Pierre DeFermat (1601-1665) from which probability theory came. In those days in France, it was common for noblemen to dabble in academic matters as a part of their social duties and pleasures. One Antoine Gombaud, the Chevalier de Méré, was such a person. He was acquainted with Pascal and many other intellectuals of his day, and he enjoyed dicing. The problem he posed to Pascal was how to fairly divide the stakes in a game which is terminated prematurely. For instance, if a player has bet that a six will appear in eight rolls of a fair die, but the game is ended after three unsuccessful trials, how much should each player take away from the total stake? Pascal reported this to Fermat, and their interaction shows a clear understanding of repeated independent trials, and the difference between conditional and unconditional probability. Pascal reasoned that the probability of winning on throw i is $(1/6)(5/6)^{i-1}$ so that adding these terms together for $i = 4, ..., 8$ gives the total probability p of winning on trials 4 through 8, which is how the stake should be divided: a proportion p to the bettor under concern and $1 - p$ to the other. But Fermat responded that this analysis is only correct if you do not know the outcome of the first three rolls. If you do, then the chance that the bettor wins on the next roll is 1/6 (note this is a conditional probability), so that the bettor should take 1/6 of the stake in exchange for quitting at trial 4, in fact at any trial. Pascal was happy with Fermat's analysis. Later on as they studied independent trials problems, they both noticed the importance of the binomial coefficient $\binom{n}{k}$ (not written in this notation in those days) to the computation of probabilities involving independent trials. It should be added, though (Cooke (p. 201)), that combinatorics itself had its origin long before. At least as early as 300 B.C., Hindu scholars were asking such questions as how many philosophical systems can be formed by taking a certain number of doctrines from a larger list of doctrines, which is clearly our basic problem of sampling without order or replacement.

The second half of the seventeenth century was a very fertile one for probability, as many of the most famous mathematicians and scientists were making contributions. Christiaan Huygens (1629-1695) was a great Dutch scholar who wrote the first book on probability: *De ratiociniis in ludo aleae*, or *On Reasoning in a Dice Game* that compiled and extended the results of Pascal and Fermat. In particular, Huygens developed ways to compute multinomial probabilities, and crystallized the idea of expectation of a game. But it was James Bernoulli (1654-1705) who wrote the seminal work *Ars Conjectandi* or *The Art of Conjecturing*, released after his death in 1713, whose work would set new directions for the burgeoning subject. In *The Art of Conjecturing*, Bernoulli set down the combinatorial underpinnings of probability in much the form as they exist today. He is not the original source of most of these results, as previously mentioned, and probably benefited in this regard by the work of his mentor Gottfried Wilhelm Leibniz (1646-1716), who made contributions to combinatorics as well as co-inventing calculus. But the most important result to the development of probability was Bernoulli's proof of a special case of the theorem that we now describe.

Probability would never be able to have a firm foundation without an unambiguous and well-accepted definition of the probability of an event. And, this probability ought to have meaningful predictive value for future performances of the experiment to which the event pertains. One needs to know that in many repetitions of the experiment, the proportion of the time that the event occurs converges to the probability of the event. This is the Law of Large Numbers (a phrase coined by Simeon-Denis Poisson (1781-1840) for whom the Poisson distribution is named).

The Law of Large Numbers actually takes two forms, one called the *Weak Law* and the other called the *Strong Law*. Here are the modern statements of the two theorems.

Theorem 1. (Weak Law of Large Numbers) Assume that $X_1, X_2, X_3, \ldots$ is a sequence of independent and identically distributed random variables. Let

$$\overline{X}_n = \tfrac{1}{n} \Sigma_{i=1}^n X_i$$

be the sample mean of the first n X's, and let μ denote the mean of the distribution of the X's. Then for any $\epsilon > 0$,

$$P[|\overline{X}_n - \mu| \geq \epsilon] \longrightarrow 0 \text{ as } n \longrightarrow \infty \tag{1}$$

Theorem 2. (Strong Law of Large Numbers) Under the assumptions of the Weak Law of Large Numbers, for all outcomes ω except possibly some in a set of probability 0,

$$\overline{X}_n(\omega) \longrightarrow \mu \text{ as } n \longrightarrow \infty \qquad (2)$$

Both theorems talk about the convergence of the sequence of sample means to the population mean, but the convergence modes in (1) and (2) are quite different. To illustrate the difference let us use simulation.

Convergence mode (1), usually called *weak convergence* or *convergence in probability*, refers to an aspect of the probability distribution of $\overline{X}$. Fixing a small ϵ > 0, as the sample size n gets larger, it is less and less likely for $\overline{X}$ to differ from μ by at least ϵ. We can show this by looking at the empirical distribution of $\overline{X}$ for larger and larger n, and observing that the relative frequencies above $\mu + \epsilon$ and below $\mu - \epsilon$ become smaller and smaller. Here is a command to simulate a desired number of sample means for a given sample size from a given distribution. You may remember that we used the same command earlier in Section 2.6. We apply the command to sample from the uniform(0,1) distribution, for which $\mu = .5$, taking $\epsilon = .04$ and sample sizes $n = 100$, 200, and 300.

```
Needs["KnoxProb`Utilities`"]
```

```
SimulateSampleMeans[nummeans_,
    distribution_, sampsize_] := Table[
    Mean[RandomArray[distribution, {sampsize}]],
    {nummeans}]
```

```
SeedRandom[18732]
list1 = SimulateSampleMeans[
    100, UniformDistribution[0, 1], 100];
list2 = SimulateSampleMeans[100,
    UniformDistribution[0, 1], 200];
list3 = SimulateSampleMeans[100,
    UniformDistribution[0, 1], 300];
Histogram[list1, 6, Endpoints → {.38, .62}];
Histogram[list2, 6, Endpoints → {.38, .62}];
Histogram[list3, 6, Endpoints → {.38, .62}];
```

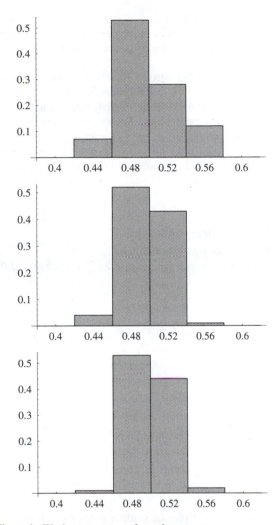

Figure 1 - Weak convergence of sample means to μ

In Figure 1 the subinterval midpoints have been chosen so that the middle two categories range from .46 to .54, that is, from $\mu - \epsilon$ to $\mu + \epsilon$. The empirical proportion of sample means that fall outside of this interval is the total height of the two outer rectangles. You see this proportion becoming closer to 0 as the sample size increases. This is what weak convergence means.

The mode of convergence characterized by (2), usually called *strong*, or *almost sure* convergence, is more straightforward. Idealize an experiment that consists of observing an infinite sequence $X_1, X_2, X_3, \ldots$ of independent and identically distributed random variables. For a particular experimental outcome ω, the partial sample means $\overline{X}_1(\omega), \overline{X}_2(\omega), \overline{X}_3(\omega), \ldots$ form an ordinary sequence of numbers. For almost every outcome, this sequence of numbers converges in the calculus sense to μ. It turns out that in general strong convergence of a sequence of random variables implies weak convergence, but weak convergence does not imply strong convergence. Both modes of convergence occur for the sequence of partial means.

We can illustrate the Strong Law using the command SimMeanSequence from Exercise 10 of Section 2.6 and Example 2 of Section 3.3, which successively simulates observations from a discrete distribution, updating $\overline{X}$ each time m new observations have been generated. So we are actually just picking out a subsequence of $(\overline{X}_n(\omega))$, which hits every mth sample mean, but that will let us plot the subsequence farther out than the whole sequence could have been plotted. The code for SimMeanSequence is repeated in the closed cell below.

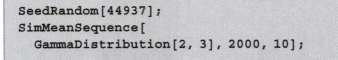

```
SeedRandom[44937];
SimMeanSequence[
   GammaDistribution[2, 3], 2000, 10];
```

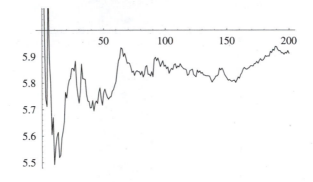

Figure 2 - Strong convergence of sample means to μ

The connected list plot of sample means in Figure 2 is the result of calling this command on the $\Gamma(2, 3)$ distribution, whose mean is $\mu = 2 \cdot 3 = 6$, to simulate every tenth member of the sequence of sample means for a total of 2000 observations. We see the sequence of means converging to 6 as the Strong Law claims; however the convergence is not necessarily rapid or smooth. Even after 500 observations ($m = 50$ groups of 10), the sample mean is about 5.7.

Activity 1 Read the code for SimulateSampleMeans and SimMeanSequence carefully to make sure you see how the algorithms relate to Theorems 1 and 2. Use these *Mathematica* commands on different probability distributions. Feel free to change the sample size and number of means. Do your experiments confirm the conclusions of the Laws of Large Numbers?

In Exercises 2 and 3 you are led through a fairly easy proof of the Weak Law of Large Numbers based on a lemma called *Chebyshev's Inequality*, which is interesting in itself. This states that for any random variable X with finite mean μ and variance σ^2,

$$P[\,|X - \mu| \geq k\sigma] \leq \tfrac{1}{k^2} \tag{3}$$

for any $k > 0$. In words, Chebyshev's inequality says that the chance that X is more than k standard deviations away from its mean is no more than $1/k^2$. The remarkable thing about the result is that it is a bound on the tail probability of the distribution that does not depend on what the distribution is.

The Russian mathematician who proved the inequality is worthy of special mention. Pafnutii L'vovich Chebyshev (1821-1894) formally introduced the concept of random variables and their expectation in their modern forms. Besides his proof of the weak law for means, he is responsible for stating the Central Limit Theorem using random variables, and he was the teacher of Andrei Andreevich Markov (1856-1922) who introduced the dependent trials analysis that was later to bear his name (see Section 6.1 on Markov chains).

James Bernoulli proved the weak law in *The Art of Conjecturing* in the following special case. Let the sequence $X_1, X_2, \dots$ be a sequence of what we call today independent Bernoulli random variables, taking the value 1 with probability p and 0 with probability $1 - p$. Then $\mu = E[X] = p$. But the sample mean $\overline{X} = \sum_{i=1}^{n} X_i / n$ is just the proportion of successes in the sample. Therefore the statement that $P[|\overline{X}_n - \mu| \geq \epsilon] \longrightarrow 0$ as $n \longrightarrow \infty$ says that as we repeat the experiment more and more often, the probability approaches 0 that the sample success proportion differs by a fixed positive amount from the true success probability p. The convergence of the sample proportion of times that an event occurs justifies the long-run frequency approach to probability, in which the probability of an event is the limit of the empirical long-run frequency. It also is a precursor to the modern theory of estimation of parameters in statistics.

The strong law is much more difficult to show than the weak law. In fact it was not until 1909 that Emile Borel (1871-1956) showed the strong law for the special case of a Bernoulli sequence described above. Finally in 1929, Andrei Nikolaevich Kolmogorov (1903-1987) proved the general theorem. Kolmogorov was one of the later members of the great Russian school of probabilists of the late 1800's and early 1900's, which included Chebyshev, Markov, and Alexsandr Mikhailovich Lyapunov (1857-1918) about whom we will hear again in the next section. That tradition has continued in the latter part of the twentieth century, as has the traditional role of the French in probability theory.

Mathematica for Section 5.1

Command	Location
RandomArray[dist, n]	Statistics` ContinuousDistributions`
UniformDistribution[a, b]	Statistics` ContinuousDistributions`
GammaDistribution[a, b]	Statistics` ContinuousDistributions`
Mean[list]	Statistics` DescriptiveStatistics`
SeedRandom[]	kernel
Histogram[list, n]	KnoxProb` Utilities`
SimulateSampleMeans[n, dist, size]	Section 2.6
SimMeanSequence[dist, n, m]	Section 3.3

Exercises 5.1

1. (*Mathematica*) The Chevalier de Méré also posed the following problem to Pascal, which he thought implied a contradiction of basic arithmetic. Pascal expeditiously proved him wrong. How many rolls of two dice are necessary such that the probability of achieving at least one double six is greater than 1/2?

2. Prove Chebyshev's inequality, formula (3), for a continuous random variable X with p.d.f. f, mean μ, and standard deviation σ. (Hint: Write a formula for the expectation of $(X - \mu)^2 / (\sigma^2 k^2)$, and split the integral into the complementary regions of x values for which $|x - \mu| \geq k\sigma$ and $|x - \mu| < k\sigma$.)

3. Use Chebyshev's inequality to prove the weak law of large numbers.

4. (*Mathematica*) To what do you think the sample variance S^2 converges strongly? Check your hypothesis by writing a command similar to SimMeanSequence and running it for several underlying probability distributions.

5. (*Mathematica*) Is the Chebyshev bound in formula (3) a tight bound? Investigate by computing the exact value of $P[|X - \mu| \geq 2\sigma]$ and comparing to the Chebyshev bound when:
(a) $X \sim \Gamma(3, 4)$; (b) $X \sim$ uniform(0, 1); (c) $X \sim N(2, 4)$.

6. Consider a sample space Ω of an infinite sequence of coin flips in which we record $X_i = \pm 1 / 2^i$ on the ith trial according to whether the ith trial is a head or tail, respectively. Show that the sequence (X_n) converges strongly to 0. Show also that it converges weakly to 0.

7. Suppose that a random variable X has a finite mean μ and we have an a priori estimate of its standard deviation $\sigma \approx 1.6$. How large a random sample is sufficient to ensure that the sample mean $\overline{X}$ will estimate μ to within a tolerance of .1 with 90% probability?

5.2 Central Limit Theorem

In Section 4.3 we observed the very important fact that the sample mean $\overline{X}$ of a random sample from a normal population is itself a normally distributed random variable with mean μ and variance σ^2/n. From this we saw that we could assess the reasonableness of hypothesized values of μ; if $\overline{X}$ is unlikely to have taken its observed value for a given μ, then we disbelieve that μ. We saw these ideas again in Section 4.5 on the t-distribution. But in each case we made the apparently restrictive assumption that the population being sampled from is normally distributed. There is no theorem more useful in statistical data analysis than the *Central Limit Theorem*, because it implies that approximate statistical inference of this kind on the mean of a distribution can be done without assuming that the underlying population is normal, as long as the sample size is large enough. So, it is our purpose to explore this theorem in this section, and we will also take the opportunity to give some of the history of the normal distribution, the Central Limit Theorem itself, and the beginnings of the subject of statistics.

It was Abraham DeMoivre (1667-1754) who, in a 1733 paper, first called attention to an analytical result on limits which enables the approximation of binomial probabilities $\binom{n}{k} p^k (1-p)^{n-k}$ for large n by an expression which turns out to be a normal integral. Although DeMoivre was born at the time when the great French probability tradition was developing, the fact that his family was Protestant made life in France uncomfortable, and even led to DeMoivre's imprisonment for two years. After he was freed in 1688 he spent the rest of his life in England where Newton's new Calculus was being worked out.

While DeMoivre had confined his work mostly to the case $p = 1/2$, Pierre de Laplace (1749-1827), motivated by an effort to statistically estimate p given the observed proportion of successes in a very large sample, generalized DeMoivre's work to arbitrary success probabilities in 1774. He essentially showed that (Katz, p. 550)

$$\lim_{n \to \infty} P[\, |\hat{p} - p| \le \epsilon \,] = \lim_{n \to \infty} \frac{2}{\sqrt{2\pi}} \int_0^{\epsilon/\sigma} e^{-u^2/2}\, du$$

where $\hat{p}$ is the sample proportion of successes and σ is the standard deviation of $\hat{p}$, namely $\sqrt{p(1-p)/n}$. Leonhard Euler had already shown that $\int_0^\infty e^{-u^2/2}\,du = \sqrt{\pi/2}$; so since σ is approaching 0, the limit above is 1. Notice that with this result Laplace has a novel proof of the Weak Law of Large Numbers by approximation, and is just inches from defining a normal density function. Laplace published in 1812 the first of several editions of his seminal work on probability, *Theorie Analytique des Probabilites*, which contained many results he had derived over a period of years. In this book he proved the addition and multiplication rules of probability, presented known limit theorems to date, and began to apply probability to statistical questions in the social sciences. But Laplace was actually more interested in celestial mechanics and the analysis of measurement errors in astronomical data than in social statistics. Because of the rise of calculus, celestial mechanics was a hot topic in European intellectual circles at that time, and the contemporary advances in probability combined with calculus formed the breeding ground for mathematical statistics.

Carl Friedrich Gauss (1777-1855), who rose from humble family beginnings in Brunswick, Germany to become one of the greatest and most versatile mathematicians of all time, took up Laplace's study of measurement errors in 1809. Gauss was interested in a function $f(x)$ which gave the probability (we would say probability density instead) of an error of magnitude x in measuring a quantity. Laplace had already given reasonable assumptions about such a function: it should be symmetric about 0 and the integral of $f(x)$ over the whole line should be 1; in particular, $f(x)$ should approach 0 as x approaches both $+\infty$ and $-\infty$. But it was Gauss who arrived at the formula for a normal density with mean zero, by considering the maximization of the joint probability density function of several errors, and making the further supposition that this maximum should occur at the sample mean error. He was able to derive a differential equation for f in this way, and to show a connection to the statistical problem of least squares curve fitting. Gauss' tremendous contributions have led to the common practice of referring to the normal distribution as the *Gaussian distribution*.

Let us return to the Central Limit Theorem, the simplest statement of which is next.

Theorem 1. (Central Limit Theorem) Let X_1, X_2, X_3, ... be a sequence of independent and identically distributed random variables with mean μ and variance σ^2. Let $\overline{X}_n$ be the mean of the first n X's. Then the cumulative distribution function F_n of the random variable

$$Z_n = \frac{\overline{X}_n - \mu}{\sigma/\sqrt{n}} \tag{1}$$

converges to the c.d.f. of the $N(0, 1)$ distribution as $n \longrightarrow \infty$.

Thus, the Central Limit Theorem allows us to say (vaguely) that the distribution of $\overline{X}_n$ is approximately $N(\mu, \sigma^2/n)$ for "large enough" n, despite the fact that the sample members $X_1, X_2, \ldots, X_n$ may not themselves be normal. How large is "large enough" is a very good question. In Figure 3 for example is a scaled histogram of 500 simulated sample means from samples of size $n = 25$ from a $\Gamma(2, 1)$ distribution, with a normal density of mean $\mu = 2 \cdot 1 = 2$ and variance $\sigma^2/n = 2 \cdot 1^2/25 = 2/25$ superimposed. (We used the command SimulateSampleMeans from the last section; you should open up the closed cell below the package loading command to see the code.)

```
Needs["KnoxProb`Utilities`"]
```

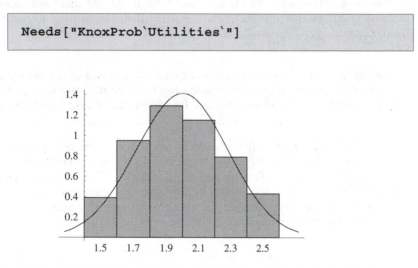

Figure 3 - Empirical and limiting distribution of $\overline{X}$, $n = 25$, population distribution $\Gamma(2, 1)$

The fit of the empirical distribution of sample means to the appropriate normal density is quite good, even for n as small as 25. You should try running this experiment with the gamma distribution again several times (without resetting the random seed), to see whether this is a consistent result. Also, do the activity below.

Activity 1   Rerun the commands that produced Figure 1 using the uniform(0, 2) distribution instead. Experiment with sample sizes 25, 20, and 15. Does the histogram begin to depart drastically from the normal density as the sample size decreases?

Your experimental runs probably indicated a good fit of the empirical distribution of $\overline{X}$ to normality even for the smallest sample size of 15 for the uniform distribution. Experience has shown that $n = 25$ is fairly safe for most population distributions, and for some of the more symmetric population distributions even fewer samples are necessary. This is because there is a general result

(Loève (p. 300)) that gives a bound on the difference between the actual c.d.f. F_n of Z_n in (1) and the standard normal c.d.f. G:

$$|F_n(x) - G(x)| \leq \frac{c}{n^{1/2}\sigma^3} \cdot E[|X|^3] \quad \text{for all } x, \tag{2}$$

where c is a constant. You can see that the maximum absolute difference decreases as n grows, as you would expect, but it increases as the third absolute moment $E[|X^3|]$ of the distribution being sampled from increases. Actually, the ratio $E[|X|^3]/\sigma^3$ can be used to measure the lack of symmetry of a distribution, called its *skewness*. Thus, we have the qualitative result that the more symmetric is the underlying distribution, the better the fit of the distribution of $\overline{X}$ to normality for fixed n, alternatively, the smaller n can be taken for an equally good fit.

Activity 2 If one distribution has a skewness that is half of the skewness of a second distribution, what is the relationship between the sample sizes required to have equal upper bounds on the absolute difference of c.d.f.'s in formula (2)?

The Central Limit Theorem can be proved by a moment-generating function argument. We will outline the proof that the moment-generating function of Z_n converges to the standard normal m.g.f. $M(t) = e^{t^2/2}$ as $n \longrightarrow \infty$, which must imply that the c.d.f. of Z_n converges to the standard normal c.d.f. at every point x. (The latter is actually a subtle theorem, whose proof we will omit.)

We can write the m.g.f. of Z_n as:

$$
\begin{aligned}
E[e^{tZ_n}] &= E\left[e^{t \cdot (\overline{X}_n - \mu)/(\sigma/\sqrt{n})}\right] \\
&= E\left[e^{\frac{t\sqrt{n}}{\sigma} \cdot \frac{1}{n} \Sigma(X_i - \mu)}\right] \\
&= E\left[e^{\frac{t}{\sigma\sqrt{n}} \cdot \Sigma(X_i - \mu)}\right] \\
&= \prod_{i=1}^{n} E\left[e^{\frac{t}{\sigma\sqrt{n}} \cdot (X_i - \mu)}\right]
\end{aligned} \tag{3}
$$

Let $M_1(t)$ be the m.g.f. of every $X_i - \mu$. Note that $M_1(0) = E[e^0] = 1$, $M_1'(0) = E[X_i - \mu] = 0$, and $M_1''(0) = E[(X_i - \mu)^2] = \text{Var}(X_i) = \sigma^2$. Expanding $M_1(t)$ in a second-order Taylor series about 0 we have for some $\xi \in (0, t)$,

$$
\begin{aligned}
M_1(t) &= M_1(0) + M_1'(0)(t - 0) + \tfrac{1}{2} M_1''(0)(t - 0)^2 \\
&\quad + \tfrac{1}{6} M_1'''(\xi)(t - 0)^3 \\
&= 1 + \tfrac{1}{2}\sigma^2 t^2 + \tfrac{1}{6} M_1'''(\xi) t^3
\end{aligned}
$$

Substituting this into (3), the m.g.f. of Z_n becomes

$$E[e^{t Z_n}] = \prod_{i=1}^{n} M_1\left(\frac{t}{\sigma \sqrt{n}}\right)$$

$$= \prod_{i=1}^{n} \left(1 + \frac{1}{2} \sigma^2 \frac{t^2}{\sigma^2 n} + \frac{1}{6} M_1'''(\xi) \frac{t^3}{\sigma^3 n^{3/2}}\right)$$

$$= \left(1 + \frac{t^2/2}{n} + \frac{1}{6} M_1'''(\xi) \frac{t^3}{\sigma^3 n^{3/2}}\right)^n, \quad \xi \in \left(0, \frac{t}{\sigma \sqrt{n}}\right)$$

If the third derivative of M_1 is a well-behaved function near zero, then the third term in parentheses goes swiftly to zero as $n \longrightarrow \infty$. The limit of the exponential expression is therefore the same as that of $\left(1 + \frac{t^2/2}{n}\right)^n$ which from calculus is well known to be $e^{t^2/2}$. Thus the m.g.f. of Z_n converges to that of the standard normal distribution.

Though the basic result was known to DeMoivre and Laplace in the 1700's, it was not until the 1900's that the theorem as we know it was proved. And throughout much of the twentieth century, some of the best probabilistic minds have set to work improving the theorem by relaxing hypotheses. A.M. Lyapunov, working with bounds such as the one in formula (2), proved the first modern version of the Central Limit Theorem in 1901, making assumptions about the third moments of the random variables. J.W. Lindeberg proved a general version in 1922 of which ours is a corollary. Important contributions were also made by three men who can without hesitation be called the towering figures in probability in this century: A.N. Kolmogorov who proved the strong law, the beloved Princeton mathematician and author Willy Feller (1906-1970), and the central figure in the modern French school of probability Paul Lévy (1886-1972). These and other probabilists have been interested in the more general question of what limiting distributions for sums of random variables are possible, if the restrictions of identical distributions, independence, and finiteness of the variance are removed.

The next example shows how our general statement involving means covers the special cases studied by DeMoivre and Laplace involving approximation of binomial probabilities.

Example 1 Suppose that we are conducting a poll to find out the proportion of the voting citizenry in a state that is in support of a proposition legalizing gambling. If we decide to sample 500 voters at random and 315 are in support, give an approximate 95% confidence interval for that proportion. How many voters should be sampled if we want to estimate the proportion to within 2% with 95% confidence?

To form a mathematical model for the problem, we suppose that there is an unknown proportion p of the voters who support the proposition. We sample $n = 500$ of them, in sequence and with replacement so that the voters who are sampled form 500 independent Bernoulli trials with success probability p. We can denote their responses by $X_1, X_2, \ldots, X_{500}$ where each $X_i = 1$ or 0, respectively, according to whether the voter supports the proposition or not. Then the sample proportion $\hat{p}$ who are in favor is the same as the sample mean $\overline{X}$ of the random variables X_i. Therefore the Central Limit Theorem applies to $\hat{p}$. By properties of the Bernoulli distribution, the common mean of the X_i's is $\mu = p$ and the variance is $\sigma^2 = p(1-p)$. Thus, the appropriate mean and standard deviation with which to standardize $\hat{p}$ in the Central Limit Theorem are p and $\sqrt{p(1-p)/n}$, respectively. We therefore have that

$$Z_n = \frac{\hat{p}-p}{\sqrt{p(1-p)/n}} \approx N(0, 1) \tag{4}$$

Remember that to construct confidence intervals, the approach is to begin with a transformed random variable whose distribution is known, pick quantiles for that distribution that are suitable for the target level of confidence, then algebraically isolate the parameter of interest in the resulting inequalities. For a 95% confidence interval, each tail needs probability .025; so here is the suitable normal quantile.

```
z = Quantile[NormalDistribution[0, 1], .975]
```

1.95996

Then,

$$.95 \approx P[-z \le Z_n \le z] = P\left[-z \le \frac{\hat{p}-p}{\sqrt{p(1-p)/n}} \le z\right]$$
$$= P\left[\hat{p} - z\sqrt{p(1-p)/n} \le p \le \hat{p} + z\sqrt{p(1-p)/n}\right]$$

Hence to give an interval estimate of p we can start with the point estimate $\hat{p}$ and then add and subtract the margin of error term $z\sqrt{p(1-p)/n}$. The only trouble with this is that the margin of error depends on the unknown p. But since our probability is only approximate anyway, we don't lose much by approximating the unknown margin of error as well, by replacing p by the sample proportion $\hat{p}$. We let *Mathematica* finish the computations for the given numerical data.

```
p̂ = 315/500

margin = z * √(p̂ (1 - p̂) / 500)
{p̂ - margin, p̂ + margin}
```

$\dfrac{63}{100}$

0.0423189

{0.587681, 0.672319}

So the sample proportion is 63% and we are 95% sure that the true proportion of all voters in support of the proposition is between about 59% and 67%.

The second question is important to the design of the poll. Our margin of error in the first part of the question was about 4%; what should we have done to cut it to 2%? Leaving n general for a moment, we want to satisfy

$$z\sqrt{p(1-p)/n} \le .02$$

We know that the quantile z is still 1.96 because the confidence level is the same as before, but prior to sampling we again do not know p, hence we cannot solve for n. There are two things that can be done: we can produce a pilot sample to estimate p by $\hat{p}$, or we can observe that the largest value that $p(1-p)$ can take on is 1/4, when $p = 1/2$. (Do the simple calculus to verify this.) Then $z\sqrt{p(1-p)/n} \le z\sqrt{1/4n}$, and as long as the right side is less than or equal to .02, then the left side will be as well. Thus,

$$\frac{z}{\sqrt{4n}} \le .02 \Longleftrightarrow \sqrt{n} \ge \frac{z}{2(.02)}$$

Squaring both sides of the last inequality gives the following threshhold value for the sample size n:

```
(  z  )²
( ——— )
( 2*.02)
```

2400.91

We would need to poll 2401 people in order to achieve the .02 margin with 95% confidence. ∎

Mathematica for Section 5.2

Command	Location
PDF[dist, x]	Statistics` ContinuousDistributions`
Quantile[dist, p]	Statistics` ContinuousDistributions`
NormalDistribution[μ, σ]	Statistics` ContinuousDistributions`
GammaDistribution[a, b]	Statistics` ContinuousDistributions`
SeedRandom[seed]	kernel
Histogram[list, n]	KnoxProb` Utilities`
SimulateSampleMeans[n, dist, m]	Section 2.6

Exercises 5.2

1. Use the Central Limit Theorem for means, Theorem 1, to formulate a similar limit theorem for the partial sums $S_n = \sum\limits_{i=1}^{n} X_i$.

2. The differential equation that Gauss arrived at for his density function of observed errors is

$$\frac{f'(x)}{x f(x)} = -k$$

where k is a positive constant. Solve this to obtain $f(x)$.

3. The DeMoivre-Laplace limit theorem says that

$$\sum_{k=a}^{b} \binom{n}{k} p^k (1-p)^{n-k} \xrightarrow[n \to \infty]{} \int_{(a-\mu)/\sigma}^{(b-\mu)/\sigma} \frac{1}{\sqrt{2\pi}} e^{-z^2/2} \, dz$$

where $\mu = np$ and $\sigma^2 = np(1-p)$. Explain how this result follows as a special case of the Central Limit Theorem.

4. The Central Limit Theorem has been extended to the case where $X_1, X_2, X_3 \ldots$ are independent but not necessarily identically distributed. Write what you think the theorem statement would be in this case.

5. (*Mathematica*) The Central Limit Theorem allows the distribution being sampled from to be discrete as well as continuous. Compare histograms of 300 sample means of samples of sizes 20, 25, and 30 to appropriate normal densities for each of these two distributions: (a) Poisson(3); (b) geometric(.5).

6. (*Mathematica*) Suppose that your probability of winning a game is .4. Use the Central Limit Theorem to approximate the probability that you win at least 150 out of 400 games.

7. (*Mathematica*) Do you find significant statistical evidence against the hypothesis that the mean of a population of SAT scores is 540, if the sample mean among 100 randomly selected scores is 550 and the sample standard deviation is 40?

CHAPTER 6
APPLICATIONS OF PROBABILITY

6.1 Markov Chains

Probability theory has ridden the wave of the 20th century technological and telecommunications revolution. The analysis of modern business systems for production, inventory, distribution, computing, and communication involves studying the effect of random inputs on complex operations, with an eye toward predicting their behavior, and improving the operations. The availability of fast computers has also made it possible to solve probabilistic problems that were impractical until now.

This chapter is meant to introduce you to a few applied topics in probability that have been the focus of much effort in the second half of the 20th century, and which will continue to flourish in the 21st century. And the work is all based on the old foundational results of probability from Chapters 1–4.

We will begin with the subject of *Markov chains*, which roughly are systems taking on various states as time progresses, subject to random influences. Here are just a few systems which could be amenable to Markov chain modeling.

1. The state of deterioration of a piece of equipment
2. The popularity of a politician
3. The inventory level of an item in a store
4. The number of jobs waiting for processing by a computer server
5. The value of a retirement investment

You may also recall the random walk that we simulated in Section 1.3, which is a special kind of Markov chain.

Activity 1 Try to think of more examples of time dependent random processes in the world similar to 1-5 above.

A precise definition of a Markov chain is below. Note that the random states of the system are modeled as random variables, and we impose an extra condition on the conditional distributions that makes future states independent of the past, given the present.

Definition 1. A sequence $X_0, X_1, X_2, \ldots$ of random variables is said to be a *Markov chain* if for every $n \geq 0$,

$$P[X_{n+1} = j \mid X_0 = i_0, X_1 = i_1, \ldots X_n = i_n] = P[X_{n+1} = j \mid X_n = i_n] \qquad (1)$$

In other words, at each time n, X_{n+1} is conditionally independent of $X_0, X_1, \ldots, X_{n-1}$ given X_n.

This conditional independence condition (1) makes the prediction of future values of the chain tractable, and yet it is not unreasonably restrictive. Consider the chain of store inventory levels in the third example problem above. The next inventory level will depend on the current level, and on random sales activity during the current time period, not on the past sequence of levels. There are many systems in which past behavior has no bearing on the future trajectory, given the present state, and for those systems, Markov chain modeling is appropriate.

Many interesting results have been derived about Markov chains. But for our purposes we will be content to introduce you to two problems: (1) finding the probability law of X_n; and (2) finding the limit as $n \longrightarrow \infty$ of this probability law, in order to predict the long term average behavior of the chain.

Now by formula (1), the probabilistic structure of a Markov chain can be represented by a graph called a *transition diagram*. Consider the chain $X_0, X_1, X_2, \ldots$ with possible states 1, 2, 3, and 4 shown as the labeled bubbles in Figure 1.

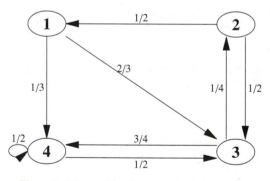

Figure 1 - The transition diagram of a Markov chain

If $X_n = 1$ at some time n, then the arrows in the transition diagram show that X_{n+1} can only take on the values 3 or 4, and it will do so with conditional probabilities 2/3 and 1/3, respectively. Because of (1), the probability law of the chain is completely determined by these single time step transition probabilities. The same transition probabilities can be assembled into a matrix called a *transition matrix*:

$$
T \doteq \text{now:}
\begin{array}{c}
\text{later:} \\
\begin{array}{c} 1 \\ 2 \\ 3 \\ 4 \end{array}
\end{array}
\begin{array}{cccc}
1 & 2 & 3 & 4 \\
\left(\begin{array}{cccc}
0 & 0 & 2/3 & 1/3 \\
1/2 & 0 & 1/2 & 0 \\
0 & 1/4 & 0 & 3/4 \\
0 & 0 & 1/2 & 1/2
\end{array} \right)
\end{array}
\tag{2}
$$

where each row refers to a possible state "now" (at time n) and each column corresponds to a possible state "later" (at time $n + 1$). For example the row 3-column 2 entry is

$$
P[X_{n+1} = 2 \mid X_n = 3] = 1/4
$$

Furthermore, the discrete probability distribution of the initial state X_0, if known, can be written as a row vector. For instance

$$
\boldsymbol{p}_0 = (1/2 \quad 1/2 \quad 0 \quad 0)
$$

is the initial distribution for which X_0 is equally likely to be in states 1 and 2. Special row vectors such as $\boldsymbol{p}_0 = (0 \quad 1 \quad 0 \quad 0)$ apply when the inital state is certain; in this case it is state 2.

Activity 2 Try multiplying each row vector $\boldsymbol{p}_0$ above on the right by the transition matrix T. Make note of not only the final answers for the entries of $\boldsymbol{p}_0 T$, but also what they are composed of. Before reading on, try to guess at the meaning of the entries of $\boldsymbol{p}_0 T$.

The Law of Total Probability is the tool to prove the main theorem about the probability distribution of X_n. The nth power of the transition matrix gives the conditional distribution of X_n given X_0, and you can pre-multiply by the initial distribution $\boldsymbol{p}_0$ to get the unconditional distribution of X_n.

Theorem 1. (a) $T^n(i, j) = P[X_n = j \mid X_0 = i]$, where T^n denotes the nth power of the transition matrix.
(b) $\boldsymbol{p}_0 \cdot T^n(j) = P[X_n = j]$, where $\boldsymbol{p}_0$ denotes the distribution of X_0 in vector form.

Proof. (a) We work by induction on the power n of the transition matrix. When $n = 1$ we have for all i and j,

$$
T^1(i, j) = P[X_1 = j \mid X_0 = i]
$$

by the construction of the transition matrix. This anchors the induction. For the inductive step we assume for a particular $n - 1$ and all states i and j that $T^{n-1}(i, j) = P[X_{n-1} = j \mid X_0 = i]$. By the Law of Total Probability,

$$P[X_n = j \mid X_0 = i] =$$
$$\sum_k P[X_{n-1} = k \mid X_0 = i] \, P[X_n = j \mid X_{n-1} = k, X_0 = i]$$

where the sum is taken over all states k. The first factor in the sum is $T^{n-1}(i, k)$ by the inductive hypothesis, and the second is $T(k, j)$ by the Markov property (1). Thus,

$$P[X_n = j \mid X_0 = i] = \sum_k T^{n-1}(i, k) \, T(k, j)$$

By the definition of matrix multiplication, the sum on the right is now the i-j component of the product $T^{n-1} \, T = T^n$, which finishes the proof of part (a).

(b) By part (a) and the Law of Total Probability,

$$\begin{aligned} \boldsymbol{p}_0 \, T^n(j) &= \sum_k p_0(k) \, T^n(k, j) \\ &= \sum_k P[X_0 = k] \, P[X_n = j \mid X_0 = k] \\ &= P[X_n = j] \end{aligned}$$

Theorem 1 says that all we need to know about the probability distribution of the state X_n of the chain at time n is contained in the nth power of the transition matrix. Try the following activity.

Activity 3   For the Markov chain whose transition matrix is as in (2), find $P[X_2 = 3 \mid X_0 = 1]$. Compare your answer to the transition diagram in Figure 1 to see that it is reasonable. If $\boldsymbol{p}_0 = (1/4 \quad 1/4 \quad 1/4 \quad 1/4)$, find the probability mass function of X_2.

Example 1   A drilling machine can be in any of five different states of alignment, labeled 1 for the best, 2 for the next best, etc., down to the worst state 5. From one week to the next, it either stays in its current state with probability .95, or moves to the next lower state with probability .05. At state 5 it is certain to remain there the next week. Let us compute the probability distribution of the state of the machine at several times n, given that it started in state 1. What is the smallest n (that is, the first time) such that it is at least 2% likely that the machine has reached its worst state at time n?

By the assumptions of the problem, the chain of machine alignment states $X_0, X_1, X_2, \dots$ has the transition matrix

```
T = {{.95, .05, 0, 0, 0},
    {0, .95, .05, 0, 0}, {0, 0, .95, .05, 0},
    {0, 0, 0, .95, .05}, {0, 0, 0, 0, 1}};
MatrixForm[
  T]
```

$$\begin{pmatrix} 0.95 & 0.05 & 0 & 0 & 0 \\ 0 & 0.95 & 0.05 & 0 & 0 \\ 0 & 0 & 0.95 & 0.05 & 0 \\ 0 & 0 & 0 & 0.95 & 0.05 \\ 0 & 0 & 0 & 0 & 1 \end{pmatrix}$$

By Theorem 1, $P[X_n = j \mid X_0 = 1]$ is row 1 of the transition matrix T raised to the nth power. The conditional distribution of X_1 given $X_0 = 1$ is of course the first row $\{.95, .05, 0, 0, 0\}$ of the transition matrix itself. We compute these first rows for powers $n = 2$, 3, and 4 below, using *Mathematica*'s MatrixPower[mat, n] function. (You could also use the period for matrix multiplication.)

```
MatrixPower[T, 2][[1]]
MatrixPower[T, 3][[1]]
MatrixPower[T, 4][[1]]
```

```
{0.9025, 0.095, 0.0025, 0., 0.}
```

```
{0.857375, 0.135375, 0.007125, 0.000125, 0.}
```

```
{0.814506, 0.171475, 0.0135375, 0.000475, 6.25×10⁻⁶}
```

For $n = 4$, we notice that the likelihood is only 6.25×10^{-6} that the worst state 5 has been reached yet. The following command allows us to inspect the 1-5 element of some of the powers T^n.

```
Table[{n, (MatrixPower[T, n])[[1, 5]]},
  {n, 20, 30}]
```

```
{{20, 0.0159015}, {21, 0.0188806},
 {22, 0.0221825}, {23, 0.0258145}, {24, 0.0297825},
 {25, 0.0340906}, {26, 0.0387414}, {27, 0.0437359},
 {28, 0.0490739}, {29, 0.0547534}, {30, 0.0607716}}
```

This output shows that it will be 22 weeks until it is at least .02 likely that the machine will be in state 5, given that it started in state 1. ∎

Our second main goal is to find, if it exists, the vector of limiting probabilities

$$\pi(j) = \lim_{n \to \infty} P[X_n = j] \tag{3}$$

The jth entry of this limiting vector π has the interpretation of the long-run proportion of time the chain spends in state j. We will not prove it here, but a sufficient condition under which this π exists is the *regularity* of the chain.

Definition 2. A Markov chain is *regular* if there is a power T^n of the transition matrix that has all non-zero entries.

So, regularity of the chain means that there is a common time n such that all states can reach all other states by a path of length n with positive probability. The two chains with diagrams in Figure 2(a) and 2(b) are <u>not</u> regular. To simplify the diagrams we have suppressed the transition probabilities. The arrows display the transitions that have positive probability. Also in (b) we have suppressed a self-loop on state 4 indicating state 4 must return to itself. In (a), state 1 can only reach state 3 by a path of length 2, or 5, or 8, etc., while state 1 can only reach itself in a path of length 3, or 6, or 9, etc. There can be no common value of n so that both 1 reaches 3 and 1 reaches itself in n steps. In (b), state 4 does not reach any of the other states at all, so the chain cannot be regular.

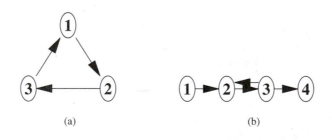

(a)   (b)

Figure 2 - Two non-regular Markov chains

The Markov chain whose transition diagram is in Figure 3 however is regular. You can check that there is a path of length 3 of positive probability from every state to every other state. (For example paths 1,3,2,1; 1,3,3,2; and 1,3,3,3 are paths of length 3 from state 1 to all of the states.)

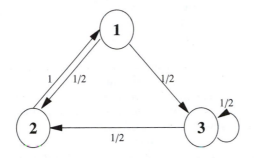

Figure 3 -A regular Markov chain

Activity 4 Use *Mathematica* to check that the Markov chain with transition matrix in (2) is regular.

Let us look at a few of the powers of the transition matrix of the chain corresponding to Figure 3.

```
T = {{0, .5, .5}, {1, 0, 0}, {0, .5, .5}};
MatrixForm[T]
```

$$\begin{pmatrix} 0 & 0.5 & 0.5 \\ 1 & 0 & 0 \\ 0 & 0.5 & 0.5 \end{pmatrix}$$

```
{MatrixPower[T, 2] // MatrixForm,
 MatrixPower[T, 3] // MatrixForm,
 MatrixPower[T, 4] // MatrixForm,
 MatrixPower[T, 5] // MatrixForm,
 MatrixPower[T, 6] // MatrixForm}
```

$$\left\{ \begin{pmatrix} 0.5 & 0.25 & 0.25 \\ 0. & 0.5 & 0.5 \\ 0.5 & 0.25 & 0.25 \end{pmatrix}, \begin{pmatrix} 0.25 & 0.375 & 0.375 \\ 0.5 & 0.25 & 0.25 \\ 0.25 & 0.375 & 0.375 \end{pmatrix}, \right.$$

$$\begin{pmatrix} 0.375 & 0.3125 & 0.3125 \\ 0.25 & 0.375 & 0.375 \\ 0.375 & 0.3125 & 0.3125 \end{pmatrix}, \begin{pmatrix} 0.3125 & 0.34375 & 0.34375 \\ 0.375 & 0.3125 & 0.3125 \\ 0.3125 & 0.34375 & 0.34375 \end{pmatrix},$$

$$\left. \begin{pmatrix} 0.34375 & 0.328125 & 0.328125 \\ 0.3125 & 0.34375 & 0.34375 \\ 0.34375 & 0.328125 & 0.328125 \end{pmatrix} \right\}$$

Rows 1 and 3 of T^n actually are identical for each n (look at the transition diagram to see why this is true), and row 2 is different from those rows. However all of the rows seem to be in closer and closer agreement as the power grows. The matrix T^n may be reaching a limit as $n \longrightarrow \infty$, which has the form of a matrix of identical rows. This would suggest that

$$\lim_{n \to \infty} T^n(i, j) = \lim_{n \to \infty} P[X_n = j \mid X_0 = i] = P[X_n = j] = \pi(j) \qquad (4)$$

independently of the initial state i. Let us find out what this limiting distribution must be, if it does exist.

If π is to be the limit in (4), then we must have

$$
\begin{aligned}
\pi(j) &= \lim_{n \to \infty} T^{n+1}(i, j) \\
&= \lim_{n \to \infty} \sum_k T^n(i, k)\, T(k, j) \\
&= \sum_k \lim_{n \to \infty} T^n(i, k)\, T(k, j) \\
&= \sum_k \pi(k)\, T(k, j)
\end{aligned}
\qquad (5)
$$

By computation (5), the limiting distribution π satisfies the system of matrix equations:

$$\pi = \pi T \quad \text{and} \quad \pi \mathbf{1} = 1 \qquad (6)$$

Here $\mathbf{1}$ refers to a column vector of 1's of the same length as π so that the equation $\pi \mathbf{1} = 1$ means that the sum of the entries of π equals 1. We require this in order for π to be a valid discrete probability mass function.

We have motivated the following theorem.

Theorem 2. If a Markov chain X_0, X_1, X_2, ... is regular, then the limiting distribution $\pi(j)$ in (4) exists and satisfies (6).

In the next example we see how Theorem 2 can be used in the long-run prediction of business revenue flow.

Example 2 A plumbing contractor services jobs of six types according to their time demands, respectively 1 to 6 hours. The fee for the service is $40 per hour. Initially the contractor believes that the durations of successive jobs are independent of one another, and they occur with the following probabilities:

1 hr: .56 ; 2 hrs: .23 ; 3 hrs: .10 ; 4 hrs: .05 ; 5 hrs: .03 ;
6 hrs: .03

However, detailed data analysis suggests that the sequence of job times may instead be a Markov chain, with slightly different column entries (next job duration probabilities) for each row (current job duration) as follows:

```
T = {{.52, .25, .11, .06, .03, .03},
     {.59, .22, .09, .03, .04, .03},
     {.50, .31, .12, .03, .02, .02},
     {.55, .24, .09, .07, .03, .02},
     {.60, .25, .08, .03, .02, .02},
     {.62, .28, .05, .03, .01, .01}};
MatrixForm[
  T]
```

$$\begin{pmatrix} 0.52 & 0.25 & 0.11 & 0.06 & 0.03 & 0.03 \\ 0.59 & 0.22 & 0.09 & 0.03 & 0.04 & 0.03 \\ 0.5 & 0.31 & 0.12 & 0.03 & 0.02 & 0.02 \\ 0.55 & 0.24 & 0.09 & 0.07 & 0.03 & 0.02 \\ 0.6 & 0.25 & 0.08 & 0.03 & 0.02 & 0.02 \\ 0.62 & 0.28 & 0.05 & 0.03 & 0.01 & 0.01 \end{pmatrix}$$

By how much does the expected long-run revenue per job change if the job times do form a Markov chain?

First, in the default case where job times are independent and identically distributed, the average revenue per job is just $\mu = E[40\,X]$, where X is a discrete random variable modeling the duration of the job, with the distribution given above. By the Strong Law of Large Numbers, with probability 1 the actual average revenue per job $\overline{X}_n$ will converge to μ. We compute:

```
μ = 40 (1 (.56) + 2 (.23) +
        3 (.10) + 4 (.05) + 5 (.03) + 6 (.03))
```

74.

Now from Theorem 2 we can compute the long-run distribution of the Markov chain, then make a similar computation of the long-run expected revenue per job. By Exercise 6, the system of equations $\pi = \pi T$ is a dependent system; therefore we can afford to discard one equation in it, and replace it with the condition that the sum of the entries of π is 1. By setting $\pi = (x_1, x_2, x_3, x_4, x_5, x_6)$ and multiplying out πT we get the following.

```
system = {x₁ ==
    .52 x₁ + .59 x₂ + .5 x₃ + .55 x₄ + .6 x₅ + .62 x₆,
  x₂ == .25 x₁ + .22 x₂ + .31 x₃ +
    .24 x₄ + .25 x₅ + .28 x₆,
  x₃ == .11 x₁ + .09 x₂ + .12 x₃ +
    .09 x₄ + .08 x₅ + .05 x₆,
  x₄ == .06 x₁ + .03 x₂ + .03 x₃ +
    .07 x₄ + .03 x₅ + .03 x₆,
  x₅ == .03 x₁ + .04 x₂ + .02 x₃ +
    .03 x₄ + .02 x₅ + .01 x₆,
  x₁ + x₂ + x₃ + x₄ + x₅ + x₆ == 1};
Solve[system, {x₁, x₂, x₃, x₄, x₅, x₆}]
```

```
{{x₁ → 0.542039, x₂ → 0.249027, x₃ → 0.102504,
  x₄ → 0.0481887, x₅ → 0.0306065, x₆ → 0.0276343}}
```

Hence the long-run average revenue per job under the Markov chain model is

```
40 (1 (.542039) + 2 (.249027) + 3 (.102504) +
    4 (.0481887) + 5 (.0306065) + 6 (.0276343))
```

74.3679

So the more sophisticated Markov chain model did not make a great change in the revenue flow prediction; it predicts only about an extra 37 cents per job. This is perhaps to be expected, because the differences between the row probability distributions and the distribution assumed for X in the i.i.d. case are small.

Mathematica for Section 6.1

(Only standard kernel operations of matrix multiplication and equation solving were used.)

Exercises 6.1

1. (*Mathematica*) For the Markov chain whose transition diagram is as in Figure 1, compute the distribution of X_3 if X_0 is equally likely to equal each of the four states. Find the limiting distribution.

2. (*Mathematica*) A store keeps a small inventory of at most four units of a large appliance, which sells rather slowly. Each week, one unit will be sold with probability .40, or no units will be sold with probability .60. When the inventory empties out, the store is immediately able to order four replacements, so that from the state of no units, one goes next week to four units if none sell, or three if one unit sells. If the store starts with four units, find the distribution of the number of units in inventory at time 2. If the Markov chain of inventory levels is regular, find its limiting distribution.

3. (*Mathematica*) Build a command in *Mathematica* which takes an initial state X_0, a transition matrix T of a Markov chain, and a final time n, and returns a simulated list of states $X_0, X_1, \ldots, X_n$.

4. Argue that an i.i.d. sequence of discrete random variables $X_0, X_1, X_2, \ldots$ is a special case of a Markov chain. What special structure would the transition matrix of this chain have?

5. (*Mathematica*) A communications system has six channels. When messages attempt to access the system when all channels are full, they are turned away. Otherwise, because messages neither arrive nor depart simultaneously, if there are free channels the next change either increases the number of channels in use by 1 (with probability .3) or decreases the number of channels in use by 1 (with probability .7). Of course when all channels are free, the next change must be an increase and when all channels are busy the next change must be a decrease. Find the transition matrix for a Markov chain that models this situation. If the chain is regular, find its limiting distribution.

6. Show that for a regular Markov chain with transition matrix T, the system $\pi = \pi T$ has infinitely many solutions. (Hint: Consider a constant times π.)

7. (*Mathematica*) A short-haul rental moving van can be borrowed from and returned to three sites in a city. If it is borrowed from site 1, then it is returned to sites 1, 2, and 3 with probabilities 2/3, 1/6, and 1/6, respectively. If it is borrowed from site 2, then it is returned to sites 1, 2, and 3 with probabilities 1/2, 1/4, and 1/4, respectively. And if it is borrowed from site 3, then it is returned to sites 1, 2, and 3 with equal probabilities. Given that the van starts at site 3, with what probability does it come to each site after four rentals? What is the long-run proportion of times that it is returned to each site?

8. A Markov chain has states 1, 2, 3, and 4, and the transition matrix below.

$$T = \begin{pmatrix} 1 & 0 & 0 & 0 \\ .3 & 0 & .7 & 0 \\ 0 & .6 & 0 & .4 \\ 0 & 0 & 0 & 1 \end{pmatrix}$$

Draw the transition diagram of the chain. Let

$$f(i) = \text{P[chain hits state 1 before state 4} \mid X_0 = i], \; i = 1, 2, 3, 4$$

Find $f(1)$ and $f(4)$ directly, then derive and solve linear equations satisfied by $f(2)$ and $f(3)$.

6.2 Queues

The word "queue" is a word used mostly in Great Britain and the former empire to refer to a waiting line. We have all been in them at stores, banks, airports and the like. We the customers come to a facility seeking some kind of service. Most often, our arrival times are not able to be predicted with certainty beforehand. There is some number of servers present, and we either join individual lines before them (as in grocery store checkout lanes) or we form a single line waiting for the next available server (as in most post office counters and amusement park rides). Some rule, most commonly first-in, first-out, is used to select the next customer to be served. Servers take some amount of time, usually random, to serve a customer. When a customer finishes, that customer may be directed to some other part of the queueing system (as in a driver's license facility where there are several stations to visit), or the customer may depart the system for good.

Predicting the behavior of human queueing systems is certainly an interesting problem, to which it is easy to relate. However, it should be noted that it is the study of very large, non-human queueing systems, especially those involved with computer and telecommunications networks, which has given probability great impetus in the latter 20th century. Because instead of human customers and servers we could just as easily consider computer jobs or messages in a queue waiting for processing by a server or communication channel.

There is a wide variety of problems in queueing theory, generated by variations in customer arrival and facility service patterns, breakdowns of servers, changes in the size of the waiting space for customers, rules other than first-in, first-out for selecting customers, etc. Our small section will just touch on the most basic of all queues: the so-called *M/M/1 queue*.

Activity 1 Think of three particular human queueing systems you have been involved in. What main features did they have; for example how many servers were there, were customer arrivals and services random, was the queue a first-in, first-out queue, etc.? Then think of three examples of non-human queueing systems.

Here are the assumptions that identify a queueing system as M/M/1. Customers arrive singly at random times $T_1, T_2, T_3, \ldots$ such that the times between arrivals

$$T_1, T_2 - T_1, T_3 - T_2, \ldots \tag{1}$$

are i.i.d. $\exp(\lambda)$ random variables. The parameter λ is called the *arrival rate* of the queue. Therefore the process that counts the total number of customer arrivals by time t is a Poisson process with rate λ. There is a single server (the 1 in the notation M/M/1), and customers form one first-come, first-served line in front of the server. Essentially infinite waiting space exists to accommodate the customers in queue. The server requires an exponentially distributed length of time with rate μ for each service, independently of the arrival process and other service times. Customers who finish service depart the system. (By the way, the two "M" symbols in the notation stand for Markov, because the memoryless nature of the exponential distribution makes the arrival and departure processes into continuous-time versions of a Markov chain.)

One of the most important quantities to compute in queueing systems is the probability distribution of the number of customers in the queue or in service, at any time t. Such knowledge can help the operators of the system to know whether it will be necessary to take action to speed up service in order to avoid congestion. From the customer's point of view, another important quantity is the distribution of the customer's waiting time in the queue. We will consider each of these questions in turn for the M/M/1 queue, producing exact results for the long-run distributions. But exact computations become difficult even for apparently simple queueing situations, so very often analysts must resort to simulating the queue many times to obtain statistical estimates of such important quantities as these. Entire languages, such as GPSS, exist to make it easy to program such simulations, but as we will show, it is not so hard to write queueing simulation programs even in a very general purpose language like *Mathematica* which has not been specifically designed for queueing simulation.

Activity 2 How would you simulate the process of customer arrivals in an M/M/1 queue? Try writing a *Mathematica* command to do this, ignoring the service issue for the moment.

Long-Run Distribution of System Size

To begin the problem of finding the long-run distribution of the number of customers in the queueing system, define M_t as the number of customers at time t, including the one being served, if any. Define also a sequence of real-valued functions of t:

$$f_k(t) = P[M_t = k], \quad k = 0, 1, 2, \ldots \tag{2}$$

If for each k, the limit as $t \longrightarrow \infty$ of $f_k(t)$ can be found, then these limits f_k form the limiting probability distribution of system size. Actually, the functions $f_k(t)$ can be computed themselves, which would give us the short-run system size distribution, but this is much more difficult than finding the limiting distribution and we will not attempt this problem here.

The unique approach to the problem of finding the limiting distribution is to construct a system of differential equations satisfied by the functions $f_k(t)$. This illustrates beautifully the power of blending apparently different areas of mathematics to solve a real problem. In the interest of space we will leave a few details out and not be completely rigorous. The interested reader can try to fill in some gaps.

We first would like to argue that the probability that two or more arrival or service events happen in a very short time interval is negligibly small. Consider for example the probability that there are two or more arrivals in a short time interval $(t, t + h]$. Because arrivals to an M/M/1 queue form a Poisson process, we can complement, use the Poisson p.m.f. and rewrite the exponential terms in a Taylor series as follows:

$$
\begin{aligned}
P[2 \text{ or more arrivals in } (t, t+h]] \ &= \ 1 - P[0 \text{ or 1 arrival in } (t, t+h]] \\
&= \ 1 - \frac{e^{-\lambda h}(\lambda h)^0}{0!} - \frac{e^{-\lambda h}(\lambda h)^1}{1!} \\
&= \ 1 - e^{-\lambda h} - \lambda h \, e^{-\lambda h} \\
&= \ 1 - (1 - \lambda h + \tfrac{1}{2}(\lambda h)^2 - \cdots) \\
& \qquad - \lambda h(1 - \lambda h + \tfrac{1}{2}(\lambda h)^2 - \cdots) \\
&= \ \tfrac{1}{2}(\lambda h)^2 + \text{higher powers of } h
\end{aligned}
$$

So this probability is a small enough function of h such that if we divide by h and then send h to zero, the limit is 0. Recall from calculus that such a function is generically called $o(h)$. We can argue similarly that if 2 or more customers are present in the system at time t, then the probability that at least 2 customers will be served during $(t, t+h]$ is an $o(h)$ function. (Try it.) The activity below leads you through another case of two queueing events in a short time.

Activity 3 Suppose there is at least one customer in the queue and we look at a time interval $(t, t+h]$ again. Write one expression for the probability of exactly one arrival in $(t, t+h]$, and another for the probability of exactly one service in $(t, t+h]$. Use them to argue that the probability of having both an arrival and a service in $(t, t+h]$ is $o(h)$.

Assuming that you are now convinced that the probability of two or more queueing events in a time interval of length h is $o(h)$, let us proceed to derive a differential equation for $f_0(t) = P[M_t = 0]$ by setting up a difference quotient

$$
\frac{f_0(t+h) - f_0(t)}{h}
$$

Now $f_0(t+h) = P[M_{t+h} = 0]$, so let us consider how the system size could have come to be 0 at time $t+h$, conditioning on the system size at time t. There are two main cases, 1 and 2 below, and a collection of others described in case 3 that involve two or more queueing events in $(t, t+h]$, which therefore have probability $o(h)$:

1. $M_t = 0$, and there were no arrivals in $(t, t+h]$;
2. $M_t = 1$, and there was one service and no new arrivals in $(t, t+h]$;
3. Neither 1 nor 2 hold, and all existing and newly arriving customers are served in $(t, t+h]$.

The total probability $P[M_{t+h} = 0]$ is the sum of the probabilities of the three cases. But the probability of no arrivals in $(t, t+h]$ is $e^{-\lambda h} = 1 - \lambda h + o(h)$ and the probability of exactly one service is $\mu h e^{-\mu h} = \mu h(1 - \mu h + o(h)) = \mu h + o(h)$. By independence, the probability of one service and no new arrivals is therefore $(1 - \lambda h + o(h))(\mu h + o(h)) = \mu h + o(h)$. Thus, by the Law of Total Probability,

$$
\begin{aligned}
f_0(t + h) &= P[M_{t+h} = 0] \\
&= P[M_t = 0](1 - \lambda h) + P[M_t = 1](\mu h) + o(h) \qquad (3) \\
&= (1 - \lambda h) f_0(t) + \mu h f_1(t) + o(h)
\end{aligned}
$$

Subtracting $f_0(t)$ from both sides, and then dividing by h and sending h to zero yields

$$
\begin{aligned}
\frac{f_0(t+h) - f_0(t)}{h} &= -\lambda f_0(t) + \mu f_1(t) + \frac{o(h)}{h} \\
\Longrightarrow f_0'(t) &= -\lambda f_0(t) + \mu f_1(t)
\end{aligned}
\qquad (4)
$$

Equation (4) is one of the equations in the infinite system of equations we seek. You are asked to show in Exercise 3 that the rest of the differential equations have the form:

$$
f_k'(t) = -(\lambda + \mu) f_k(t) + \lambda f_{k-1}(t) + \mu f_{k+1}(t), \ k \geq 1 \qquad (5)
$$

Intuitively, the first term on the right side of (5) represents the case of neither an arrival nor a service during $(t, t + h]$, the second term is for the case where there were $k - 1$ customers at time t and one new arrival appeared in $(t, t + h]$, and the third term corresponds to the case where there were $k + 1$ customers at time t, and one customer was served in $(t, t + h]$.

The system of differential equations (4) and (5) can be solved with a lot of effort (see [Gross and Harris, p. 129]), but we will just use the system to derive and solve an ordinary system of linear equations for the limiting probabilities $f_k = \lim_{t \to \infty} f_k(t) = \lim_{t \to \infty} P[M_t = k]$.

Observe that if the $f_k(t)$ functions are approaching a limit, there is reason to believe that the derivatives $f_k'(t)$ approach zero. Presuming that this is so, we can send t to ∞ in (4) and (5) to obtain the system of linear equations:

$$
\begin{cases}
0 = -\lambda f_0 + \mu f_1 \\
0 = -(\lambda + \mu) f_k + \lambda f_{k-1} + \mu f_{k+1}, \ k \geq 1
\end{cases}
\qquad (6)
$$

We can adjoin to this system the condition that $f_0 + f_1 + f_2 + \cdots = 1$ in order to force the f_k's to form a valid probability distribution. Now the top equation in (6) implies

$$
f_1 = \frac{\lambda}{\mu} f_0
$$

When $k = 1$, the bottom equation in (6) implies

$$
\mu f_2 = (\lambda + \mu) f_1 - \lambda f_0 = (\lambda + \mu) \frac{\lambda}{\mu} f_0 - \lambda f_0 = \frac{\lambda^2}{\mu} f_0
$$

$$
\Longrightarrow f_2 = \frac{\lambda^2}{\mu^2} f_0
$$

Activity 4 Check that $f_3 = \frac{\lambda^3}{\mu^3} f_0$.

In Exercise 4 we ask you to show inductively that for each $k \geq 1$,

$$f_k = \frac{\lambda^k}{\mu^k} f_0 \tag{7}$$

Clearly though, in order for the sum of the f_k's to be 1, or to be finite at all, the ratio $\rho = \lambda/\mu$ must be less than 1. This ratio ρ is called the *traffic intensity* of the M/M/1 queue, a good name because it compares the arrival rate to the service rate. If $\rho < 1$, then $\mu > \lambda$, and the server works fast enough to keep up with the arrivals. I will leave it to you to show that under the condition $\sum_{k=0}^{\infty} f_k = 1$, the limiting M/M/1 queue system size probabilities are

$$f_k = \lim_{t \to \infty} P[M_t = k] = \rho^k (1 - \rho), \ k = 0, 1, 2, \ ... \tag{8}$$

So we see that the limiting system size has the geometric distribution with parameter $1 - \rho$ (see formula (1) of Section 2.3). An interesting consequence is that the average number of customers in the system is

$$L = \frac{\rho}{1-\rho} \tag{9}$$

From this expression, notice that as the traffic intensity ρ increases toward 1, the average system size increases to ∞, which is intuitively reasonable.

Example 1 At a shopping mall amusement area, customers wanting to buy tokens for the machines form a first-in, first-out line in front of a booth staffed by a single clerk. The customers arrive according to a Poisson process with rate 3/min., and the server can serve at an average time of 15 sec. per customer. Find the distribution and average value of the number of customers awaiting service. Suppose that replacing the booth clerk by a machine costs $2000 for the machine itself plus an estimated $.50 per hour for power and maintenance, and with this self-service plan customers could serve themselves at an average time of 12 sec. apiece. The human server is paid $5/hr. Also, an estimate of the intangible cost of inconvenience to customers is $.20 per waiting customer per minute. Is it worthwhile to replace the clerk by the machine?

The given information lets us assume that the times between customer arrivals are exponentially distributed with rate parameter $\lambda = 3$/min. If we are willing to assume that service times are independent exponential random variables, which are furthermore independent of arrivals, then the M/M/1 assumptions are met. The service rate is the reciprocal of the mean service time, which is $1/(1/4$ min.$) = 4$/min. Thus the queue traffic intensity is $\rho = \lambda/\mu = 3/4$. By formula (8) the long-run probability of k customers in the system is

$$f_k = \tfrac{1}{4} \left(\tfrac{3}{4}\right)^k, \ k = 0, 1, 2, \ ...$$

and by formula (9) the average number of customers in the system is

$$L = \frac{3/4}{1-3/4} = \frac{3/4}{1/4} = 3$$

To answer the second question, we must compute the average cost per minute C_h for the human server system and the average cost per minute of the machine system C_m. Note that the total cost for x minutes of operations under the human system is just $C_h\, x$, but because of the fixed purchase cost of the machine, the total cost for x minutes is $C_m\, x + 2000$ under the machine scenario. For consistency let us express all monetary units in dollars and all times in minutes. Then for instance the clerk's pay rate is ($5/hr.)/(60 min./hr.) = $(1/12)/min., and the cost of operation of the machine is ($.50/hr.)/(60 min./hr) = $(1/120)/min. Now in calculating C_h and C_m there are two aspects of cost to consider: operating cost and the overall cost to waiting customers. The latter would be the average number of customers waiting, times the cost rate of $.20/min. per customer. For the human server system, the overall cost per minute would be

$C_h =$
$\$(1/12)/\text{min.} + (3\text{ customers}) \times (\$.20/\text{customer})/\text{min.} = \$.683/\text{min.}$

Since the machine average service time is 12 sec. = 1/5 min., its service rate is 5/min. and the resulting queue traffic intensity is $\rho = \lambda/\mu = 3/5$. The average number of customers in the system under the machine scenario is therefore

$$L = \frac{3/5}{1-3/5} = \frac{3/5}{2/5} = \frac{3}{2}$$

Thus the overall cost per minute for the machine system is

$C_h = \$(1/120)/\text{min.} + (3/2\text{ customers}) \times (\$.20/\text{customer})/\text{min.} =$
   $\$.308/\text{min.}$

The cost rate for the machine is less than half that of the human server. And notice that the bigger factor in the improvement was not the running cost but the savings in customer waiting cost. Taking into account the purchase cost, the use of the machine becomes economical after it has been used for at least x minutes, where x satisfies

$$C_h\, x = C_m\, x + 2000 \Longrightarrow .683\, x = .308\, x + 2000 \Longrightarrow .375\, x = 2000$$

The solution value for x turns out to be about 5333 minutes, or about 89 hours. If the arcade is open for 10/hrs. per day, then before 9 working days have elapsed the machine will have begun to be less costly than the human server. ∎

Waiting Time Distribution

The second major question that we will address is the distribution of the waiting time of a customer who arrives to an M/M/1 queue at some time t. There are two things that we might mean by this: the time W_q that a customer spends in queue prior to entering service, or the time W spent in the system in total, including the customer's time in service. We will work on W, saving the W_q problem for the exercises.

Assume that the queue has reached equilibrium, so that we are really looking at the limiting distribution as $t \longrightarrow \infty$ of the waiting time W of a customer who arrives at time t. This allows us to make use of the limiting system size probabilities in formula (8).

When our customer arrives, there is some number $n \geq 0$ of customers ahead. Because of the memoryless property of the exponential service time distribution, the distribution of the waiting time is the same as if the customer currently in service, if any, just began that service. So if n customers are ahead, the waiting time in system of the new customer will be the total $U_1 + U_2 + \cdots + U_{n+1}$ of $n + 1$ independent $\exp(\mu)$ distributed service times (including the new customer as well as the n customers ahead). Remember that the moment-generating function method implies that the distribution of the sum of $n + 1$ independent $\exp(\mu) = \Gamma(1, 1/\mu)$ random variables is $\Gamma(n + 1, 1/\mu)$. So, conditioned on the event that there is a particular number $n \geq 0$ of customers in the system upon the arrival of the new customer, $W \sim \Gamma(n + 1, 1/\mu)$. By the Law of Total Probability we can write the limiting c.d.f. of W as

$$
\begin{aligned}
F(s) = P[W \leq s] \;&=\; \sum_{n=0}^{\infty} P[W \leq s \mid n \text{ in system}]\, P[n \text{ in system}] \\[2mm]
&=\; \sum_{n=0}^{\infty} \left(\int_0^s \frac{\mu^{n+1}}{\Gamma(n+1)}\, x^n\, e^{-\mu x}\, dx \right) \rho^n (1 - \rho) \\[2mm]
&=\; \mu(1 - \rho) \int_0^s e^{-\mu x} \left(\sum_{n=0}^{\infty} \frac{(\mu x \rho)^n}{n!} \right) dx \\[2mm]
&=\; \mu(1 - \rho) \int_0^s e^{-\mu x}\, e^{\mu x \rho}\, dx \\[2mm]
&=\; \mu(1 - \rho) \int_0^s e^{-\mu x (1 - \rho)}\, dx \\[2mm]
&=\; 1 - e^{-\mu(1-\rho)s}
\end{aligned}
\tag{10}
$$

Hence *W* has density function

$$f(s) = F'(s) = \mu(1 - \rho)\,e^{-\mu(1-\rho)s} \tag{11}$$

which we recognize as the $\exp(\mu(1 - \rho))$ distribution. It follows that the long-run mean waiting time is $E[W] = 1/(\mu(1 - \rho))$, which is intuitively reasonable. As the traffic intensity ρ increases toward 1, the equilibrium queue length increases, thus increasing the average waiting time of the new customer. Also, as the service rate μ increases, the average waiting time decreases. Note also that $\mu(1 - \rho) = \mu(1 - \lambda/\mu) = \mu - \lambda$, so that the closer λ is to μ, the longer is the expected wait.

Example 2 Suppose that a warehouse processes orders for shipment one at a time. The orders come in at a rate of 4 per hour, forming a Poisson process. Under normal working conditions, a team of order fillers working on preparing the order can complete an order in an exponential amount of time with rate 4.2 per hour. Hence the team can just barely keep up with input. What is the average time for a newly arriving order to be completed? If more people can be assigned to the team, the service rate can be increased. To what must the service rate be increased so that orders require no more than one hour of waiting on average from start to finish?

We are given that $\lambda = 4.0$/hr., and the default service rate is $\mu = 4.2$/hr. Hence the traffic intensity is $\rho = 4.0/4.2 = .952$, and the expected waiting time in equilibrium is $1/(4.2\,(1 - .952)) \approx 4.96$. This is an excessive amount of time for an order. To answer the second question, we need to choose μ such that

$$E[W] = \tfrac{1}{\mu - \lambda} = \tfrac{1}{\mu - 4} \le 1 \Longrightarrow \mu - 4 \ge 1 \Longrightarrow \mu \ge 5$$

So an increase in service rate from 4.2 to just 5 per hour is enough to cut the average waiting time from about 5 hours to 1 hour. ∎

Simulating Queues

Consider a single-server first-in, first-out queue as before except that we generalize: (1) the distribution of customer interarrival times; and (2) the distribution of customer service times. We have so far treated only the case where both of these distributions are exponential. It is much harder, and sometimes impossible, to get analytical results about waiting times and limiting distributions for general interarrival and service distributions; so it is important to be able to estimate these quantities by observing the results of simulation. Simulation can also be used in more complicated cases than single-server first-in, first-out queues, but we will not treat such problems here.

It is very easy in *Mathematica* to simulate a sequence of arrival times with a given interarrival distribution. Notice that the *n*th arrival time T_n is the sum of the first *n* interarrival times. The following utility function takes advantage of this observation by keeping a running sum of interarrival times, simulating the next one, and adjoining the updated running sum to the end of a list of arrival times. The arguments to the function are the interarrival distribution and the number of arrivals to simulate, and the output is the list of arrival times.

```
Needs["Statistics`ContinuousDistributions`"]
```

```
SimArrivals[arrdist_, numarrs_] :=
 Module[
   {arrtimelist, currtime, nextinterarr},
   arrtimelist = {};
   currtime = 0;
   Do[nextinterarr = Random[arrdist];
      currtime = currtime + nextinterarr;
      AppendTo[arrtimelist, currtime],
    {numarrs}];
   arrtimelist]
```

```
SeedRandom[787632]
```

```
SimArrivals[ExponentialDistribution[3], 4]
```

{0.783599, 1.85244, 2.4792, 2.51363}

It is even easier to simulate a list of service times for those customers. We just simulate a table of the desired number of observations from the desired distribution, as in the next command.

```
SimServiceTimes[servicedist_, numservices_] :=
 RandomArray[servicedist, {numservices}]
```

```
SimServiceTimes[ExponentialDistribution[4], 4]
```

{0.522037, 0.586574, 0.37423, 0.214739}

All of the behavior of the queue is determined by these lists of arrival and service times, but there is still the considerable problem of extracting information from the lists. As an example let us consider the relatively easy problem of finding from a simulated queueing process a list of customer waiting times.

Think of the queueing behavior for the simulated output above. Customer 1 arrives at time .783599. It takes .522037 time units to serve this customer, who therefore departs at time .783599 + .522037 = 1.30564. Customer 2 does not arrive until time 1.85244, at which time no one is in the queue, so he immediately enters service. Since his service time was .586574, he departs at time 1.85244 + .586574 = 2.43901. Customer 3 arrives shortly afterward at time 2.4792 and requires .37423 time units to serve, so this customer departs at time 2.4792 + .37423 = 2.85343. Meanwhile, customer 4 has come in at time 2.51363, so he has to wait until time 2.85343 to begin service. The departure time of customer 4 is then 2.85343 + .214739 = 3.06817.

Activity 5 If five successive customer arrival times are: 2.1, 3.4, 4.2, 6.8, 7.4 and their service times are: 1.0, 2.3, 1.7, 3.4, 0.6, find the departure times of each customer, and find the amount of time each customer is in the system. What is the longest queue length?

In this example we had three customers who immediately on arrival went into service, so that their waiting times were their service times, and one who had to wait in queue. But notice that any customer's waiting time is just his departure time minus his arrival time, so that producing a list of waiting times will be simple if we can produce a list of departure times. This too is not hard, because a customer's departure time is his service time plus either his arrival time, or the departure time of the previous customer, whichever is later. We can generate this list of departure times iteratively by this observation, beginning with the first customer, whose departure time is just his arrival time plus his service time. The next function does this. It takes the arrival time and service time lists as inputs, and returns the list of departure times using the strategy we have discussed.

```
DepartureTimes[arrtimelist_, servtimelist_] :=
  Module[{numcustomers,
     deptimelist, customer, nextdeptime},
   numcustomers = Length[servtimelist];
   deptimelist =
    {arrtimelist[[1]] + servtimelist[[1]]};
   Do[nextdeptime = servtimelist[[customer]] +
       Max[deptimelist[[customer - 1]],
        arrtimelist[[customer]]];
     AppendTo[deptimelist, nextdeptime],
     {customer, 2, numcustomers}];
   deptimelist]
```

Here we apply the function to the arrival and service lists in the example above.

```
arrlist = {0.783599, 1.85244, 2.4792, 2.51363};
servlist =
   {0.522037, 0.586574, 0.37423, 0.214739};
DepartureTimes[arrlist, servlist]
```

{1.30564, 2.43901, 2.85343, 3.06817}

Now waiting times are departure times minus arrival times; so we can build a function which calls on the functions above, plots a histogram of waiting times, and computes the mean and variance.

```
Needs["KnoxProb`Utilities`"];
```

```
SimulateWaitingTimes[arrdist_,
  servdist_, numcustomers_] :=
 Module[{arrtimelist, servtimelist,
   deptimelist, waitimelist},
  arrtimelist = SimArrivals[
    arrdist, numcustomers];
  servtimelist = SimServiceTimes[
    servdist, numcustomers];
  deptimelist = DepartureTimes[
    arrtimelist, servtimelist];
  waitimelist = deptimelist - arrtimelist;
  Print["Mean waiting time: ",
   N[Mean[waitimelist]],
   " Variance of waiting times: ",
   N[Variance[waitimelist]]];
  Histogram[waitimelist, 6]]
```

Here is an example run in a setting that we already know about: an M/M/1 queue with arrival rate $\lambda = 3$ and service rate $\mu = 4$. The distribution suggested by the histogram in Figure 4(a) does look exponential, and the mean waiting time of 1.10157 did come out near the theoretical value of $1/(\mu - \lambda) = 1$ in this run. You should try running the command several more times to get an idea of the variability. Also try both increasing and decreasing the number of customers. What do you expect to happen as the number of customers gets larger and larger?

```
SeedRandom[9689321]
```

```
g1 = SimulateWaitingTimes[
    ExponentialDistribution[3],
    ExponentialDistribution[4], 1000];
g2 = SimulateWaitingTimes[UniformDistribution[
    0, 2], UniformDistribution[0, 1.5], 1000];
Show[GraphicsArray[{g1, g2}]];
```

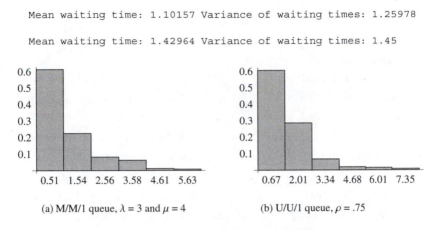

(a) M/M/1 queue, $\lambda = 3$ and $\mu = 4$ (b) U/U/1 queue, $\rho = .75$

Figure 4 - Histograms of 1000 waiting times

Our command allows us to use any interarrival and service distribution we choose. To see another example, we repeat the simulation for a uniform[0, 2] interarrival distribution and a uniform[0, 1.5] service distribution. The result is shown in Figure 4(b). Note that the service rate is the reciprocal of the mean time 1/.75, and the arrival rate is the reciprocal of the interarrival mean 1/1, so the traffic intensity is $\rho = .75$. The distribution of waiting times is rather right-skewed, reminiscent of a gamma distribution, and the mean time in system is around 1.43; in particular it is more than the mean time in the M/M/1 simulation even though the traffic intensity is the same.

Mathematica for Section 6.2

Command	Location
SeedRandom[seed]	kernel
Random[dist]	Statistics` ContinuousDistributions`
RandomArray[dist, n]	Statistics` ContinuousDistributions`
ExponentialDistribution[λ]	Statistics` ContinuousDistributions`
UniformDistribution[a, b]	Statistics` ContinuousDistributions`
Mean[datalist]	Statistics` DescriptiveStatistics`
Variance[datalist]	Statistics` DescriptiveStatistics`
Histogram[datalist, numrectangles]	KnoxProb` Utilities`
SimArrivals[arrdist, numarrs]	Section 6.2
SimServiceTimes[serv, services]	Section 6.2
DepartureTimes[arrlist, servlist]	Section 6.2
SimulateWaitingTimes[arr, serv, n]	Section 6.2

Exercises 6.2

1. Customers waiting in a post office line experience a frustration cost of 10 cents a minute per customer while they wait. What target traffic intensity should the postal supervisors aim for so that the average frustration cost per unit time of the whole queue is 20 cents a minute?

2. Customers arrive to a department store service desk according to a Poisson process with rate 8 per hour. If it takes 5 minutes on average to serve a customer, what is the limiting probability of having at least 2 people in the system? What is the limiting probability that an arriving customer can go directly into service?

3. Derive the differential equation (5) for the short-run M/M/1 system size probabilities.

4. Verify by mathematical induction expression (7) for the limiting M/M/1 system size probabilities.

5. What in general should be the relationship between the expected waiting time in the system E[W] and the expected waiting time in the queue E[W_q]?

6. Derive the long-run probability distribution of the time W_q that a customer spends in queue prior to entering service. Be careful: this random variable has a mixture of a discrete and continuous distribution, because there is non-zero probability that W_q equals 0 exactly, but for states greater than zero there is a sub-density,

which when integrated from 0 to ∞ gives the complementary probability to that of the event $\{W_q = 0\}$.

7. People come to a street vendor selling hot pretzels at a rate of 1 every 2 minutes. The vendor serves them at an average time of 1 minute and 30 seconds. Find the long–run probability that a customer will take at least 2 minutes to get a pretzel.

8. The operator of a queueing system must solve an optimization problem in which he must balance the competing costs of speeding up service, and making his customers wait. There is a fixed arrival rate of λ, operating costs are proportional to the service rate, and customer inconvenience cost is proportional to the expected waiting time of a customer. Find the optimal service rate in terms of λ and the proportionality constants.

9. (*Mathematica*) Use the commands of the section to study the long–run waiting time distribution and average waiting time when the interarrival and service time distributions are approximately normal. What do you have to be careful of when you choose the parameters of your normal distributions?

10. (*Mathematica*) Build a command that takes a list of arrival and service times and a fixed time t, and returns the number of people in the system at time t. Combine your command with the arrival and service time simulators of this section to produce a command that simulates and plots the system size as a function of t.

11. One case in which it is possible to derive the short-run distribution of system size is the M/M/1/1 queue. The extra 1 in the notation stands for 1 waiting space for the customer in service; hence any new customer arriving when someone is already being served is turned away. Therefore the system size M_t can only take on the values 0 or 1. Derive and solve a differential equation for $f_1(t) = P[M_t = 1] = 1 - f_0(t) = 1 - P[M_t = 0]$. (You will need to make the assumption that the system is empty at time 0.)

12. A self-service queueing system, such as a parking garage or a self-service drink area in a restaurant, can be thought of as an M/M/∞ queue, i.e., a queue with an unlimited number of servers. All customers therefore go into service immediately on arrival. If n customers are now in such a queue, what is the distribution of the time until the next customer is served? (Assume customers serve themselves at the same rate μ.) Obtain a system of differential equations analogous to (5) (case $k \geq 1$ only) for the short-run system size probabilities. You need not attempt to solve either the differential equations or the corresponding linear equations for the limiting probabilities.

6.3 Mathematical Finance

The comments that we made at the start of Section 6.2 about queueing theory and its effect on the rapid development of probability could also be made about the area of mathematical finance. Increasingly, economists have studied aspects of financial problems that are subject to random influences. Since time is usually a factor, the study of random processes like the Poisson process, Markov chains and others has become a key element in the education of finance specialists. Several Nobel prizes in economics have been awarded to people who have made key contributions to mathematical finance. Here are just a few example problems in which probabilistic analysis is important.

1. An investor must decide how to allocate wealth between a risky asset such as a stock, and a non-risky asset such as a bond.

2. An *option* on a stock is a contract that can be taken out which lets the option owner buy a share of the stock at a fixed price at a later time. What is the fair value of this option?

3. A firm must decide how to balance its equity (money raised by issuance of stock to buyers, who then own a part of the company) and debt (money raised by selling bonds which are essentially loans made by the bond-holders to the company). The main considerations are the tax-deductibility of the interest on the bonds, and the chance of default penalties if the company is unable to make payments on the bonds. The goal is to maximize the total value of the money raised.

4. How should an individual consume money from a fund which appreciates with interest rates that change randomly as time progresses, if the person wants a high probability of consumption of at least a specified amount for a specified number of years?

Activity 1   The insurance industry has always been filled with financial problems in which randomness is a factor. Write descriptions of at least two problems of this kind.

In the interest of brevity we will just look at problems 1 and 2 above, on finding an optimal combination of two assets in a portfolio, and pricing a stock option.

Optimal Portfolios

All other things being equal, you would surely prefer an investment whose expected return is 6% to one whose expected return is 4%. But all other things are rarely equal. If you knew that the standard deviation of the return on the first asset was 4% and that of the second was 1% it might change your decision. Chebyshev's inequality says that most of the time, random variables take values within 2 standard deviations of their means, so that the first asset has a range of returns in $[-2\%, 14\%]$ most of the time, and the second asset will give a return in $[2\%, 6\%]$ most of the time. The more risk averse you are, the more you would tend to prefer the surer, but smaller, return given by the second asset to the more variable return on the first.

Let us pose a simple mathematical model for investment. There are just two times under consideration: "now" and "later". The investor will make a decision now about what proportion of wealth to devote to each of two potential holdings, and then a return will be collected later. One of these two holdings is a bond with a certain rate of return r_1, meaning that the total dollars earned in this single period is certain to be r_1 times the dollars invested in the bond. The other potential investment is a risky stock whose rate of return R_2 is a random variable with known mean μ_2 and standard deviation σ_2. The investor starts with wealth W and is to choose the proportion q of that wealth to invest in the stock, leaving the proportion $1 - q$ in the bond. The dollar return on this portfolio of assets is a combination of a deterministic part and a random part:

$$W(1 - q)\, r_1 + W\, q\, R_2$$

Hence the rate of return, i.e., the dollar return divided by the number of dollars W invested, is

$$R = (1 - q)\, r_1 + q\, R_2 \tag{1}$$

Because R_2 is random, the investor cannot tell in advance what values this portfolio return R will take. So, it is not a well-specified problem to optimize the rate of return given by (1). The investor might decide to optimize expected return, but this is likely to be a trivial problem, because the risky stock will probably have a higher expected return $\mu_2 = E[R_2]$ than the return r_1 on the bond; so the investor should simply put everything into the stock if this criterion is used. Moreover, this optimization criterion takes no account of the riskiness of the stock, as measured by $\sigma_2{}^2$, the variance of the rate of return.

There are various reasonable ways to construct a function to be optimized which penalizes variability according to the degree of the investor's aversion to risk. One easy way is based on the following idea. Suppose we are able to query the investor and determine how many units of extra expected rate of return he would demand in exchange for an extra unit of risk on the rate of return. Suppose that number of units of extra return is a (for aversion). Then the investor would be indifferent between a portfolio with mean rate of return μ and variance σ^2 and another with mean $\mu + a$ and variance $\sigma^2 + 1$, or another with mean $\mu + 2a$ and variance $\sigma^2 + 2$, etc. Such risk averse behavior follows if the investor attempts to maximize

$$\mu_p - a\,\sigma_p{}^2 \tag{2}$$

where μ_p is the mean portfolio rate of return and σ_p^2 is the variance of the rate of return. This is because with objective (2), the value to this investor of a portfolio with mean $\mu + k\,a$ and variance $\sigma^2 + k$ is

$$(\mu + k\,a) - a(\sigma^2 + k) = \mu - a\,\sigma^2$$

which is the value of the portfolio of mean μ and variance σ^2. So we will take formula (2) as our objective function.

From formula (1), the mean portfolio rate of return as a function of q is

$$\mu_p = E[(1-q)\,r_1 + q\,R_2] = (1-q)\,r_1 + q\,\mu_2 \tag{3}$$

and the variance of R is

$$\sigma_p^2 = \mathrm{Var}((1-q)\,r_1 + q\,R_2) = q^2\,\mathrm{Var}(R_2) = q^2\,\sigma_2^2 \tag{4}$$

Combining (2), (3), and (4), an investor with risk aversion constant a will maximize as a function of q:

$$f(q) = r_1 + (\mu_2 - r_1)\,q - a\,\sigma_2^2\,q^2 \tag{5}$$

By simple calculus, the optimal value q^* for this objective is

$$q^* = \frac{\mu_2 - r_1}{2\,a\,\sigma_2^2} \tag{6}$$

Activity 2 Verify the formula for q^* in (6).

Notice from (6) the intuitively satisfying implications that as either the risk aversion a or the stock variance σ_2^2 increase, the fraction q^* of wealth invested in the stock decreases, and as the gap $\mu_2 - r_1$ between the expected rate of return on the stock and the certain rate of return on the bond increases, q increases.

Example 1 Suppose that the riskless rate of return is $r_1 = .04$, and the investor has the opportunity of investing in a stock whose expected return is $\mu_2 = .07$, and whose standard deviation is .03. Let us look at the behavior of the optimal proportion q^* as a function of the risk aversion a.

First, here is a *Mathematica* function that gives q^* in terms of all problem parameters.

```
qstar[a_, r1_, μ2_, σ2_] := μ2 - r1
                            ─────────
                            2 * a * σ2²
```

The plot of this function in Figure 5 shows something rather surprising.   The optimal proportion q^* does not sink below $1 = 100\%$ until the risk aversion becomes about 17 or greater. (Try zooming in on the graph to find more precisely where q^* intersects the horizontal line through 1.)   There are two ways to make sense of this: investors whose risk aversions are less than 17 hold all of their wealth in the stock and none in the bond, or if it is possible to borrow cash (which is called holding a *short position* in the bond), then an investor with low risk aversion borrows an amount $|1 - q^*|\,W$ of cash and places his original wealth plus this borrowed amount in the stock.   His overall wealth is still W, because he owns $W + (q^* - 1)\,W$ of stock and $-(q^* - 1)\,W$ of cash.   He hopes that the stock grows enough to let him pay off his cash debt, and still profit by the growth of his holding in stock.

```
Plot[{qstar[a, .04, .07, .03], 1}, {a, 2, 20},
    PlotStyle → {RGBColor[0, 0, 0], GrayLevel[.7]},
    AxesOrigin → {0, 0},
    DefaultFont → {"Times-Roman", 8},
    AxesLabel → {"a", "q*"}, PlotRange → {0, 5}];
```

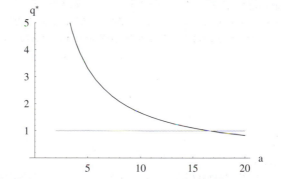

Figure 5 - Optimal proportion of wealth in risky asset as a function of risk aversion

Let us simulate the experiment of making this investment 200 times, and keep track of the investor's profit under the optimal choice q^*, and then under a less than optimal choice. This will give us numerical evidence that this borrowing is a good thing for the investor. We will plot a histogram, and compute a sample mean and standard deviation of profit in each case. The distribution of the stock rate of return will be assumed normal. The rate of return information will be as described in the problem, and we will take an initial \$100 investment and an investor with risk aversion parameter $a = 10$. First we define a simulator of total returns given initial wealth W, portfolio coefficient q, the rates of return r_1 and μ_1, the standard deviation of the stock return σ_2, and the number of simulations desired.

```
Needs["KnoxProb`Utilities`"]
```

```
TotalReturn[W_, q_, r1_, μ2_, σ2_, numreps_] :=
  Table[W * ((1 - q) r1 + q * Random[
        NormalDistribution[μ2, σ2]]), {numreps}]
```

```
SeedRandom[6786389]
```

Here is a typical result using the optimal strategy. The first output is the optimal q^*, and the sample mean and standard deviation of total returns corresponding to it. Then we let *Mathematica* compute the sample mean and standard deviation for the portfolio which puts 100% of wealth in the stock, without borrowing. Beneath these are histograms of total returns for each simulation.

```
q = qstar[10, .04, .07, .03]
returns1 =
  TotalReturn[100, q, .04, .07, .03, 200];
{Mean[returns1], StandardDeviation[returns1]}
```

1.66667

{8.72528, 4.61509}

```
q = 1
returns2 =
  TotalReturn[100, q, .04, .07, .03, 200];
{Mean[returns2], StandardDeviation[returns2]}
```

1

{6.61308, 3.132}

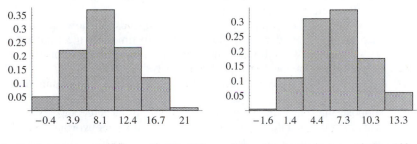

(a) optimal strategy $q = 1.6667$, $a = 10$, $W = 100$ (b) $q = 1$(no borrowing), $a = 10$, $W = 100$

Figure 6 - Distribution of profits under two strategies

Since the first run began with more money in the stock, it is to be expected that the mean total return of about 8.7 exceeds the mean total return for the second scenario of about 6.6. It is also to be expected that the standard deviation of the return is a little higher in the first case than the second. But the distribution of returns in Figure 6 is noticeably more desirable in the optimal case than in the $q = 1$ case, since the center of the distribution is higher, and its lowest and highest categories are both higher than those for the non-optimal case. ■

Activity 3   Compare the distribution, mean, and standard deviation in the optimal case to some suboptimal cases for an investor with high risk aversion of 30.

    Extensions of this basic problem are possible. Exercise 4 investigates what happens when there are two stocks of different means and variances. An important result in the multiple stock case is the *Separation Theorem*, which says that risk aversion affects only the balance between the bond and the group of all stocks; all investors hold the stocks themselves in the same relative proportions. One can also consider multiple time periods over which the portfolio is allowed to move. Mathematical economists have also studied different objective criteria than the mean-variance criterion we have used. One of the main alternatives is to maximize a utility function $U(R)$ of the rate of return on the portfolio, where U is designed to capture both an investor's preference for high return, and aversion to risk.

Option Pricing

    Our second problem is the option pricing problem. An asset called a *call option of European type* can be purchased, which is a contract that gives its owner the right to buy a share (actually, it is more commonly 100 shares) of another risky asset at a fixed price E per share at a particular time T in the future. The price E is called the *exercise price* of the option. The question is, what is the present value of such a call option contract? (The "European" designation means that it is not possible to exercise the option until time T, while options of *American type* permit the option to be exercised at any time up to T.)

The value of the European call option depends on how the risky asset, whose price at time t we denote by $P(t)$, moves in time. If the final price $P(T)$ exceeds E, the owner of the option can exercise it, buy the share at price E, then immediately resell it at the higher price $P(T)$ for a profit of $P(T) - E$. But if $P(T) < E$ then it does not pay to exercise the option to buy, so the option becomes valueless.

Suppose that the value of money discounts by a percentage of r per time period. This means that a dollar now is worth $1 + r$ times the worth of a dollar next time period; equivalently a dollar in the next time period is worth only $(1 + r)^{-1}$ times a dollar in present value terms. Similarly, a dollar at time T is only worth $(1 + r)^{-T}$ times a dollar now. Therefore the present value of the call option is the expectation

$$E\left[\frac{1}{(1+r)^T} \max(P(T) - E, 0) \right] \tag{7}$$

Activity 4   A *European put option* is a contract enabling the owner to sell a share of a risky asset for a price E at time T. Write an expression similar to (7) for the value of a put option.

To proceed, we need to make some assumptions about the risky asset price process $P(t)$. Nobel winners Fischer Black and Myron Scholes in their seminal 1973 paper chose a continuous time model in which $\log(P(t))$ has a normal distribution with known parameters, implying that $P(t)$ has a distribution called the *log-normal distribution*. The Black-Scholes solution, which has dominated financial theory since its inception, is expressed in terms of the standard normal c.d.f. and some parameters which govern the stock price dynamics. But there is quite a lot of machinery involved in setting up the model process, and finding the expectation in (7) for this continuous time problem. So we will use the simpler Cox, Ross, and Rubinstein model (see Lamberton and LaPeyre, p. 12) in discrete time. It turns out that, in the limit as the number of time periods approaches infinity while the length of a period goes to 0, the Cox, Ross, and Rubinstein call option value converges to the Black-Scholes call option value.

In the Cox, Ross, and Rubinstein framework, we assume that from one time to the next the asset price changes by:

$$P(t + 1) = \begin{cases} (1 + a)\, P(t) & \text{with probability } p \\ (1 + b)\, P(t) & \text{with probability } 1 - p \end{cases} \tag{8}$$

where $a < b$. So there are two possible growth rates for the price in one time period: a or b, and these occur with probabilities p and $(1 - p)$. By the way, a need not be positive; in fact, it is probably useful to think of the a transition as a "down" movement of the price and the b transition as an "up" price movement. One other assumption is that the ratios $Y(t) = P(t)/P(t-1)$ are independent random variables. By (8),

$$Y(t) = \begin{cases} (1+b) & \text{with probability } p \\ (1+a) & \text{with probability } 1-p \end{cases} \quad \text{and} \quad P(t) = P(0)\, Y_1\, Y_2 \cdots Y_t \qquad (9)$$

If the initial price of the risky asset is x, then in the time periods 1, 2, ... , T the price will undergo some number j of "ups" and consequently $T - j$ "downs" to produce a final price of $x(1+b)^j (1+a)^{T-j}$. By our assumptions, the number of "ups" has the binomial distribution with parameters T and p. Therefore, the expectation in (7), i.e., the present value of the call option, is a function of the initial price x, the exercise price E, and the exercise time T by

$$V(x, T, E) = \tfrac{1}{(1+r)^T} \sum_{j=0}^{T} \binom{T}{j} p^j (1-p)^{T-j} \max(x(1+b)^j (1+a) \qquad (10)$$

It appears that formula (10) solves the option valuation problem, but there is a little more to the story. Economic restrictions against the ability of investors to arbitrage, i.e., to earn unlimited risk-free returns at rates above the standard risk-free rate r, imply that we may assume not only that $a < r < b$, but also that

$$E\left[\tfrac{P(t)}{P(t-1)}\right] = E[Y_t] = 1 + r$$

The argument that justifies this is actually rather subtle, and can be found in most finance books, including Lamberton and LaPeyre mentioned above. It hinges on the fact that if the option were valued in any other way, a portfolio of the stock and the option could be constructed which would hedge away all risk, and yet would return more than the riskless rate. We will omit the details. By construction of Y_t,

$$(1+b)\,p + (1+a)(1-p) = 1 + r \Longrightarrow p = \tfrac{r-a}{b-a} \qquad (11)$$

after some easy algebra. So with the "up" and "down" percentages b and a, and the riskless rate r as given parameters, equation (10) is now a complete description of the fair option value.

Also, the same reasoning as before applies to the problem of finding the present value at time t of the same option at the time t between 0 and T. All of the T's in formula (10) would be replaced by $T - t$, which is the remaining number of time steps.

Example 2 Let us define the option price function V as a *Mathematica* function that we call OptionValue, and examine some of its characteristics. In addition to current time t, current price x, T, and E, we will also put b, a, and r as arguments, and let the function compute p. We use variable name EP instead of E for the exercise price, since E is a reserved name.

```
OptionValue[t_, x_, T_, EP_, b_, a_, r_] :=
 Module[{ p, V },
        r - a
   p = ——————— ;
        b - a
            1
   V = —————————— *
        (1 + r)^(T-t)

      NSum[Binomial[T - t, j] * p^j * (1 - p)^(T-t-j) *
        Max[0, x * (1 + b)^j * (1 + a)^(T-t-j) - EP],
        {j, 0, T - t}];
   V]
```

Here is one example run. Think carefully about the question in the activity that follows.

```
OptionValue[0, 22, 3, 18, .07, .03, .05]
```

6.45092

Activity 5   In the output above the difference between the stock price and the option exercise price is $P(0) - E = \$22 - \$18 = \$4$. Why is $6.45 a reasonable answer for the value of the option?

Figure 7(a) shows the graph of option price against stock price, for fixed time $t = 0$, time horizon $T = 8$, and exercise price $E = 18$. We take $b = .09$, $a = .03$, and $r = .05$.   We observe that the option is worthless until the stock price is larger than 11 or so, which is to be expected because if the stock price is so much lower than E, and can only rise at best by a little less than 10% for each of 8 time periods, it has little if any chance of exceeding the exercise price by time T. As the initial stock price rises, the price has a much better chance of exceeding 18 at time $T = 8$, so the amount the buyer expects to receive increases, and the price of the option therefore increases.

```
g1 = Plot[OptionValue[0, x, 8, 18, .09, .03, .05],
    {x, 5, 22}, DisplayFunction → Identity,
    AxesOrigin → {5, -.5}, PlotRange → All,
    AxesLabel → {"stock", "option"},
    DefaultFont → {"Times-Roman", 8}];
g2 = Plot[OptionValue[0, 22, 8, EP,
    .09, .03, .05], {EP, 15, 35},
    DisplayFunction → Identity,
    AxesLabel → {"E", "option"},
    DefaultFont → {"Times-Roman", 8}];
Show[GraphicsArray[{g1, g2}],
    DisplayFunction → $DisplayFunction];
```

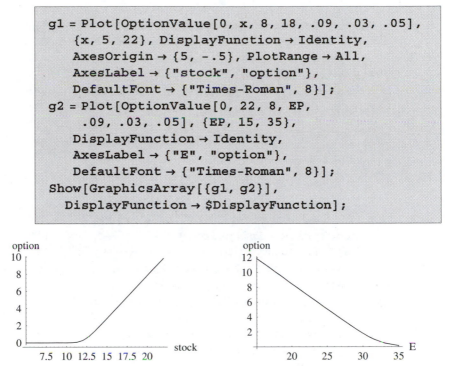

(a) $T = 8$, $E = 18$, $b = .09$, $a = .03$, $r = .05$ (b) $T = 8$, $b = .09$, $a = .03$, $r = .05$, stock price \$22.

Figure 7 - (a) Option price at time 0 as a function of stock price; (b) Option price at time 0 as a function of exercise price

We see the same phenomenon in a different way in Figure 7(b), in which the option price is plotted against the exercise price E, for fixed initial stock price \$22 (the other parameters have the same values as before). As the exercise price increases, it becomes more likely that the stock price will not exceed it, lowering the price of the option. Thus, the price of the option goes down as the exercise price increases. To complete the study, in Exercise 6 you are asked to study the effect of lengthening the time horizon on the option price. ∎

We can use the OptionValue function in a simple way to simulate a stock price and its option price together. After loading the Graphics`MultipleListPlot` package in order to gain access to the MultipleListPlot command, we define the function SimStockAndOption. This function simulates a series of "ups" and "downs" of the stock price, and updates and returns lists of stock prices and corresponding option prices at each time up to T.

```
Needs["Graphics`MultipleListPlot`"]
```

```
SimStockAndOption[
  initprice_, T_, EP_, b_, a_, r_] :=
 Module[{ p, stockprice, stocklist,
   optionprice, optionlist},
          r - a
     p = ───── ;
          b - a
    stocklist = {{0, initprice}};
    optionlist = {{0, OptionValue[
       0, initprice, T, EP, b, a, r]}};
   stockprice = initprice;
  Do[If[Random[] ≤ p,
    stockprice = (1 + b) stockprice,
    stockprice = (1 + a) stockprice];
   AppendTo[stocklist, {t, stockprice}];
   optionprice =
    OptionValue[t, stockprice, T, EP, b, a, r];
   AppendTo[optionlist, {t, optionprice}],
   {t, 1, T}];
  MultipleListPlot[stocklist,
   optionlist, PlotJoined → True,
   PlotStyle → {GrayLevel[.7], Thickness[.01]},
   DefaultFont → {"Times-Roman", 8}];
  {stocklist, optionlist}]
```

Here is an example simulation, with a starting stock price of \$15, exercise time $T = 10$, exercise price $E = \$18$, multipliers $b = .08$, $a = -.02$, and riskless rate $r = .04$. Feel free to run the command several times with this choice of parameters, and then to change the parameters to find patterns. One thing that you will always see is that the difference between the stock price and the option price is the exercise price at the final time (Why?).

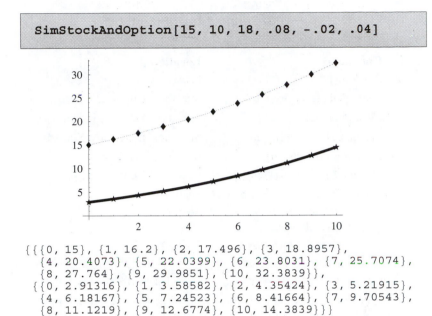

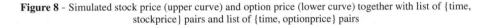

Figure 8 - Simulated stock price (upper curve) and option price (lower curve) together with list of {time, stockprice} pairs and list of {time, optionprice} pairs

This chapter has only scratched the surface of the many interesting extensions and applications of probability. Other important areas, such as reliability theory, stochastic optimization, information theory, the analysis of algorithms, and inventory theory, await you. Or, you might decide you want to study Markov chains and other random processes, queues, or finance more deeply. Probably the most important next step is the subject of statistics, which we have previewed a number of times. It is my hope that your experience with this book has prepared you well for all of these studies, as well as enhancing your problem solving abilities, and giving you a new view of a world fraught with uncertainties.

Mathematica for Section 6.3

Command	Location
Random[]	kernel
NormalDistribution[μ, σ]	Statistics` ContinuousDistributions`
Mean[datalist]	Statistics` DescriptiveStatistics`
StandardDeviation[datalist]	Statistics` DescriptiveStatistics`
Histogram[datalist, n]	KnoxProb` Utilities`
MultipleListPlot[list1, list2]	Graphics` MultipleListPlot`
OptionValue[t, x, T, EP, b, a, r]	Section 6.3
SimStockAndOption[init, T, EP, b, a, r]	Section 6.3

Exercises 6.3

1. If your investment strategy is to pick just one stock from among the following three to invest in, and your risk aversion is $a = 5$, which stock would you pick?
    Stock 1: $\mu_1 = .05$, $\sigma_1^2 = .008$; Stock 2: $\mu_2 = .06$, $\sigma_1^2 = .01$;
    Stock 3: $\mu_1 = .09$, $\sigma_1^2 = .02$

2. Find the optimal portfolio of one stock and one bond for an investor with risk aversion $a = 12$, if the non-risky rate of return is 4%, and the stock rate of return is random with mean 5% and standard deviation 4%.

3. (*Mathematica*) Simulate the total returns for 200 replications of the investment problem in Exercise 2, for the optimal portfolio, for a portfolio consisting of 100% stock, and for a portfolio consisting of 50% stock and 50% bond. Assume an initial wealth of $1000. Produce and compare histograms and summary statistics of the simulated returns.

4. Suppose that two independent stocks with mean rates of return μ_1 and μ_2 and standard deviations σ_1 and σ_2 are available, as well as the non-risky bond with rate of return r. Find expressions for the optimal proportions of wealth that an investor with risk aversion a is to hold in each asset. Does the relative balance between the two risky assets depend on the risk aversion?

5. Suppose that the non-risky rate of return is 5%, and we are interested in a two period portfolio problem in which we collect our profit at the end of the second time period. If the stock follows a binomial model as described in the second subsection, with $b = .10, a = 0$, and $p = .4$, and the investor's risk aversion is 15, find the optimal portfolio.

6. (*Mathematica*) In Example 2, produce a graph of the option price at time 0 as a function of the exercise time T, with the initial stock price = $14, and other parameters as in the example: $E = 18$, $b = .09$, $a = .03$, $r = .05$. Then see how the graph changes as you vary the initial stock price.

7. A stock follows the binomial model with initial price $30, $b = .08$, $a = 0$, and the riskless rate is $r = .04$. Find (by hand) the probability p, the distribution of the stock price two time units later, and the value of an option whose exercise time is 2 and whose exercise price is $32.

8. Express the value of the European call option (10) in terms of the binomial cumulative distribution function.

9. An *American call option* is one that can be exercised at any time prior to the termination time T of the contract. Argue that a lower bound for the value of an American call option $V(S, t, E)$ is $\max(S - E, 0)$. Sketch the graph of this lower bound as a function of S. (Incidentally, it can be shown that the value of an American call is the same as that of a European call. So this lower bound holds for the European call option as well.)

10. Consider a single time period option model with a general binomial model expressed by (8) holding for the stock upon which the option is based. Again let r be the non–risky rate. Show that the investment action of holding one option at time 0 can be *replicated* by holding a certain number N of shares of the stock, and borrowing an amount B at the non-risky rate, i.e., whether the stock goes up or down in the single time period, this replicating portfolio performs exactly as the option would perform. Find expressions for the N and B that do this replication.

Appendices

Appendix A

Short Answers to Selected Exercises

Section 1.1

1. Approach 1: sample space {000, 001, ..., 999}. A typical simple event is a three digit number with leading 0's, such as 027 or 635. The probability of an event is its cardinality divided by 1000. Approach 2: sample space {−1, 499}. Give outcome −1 probability 999/1000 and outcome 499 probability 1/1000.

2. Sample space: {{A,B}, {A,C}, {A,D}, {B,C}, {B,D}, {C,D}}. P[A serves] = 3/6.

3. (a) P[25 ACT and B] = 10/179; P[C] = 56/179; P[23 ACT] = 57/179.
(b) P[A given 24 ACT] = 6/43. (c) P[23 ACT given B] = 5/38 ≠ 57/179.

5. P[king on 1st] = 4/52. P[queen on 2nd] = 4/52.

6. $P[X = -1]$ = 999/1000 and $P[X = 499]$ = 1/1000. $E[X] = -1/2$.

7. $P[X = 0]$ = 4/6 and $P[X = 1]$=2/6. $E[X]$ = 2/6.

8. $P[X = 0]$ = 1/4, $P[X = 1]$ = 2/4, $P[X = 2]$ = 1/4. $E[X]$ = 1.

9. $P[X = 23]$ = 57/179, $P[X = 24]$ = 43/179, $P[X = 25]$ = 39/179, $P[X = 26]$ = 40/179. $E[X]$ = 4358/179 ≈ 24.3.

Section 1.2

2. The probability of a hit of any kind is $\frac{59}{216}$ ≈ .273. This batter should have home run symbols in a set of squares whose probability totals to 11/216; for instance in squares red=1, whitesum=6 and red=3, whitesum=8, and red=5, whitesum=2 .

3. (a) $\Omega = \{(r_1, r_2, r_3)$ where each $r_i \in$ {a, b, c, d}}. Examples: (a,c,d) (c,b,b).
(c) P[at least 2 right] = 10/64.

4. (c) P[Bubba or Erma is in the sample] = 9/10.

6. P[10 is not in the sample] = 72/90.

7. (a) P[less than 20] ≈ .335; P[between 20 and 30] ≈ .152; P[between 30 and 40] ≈ .121; P[between 40 and 100] ≈ .313; P[over 100] = $\frac{10.5}{133.9}$ ≈ .078.
P[at least 30] ≈ .512.
(b) The probability that at least one of the two earned more than $100,000 is roughly .14976.

8. (b) P[3rd quadrant] = 1/4; (c) P[not in circle radius 2] = $1 - \pi/4$;
(d) P[either 3rd quadrant or not in circle radius 2] = $1 - 3\pi/16$.

9. $P[(A \cup B)^c]$ = .26.

13. $P[\{\omega_1\}]$ = .2, $P[\{\omega_2\}]$ = .4, $P[\{\omega_3\}]$ = .4.

15. 15/24.

Section 1.3

1. P[(c, d)] = $(d - c)/(b - a)$ = P[[c, d]] and P[{d}] = 0.
2. Notice that because the first few seed values are small in comparison to the modulus, the first few random numbers are small.
9. The area under $y = 1/x$ is infinite and also there is no bounded rectangle in which to enclose the region. For $y = 1/\sqrt{x}$ the second problem remains but at least the area is finite.
10. One sees a rather symmetric hill shaped frequency distribution.
12. One usually gets a mean lifetime of about 13, and an asymmetrical histogram with most of its weight on the left and a peak around 12 or 13.

Section 1.4

1.(a) There are 5! lineups. (b) The probability is $1/5$. (c) The probability is 4/5.
2. $\frac{544}{625}$.
4.

15.	0.252901
16.	0.283604
17.	0.315008
18.	0.346911
19.	0.379119
20.	0.411438
21.	0.443688
22.	0.475695
23.	0.507297
24.	0.538344
25.	0.5687
26.	0.598241
27.	0.626859
28.	0.654461
29.	0.680969
30.	0.706316
31.	0.730455
32.	0.753348
33.	0.774972
34.	0.795317
35.	0.814383

7. 2^n.
8. $C_{n,3} = \frac{1}{6} \cdot n \cdot (n - 1) \cdot (n - 2)$.
10. The probability of a full house is about .00144. The probability of a flush is about .00198. The full house is more valuable.
12. (a) 243/490. (b) 10 bad packages suffices. (c) $n = 19$ is the smallest such n.
13. The probability that subject 1 is in group 1 is n_1/n.
14. 10080.

16. P[3 or fewer runs] = 1/429.
17. P[individual 1 in the first substrata of the first stratum] = 16/180.

Section 1.5

1. (a) 126/218. (b) 25/114. (c) 295/380.
3. .11425.
4. .34.
5. (a) P[all Republicans] = $\frac{4}{35}$. (b) P[at least 2 Rep | at least 1 Rep] = $\frac{132}{204}$.
8. 5/12.
9. 9/64.
10. The proportion should approach 1/4.
12. P[lands in 1] = $\frac{1}{8}$; P[lands in 2] = $\frac{3}{8}$; P[lands in 3] = $\frac{3}{8}$; P[lands in 4] = $\frac{1}{8}$.
14. P[at least C | score between 10 and 20] = $\frac{68}{74}$.

Section 1.6

1. P[no more than 1 right] $\approx$.376.
2. The two events are independent.
4. A and B cannot be independent.
8. It is impossible to make A and B independent.
9. (b) P[HHH] = $(.49)^3$, P[HTH] = $(.49)^2(.51)$ = P[THH] = P[HHT],
P[TTH] = $(.51)^2(.49)$ = P[THT] = P[HTT], P[TTT] = $(.51)^3$.
10. P[current flows from A to B] = $(1 - (1 - p)^2)(1 - (1 - q)^3)$.
11. P[system fails by time 15] = .125.

Section 2.1

1. The values of $f_3(x)$ sum to 1 exactly; hence it is valid. The c.d.f. of this distribution is

$$F_3(x) = \begin{cases} 0 & \text{if } x < 1 \\ .16 & \text{if } 1 \le x < 2 \\ .47 & \text{if } 2 \le x < 3 \\ .65 & \text{if } 3 \le x < 4 \\ .75 & \text{if } 4 \le x < 5 \\ 1 & \text{if } x \ge 5 \end{cases}$$

2. (a) The table below gives the p.m.f.

x	2	3	4	5	6	7	8	9	10	11	12
$f(x)$	$\frac{1}{36}$	$\frac{2}{36}$	$\frac{3}{36}$	$\frac{4}{36}$	$\frac{5}{36}$	$\frac{6}{36}$	$\frac{5}{36}$	$\frac{4}{36}$	$\frac{3}{36}$	$\frac{2}{36}$	$\frac{1}{36}$

(b) The p.m.f. of the maximum random variable is

x	1	2	3	4	5	6
$f(x)$	$\frac{1}{36}$	$\frac{3}{36}$	$\frac{5}{36}$	$\frac{7}{36}$	$\frac{9}{36}$	$\frac{11}{36}$

5. If $m \in \{1, 2, ..., 20\}$, $F(m) = 1 - (1 - \frac{m}{20})^5$ and $f(m) = (1 - \frac{m-1}{20})^5 - (1 - \frac{m}{20})^5$.

7. For m in $\{5, 6, ..., 20\}$,

$$P[X = m] = \frac{\binom{m-1}{4}}{\binom{20}{5}}.$$

9. $\mu = 2.4$, $\sigma^2 = .64$, third moment $= .168$.

10. $\sigma_1^2 = .50$, $\sigma_2^2 = .25$.

11. For the distribution on the left, the third moment is 0, and for the distribution on the right, it is 2.25.

13. $E[X^3] - 3\,\mu E[X^2] + 2\,\mu^3$.

16. $E[X] = 0$; $\text{Var}(X) = n(n+1)/3$.

17. $E[2X - 6] = 7$, $\text{Var}(2X - 6) = 31$.

19. The most likely population size is 26.

20. The mean number defective is .7. The variance is about .5918.

Section 2.2

3. $E[(X - p)^3] = 2\,p^3 - 3\,p^2 + p$.

4. .382281.

5. $E[X^2] = np(1 - p) + (np)^2$.

6. The expected number of cabs is 36, and the probability that at least 30 cabs arrive is .889029.

7. P[at least 300 advances] = .999999. The mean and variance are 350 and 105 respectively.

10. The expected number of hits is 150. The standard deviation of batting average is .0205.

12. The probability that the promoters must issue at least one refund is .0774339. The expected total refund is $10.21.

14. $\binom{10}{x_1\ x_2\ x_3\ x_4} (\frac{1}{4})^{x_1} (\frac{1}{2})^{x_2} (\frac{1}{8})^{x_3} (\frac{1}{8})^{x_4}$.

Section 2.3

1. If p is approximately .0001.

2. $F(x) = P[X \le x] = 1 - (1 - p)^{x+1}$.

3. $P[X > m + n \mid X > n] = (1 - p)^m$.

4. .23999.

6. (a) .537441 (b) .518249 (c) .38064 (d) mean = 30, standard deviation = $\sqrt{420}$ ≈ 20.5.

8. p is about .285.

11. expected time = $16\frac{2}{3}$, variance = 38.9, probability of between 15 and 30 jumps around .554.

14. The expected number of games played is 70, the variance is 46.66... and the standard deviation is 6.83.

Section 2.4

1. (a) .0888353; (b) .440907; (c) .686714.

2.

	F(x;1)	F(x;2)	F(x;3)	F(x;4)
0	0.367879	0.135335	0.0497871	0.0183156
1	0.735759	0.406006	0.199148	0.0915782
2	0.919699	0.676676	0.42319	0.238103
3	0.981012	0.857123	0.647232	0.43347
4	0.99634	0.947347	0.815263	0.628837
5	0.999406	0.983436	0.916082	0.78513
6	0.999917	0.995466	0.966491	0.889326
7	0.99999	0.998903	0.988095	0.948866
8	0.999999	0.999763	0.996197	0.978637
9	1.	0.999954	0.998898	0.991868
10	1.	0.999992	0.999708	0.99716

3. μ = sample mean = 3.625.

4. $E[X^3] = \mu^3 + 3\mu^2 + \mu$.

5. $E[X(X - 1)(X - 2) \cdots (X - k)] = \mu^{k+1}$.

8. The probability that a randomly selected piece has either 1 or 2 defects is about .578624.

9. .0426209.

10. (a) $P[X > 12] = .424035$; (b) $P[X > 15 \mid X > 12] = .366914$.

11. $P[X_1 = 2, X_2 = 3, X_3 = 5] = .0124468$.

12. $P[X_s = j \cap X_{t+s} = k] = \frac{e^{-\lambda s} (\lambda s)^j}{j!} \frac{e^{-\lambda t} (\lambda t)^{k-j}}{(k-j)!}$.

13. (a) $P[T_1 > s] = e^{-\lambda s}$; (b) $P[T_1 > s+t \mid T_1 > s] = e^{-\lambda t}$.

14. $\lambda = 3$ makes them most likely.
15. The smallest such λ is about 5.6.

Section 2.5

1. $q(0\,|\,0) = 4/31, q(1\,|\,0) = 6/31, q(2\,|\,0) = 10/31, q(3\,|\,0) = 8/31, q(4\,|\,0) = 3/31$
$p(0\,|\,2) = 10/41,\ p(1\,|\,2) = 7/41,$
$p(2\,|\,2) = 10/41,\ p(3\,|\,2) = 9/41,\ p(4\,|\,2) = 5/41$
2. The X marginal is $p(1) = 4/16,\ p(2) = 3/16,\ p(3) = 3/16,\ p(4) = 6/16$.
The Y marginal is $q(1) = 9/32,\ q(2) = 9/32,\ q(3) = 7/32,\ q(4) = 7/32$.
4. (a) $P[X = 0] = 4/10,\ P[X = 1] = 3/10,\ P[X = 2] = 2/10,\ P[X = 3] = 1/10$.
(b) $P[Y = 0] = 4/10,\ P[Y = 1] = 3/10,\ P[Y = 2] = 2/10,\ P[Y = 3] = 1/10$.
(c) $P[X = 0\,|\,Y = 1] = \frac{1}{3}, P[X = 1\,|\,Y = 1] = \frac{1}{3},\ P[X = 2\,|\,Y = 1] = \frac{1}{3}$.
(d) $P[Y = 0\,|\,X = 1] = \frac{1}{3}, P[Y = 1\,|\,X = 1] = \frac{1}{3},\ P[Y = 2\,|\,X = 1] = \frac{1}{3}$.
5. $p(y\,|\,x = 3) = \binom{30}{y}\binom{25}{5-y}\Big/\binom{55}{5}$.
10. P[hits safely] = .264.
11. $P[X = 1\,|\,Y = 1] = .993,\ P[X = 0\,|\,Y = 1] = .007$;
    $P[X = 1\,|\,Y = 0] = .486,\ P[X = 0\,|\,Y = 0] = .514$.
14. $f(x_1,\ x_2) = 1/4$. Similarly the other two joint marginals $f(x_1,\ x_3)$ and
$f(x_2,\ x_3)$ are equal to 1/4 for all pairs. $f(x_1) = 1/2$ and the other two marginals
are identical.

Section 2.6

2. (a) $E[X + Y] = 3$; (b) $E[2X - Y] = 3$.
3. $E[X_1 + X_2 + X_3] = 8$; $E[X_1 + X_2 + X_3 + X_4 + X_5] = 20$.
4. mean = 24; variance = 144.
6. When the random variables are independent, $\text{Var}(\sum_{i=1}^{n} c_i X_i) = \sum_{i=1}^{n} c_i^2\,\text{Var}(X_i)$.
8. Both the covariance and correlation equal 1/2.
9. $\text{Cov}(X,\ Y) = n\,p(1 - p)$; correlation = 1.
13. A sample size of 217 is sufficient.
15. 113/16.
17. $E[Y\,|\,X = 1] = 10/4$; $E[Y\,|\,X = 2] = 17/6$.
18. $h(0) = 0,\ h(1) = 1/2,\ h(2) = 5/3,\ h(3) = 14/4$; $E[Y^2] = 2$.

Section 3.1

3. $\mathcal{H} = \{\,\{x\},\ \Omega - \{x\},\ \Omega,\ \emptyset\,\}$.
6. $c = 1$. $P[\,(2, 4) \cup (8, \infty)\,] = \frac{3}{8}$.
7. $P[\,[0, \infty)\,] = 1/2$. $P[\,[-1, 1]\,] = 1/2$.
8. $c = 1$. $P[\,[2, e]\,] = 2 - \log(4)$. $P[\,[1, 2]\,] = \log(4) - 1$.
9. $c = 81/8$. $P[\,(3, 4]^c\,] \approx .094$.

Section 3.2

1. $C = k$, $F(x) = 1 - e^{-kx}$, $x \geq 0$.
2. For $x \in [a, b]$, $F(x) = \frac{x-a}{b-a}$; $P[X > \frac{a+b}{2}] = \frac{1}{2}$.
4. $P[T > 10] = 2/e$.
5. $P[1.5 \leq X \leq 3.5] = 3/4$.
7. $x = \beta(\frac{\alpha-1}{\alpha})^{1/\alpha}$.
8. Y is uniformly distributed on $[-1, 1]$.
9. .752796; $30111.80
10. $P[V \geq 40] = .622146$; v^* is about 54.6.
11. 1/48.
12. $P[V - U \geq \frac{1}{4}] = 9/16$.
14. $P[Y > X] = .432332$.
16. $P[V > \frac{3}{4} \mid U = \frac{1}{2}] = 1/2$.
18. $P[Y > 5/4 \mid X = 3/2] = .716737$; $P[X < 3/2 \mid Y = 6/5] = .546565$.
19. (a) 2/27; (b) 8/27.

Section 3.3

1. $\text{Var}(X) = \frac{(b-a)^2}{12}$.
3. $\mu = 0$; $\text{Var}(X) = 1/6$.
5. $E[T] = 20$, $E[T^2] = 600$, $E[T^3] = 24000$.
6. (a) $P[\,|X - \mu| < \sigma] = .5774$; (b) $P[\,|X - \mu| < 2\sigma] = 1$;
(c) $P[\,|X - \mu| < 3\sigma] = 1$
8. The expected profit is $2387.
9. The covariance is $-9/100$, the correlation is $-9/31$.
11. 1/36
12. $E[X + 6Y] = 7/(2\log(1024/729))$.
16. $E[2XY] = 1/3$.
19. $E[Y \mid X = x] = \frac{x^2/2 + 1/4}{(x^2 + \frac{1}{3})}$.

Section 4.1

5. $1/(\sigma\sqrt{2\pi})$.
7. (a) .841345; (b) .97725; (c) .99865.
8. About .026.
9. About .0001.
11. (a) .182; (b) .8680; (c) .0069; (d) .1251.
12. .1111.
14. (a) About .32; (b) About .95.
15. -1.28155, -0.841621, -0.524401, -0.253347, .
0., 0.253347, 0.524401, 0.841621, 1.28155

Section 4.2

1. The individual variables look fairly normal, except for a few high price observations, and the point cloud looks elliptical with a rather tight dependence of logtax on logprice. So a bivariate normal model for the logged variables is reasonable. The sample means for the logged prices and taxes respectively are 6.92895 and 6.60455, the standard deviations are .31706 and .385655 and the sample correlation is about .86. The predicted log tax is 6.47204, and the predicted tax is the exponential of this, 646.801.

2. The probability is roughly .218.

3. Contrary to conventional wisdom, yards gained and percentage actually seem to have a positive relationship, though a loose one, with one another. The parameter estimates are: for completion percentage mean 57.55, standard deviation 4.12842, for yards gained mean 6.80125, standard deviation .940109, correlation 0.627763. (a) the probability of a percentage of at least 60 is roughly .272481; (b) the probability of a yardage of at least 7.2 is .335224; (c) the predicted percentage given yards 7.0 is about 58.0979.

4. There seems to be very little relationship, and not enough data to make a good judgement about bivariate normality. The best linear predictive relationship available is $y = 1892.92 - .0205824\,(x - 20270.5)$. This equals 1939.65 when $x = 18000$.

5. There does appear to be a good linear relationship between the logged variables. For the logged SO2 variable the mean estimate is 3.19721 and the standard deviation is 1.49761. For the logged mortality variable the mean estimate is 6.84411 and the standard deviation is 0.0662898. The correlation estimate is 0.414563. The conditional variance and standard deviation come out to be .00363912 and .0603251 respectively.

7. The relationship between variables is a loose one, but even with limited data it seems as if a bivariate normal model is appropriate. Using the estimated ACT parameters, we compute the probability of an ACT of at least 25 (i.e., not 24 or below) as .589847. Using the estimated placement parameters, we compute the probability of a placement score of at least 10 (i.e., not 9 or below) as .430556. Conditional on an ACT of 25 the distribution of placement score is normal. The conditional probability that the placement score is at least 9 is about .57.

10. If you divide $f(x, y)$ by $p(x)$ for each fixed x the resulting function will integrate to 1.

11. ρ.

13. $\dfrac{1}{2\pi\sqrt{1-\rho^2}}\exp\!\left[\dfrac{-1}{2(1-\rho^2)}\,(z^2 - 2\rho z w + w^2)\right].$

Section 4.3

1. (a) The density of Y is $g_Y(y) = \frac{3}{2} y^{1/2}$, $y \in [0, 1]$.   (b) The density of Y is $g_Y(y) = 9 y^8$, $y \in [0, 1]$.

2. $g_U(u) = \frac{1}{\sqrt{2\pi}} u^{-1/2} e^{-u/2}$, $u \geq 0$.

6. The probability that C exceeds 20 is about .655.

7. $M(t) = \frac{e^{tb} - e^{ta}}{t(b-a)}$.

8. $M(t) = p \frac{1}{1-(1-p)e^t}$, $t < -\log(1 - p)$.

10. The Bernoulli m.g.f. is $M(t) = p e^t + (1 - p)$.

11. The m.g.f. of the Poisson(μ) distribution is $M(t) = e^{\mu(e^t - 1)}$.

12. $P[\overline{X} < .250] \approx .0099$; $P[\overline{X} > .280] \approx .193$.

13. (a) The probability that the average time is more than 1.7 is about .057; (b) the probability that the total time is less than 16 is about .785.

Section 4.4

1. (a) .65697; (b) .135353; (c) .125731.

4. The probability $P[T > .5]$ is .342547.

5. The probability that the project will not be finished is .199689.

7. $E[T] = 10$, and $P[T \geq 10] = e^{-1}$.

8. (a) .151204; (b) .00229179.

9. $F(x) = 1 - e^{-\lambda x} - \lambda x e^{-\lambda x}$.

10. $E[X_1 + X_2] \approx 4.55$, and $\text{Var}(X_1 + X_2) \approx 4.13$.

11. β is around .748.

12. (a) .0464436; (b) .0656425.

14. The theoretical probabilities of 0, 1, and 2 arrivals are .135335, .270671, and .270671 respectively.

15. λ is around 2.72.

16. 5/6.

Section 4.5

1. (a) .34378; (b) .132062; (c) 6.7372; (d) $a = 3.24697$, $b = 20.4832$.

2. The m.g.f. of a $\chi^2(r)$ random variable is $M(t) = (1 - 2t)^{-r/2}$.

3. Let a and b be points for the $\chi^2(n - 1)$ distribution for which $P[X < a] = P[X > b] = .025$. The confidence interval is $\frac{(n-1)S^2}{b} \leq \sigma^2 \leq \frac{(n-1)S^2}{a}$.

4. [.0037102, .00836915].

5. It is nearly certain that the standard deviation exceeds .2.

6. .228418.

7. If $\mu = 3$ is true, it is only about .003 likely to observe such a large sample mean; so we have strong evidence that the true μ is larger than 3.

9. (a) .0234488; (b) .0616232; (c) $t = 1.31946$.

10. The data are not sufficiently unusual to cast doubt on the hypothesis $\mu = .75$.

11. Write $t_{.05}$ for the 95th percentile of the $t(n - 1)$ distribution, i.e., the point such that $P[T \le t_{.05}] = .95$. The confidence interval is $\overline{X} \pm t_{.05} \frac{s}{\sqrt{n}}$.

15. Some graphical experimentation produces $r = 14$.

16. (a) .49425; (b) .467569; (c) $a = .379201$, $b = 2.53424$.

17. Since the probability that F is as low or lower than the observed value is around 28% if the two variances are indeed equal, we do not have significant evidence against equality.

20. The 90% confidence interval is $[.563675, 2.2028]$.

Section 5.1

1. We find that $n = 25$ is the smallest number of rolls such that the desired probability exceeds $1/2$.

5. (a) In all three cases, the Chebyshev upper bound is $1/4$. When $X \sim \Gamma(3, 4)$, the actual probability is about .44; (b) When $X \sim$ uniform$(0, 1)$ the actual probability is 0; (c) When $X \sim N(2, 4)$ the actual probability is about .0455.

7. A sample of size 2560 suffices.

Section 5.2

2. $f(x) = \frac{\sqrt{k}}{\sqrt{2\pi}} e^{-kx^2/2}$.

6. Approximately .846.

7. The probability that the sample mean could have come out as large as 550 or larger is about .006. Since this probability is so small, we have significant evidence against $\mu = 540$.

Section 6.1

1. The distribution of X_3 is $\{5/96, 5/64, 67/192, 25/48\}$. The limiting distribution is $\{3/71, 6/71, 24/71, 38/71\}$.

2. The distribution of X_2 given $X_0 = 4$ is $\{0, 0, .16, .48, .36\}$. The limiting distribution is $\{.1, .25, .25, .25, .15\}$.

5. The chain is not regular.

7. The conditional distribution of X_4 given $X_0 = 3$ is $\{53/96, 43/192, 43/192\}$. The limiting distribution is $\{5/9, 2/9, 2/9\}$.

8. The solutions are $f(2) = .588235$, $f(3) = .411765$.

Section 6.2

1. $\rho = 2/3$.

2. The probability of at least 2 in the system is 4/9. The probability that a customer goes directly into service is 1/3.

5. $E[W] = E[W_q] + 1/\mu$.

6. The c.d.f. of W_q is $F_q(s) = 1 - e^{-\mu(1-\rho)s} + (1 - \rho)e^{-\mu s}$.

7. The long-run probability that a customer will take at least 2 minutes is around .716.

8. If C is the weighting factor on service rate, and D is the weighting factor on waiting time, then the optimal $\mu = \lambda + \sqrt{\frac{D}{C}}$.

11. $f_1(t) = \frac{\lambda}{\lambda+\mu} - \frac{\lambda}{\lambda+\mu} e^{-(\lambda+\mu)t}$.

12. The distribution of the time until the next customer is served is $\exp(n\,\mu)$. The differential equation is $f_k'(t) = -(\lambda + \mu\,k)\,f_k(t) + \lambda\,f_{k-1}(t) + \mu\,(k+1)\,f_{k+1}(t),\ \ k \geq 1$.

Section 6.3

1. You would be indifferent between the first two stocks.

2. The optimal proportion q^* in the stock is around 26%.

4. The optimal solutions are

$$q_1{}^* = \frac{\mu_1 - r}{2\,a\,\sigma_1{}^2}, \quad q_2{}^* = \frac{\mu_2 - r}{2\,a\,\sigma_2{}^2}$$

5. The optimal proportion in the stock is about -13%; so we see in this case that because the non-risky asset performs so much better, it is wise to borrow on the stock and invest all proceeds in that asset.

7. $p = .5$. In two time units, the stock will either go up to 34.99 with probability 1/4, or go to 32.40 with probability 1/2, or it will stay at 30 with the remaining probability 1/4. The option value is about 87.6 cents.

10. The portfolio weights are

$$N = \frac{\max\{(1+b)\,S - E, 0\} - \max\{(1+a)\,S - E, 0\}}{S(b-a)}, \quad B = \frac{N\,S(1+b) - \max\{(1+b)\,S - E, 0\}}{(1+r)}$$

Appendix B

Glossary of *Mathematica* Commands for Probability

Command Name	Location

AbsPctAtM[p, n, M, numreps]   Section 1.3
(* simulates numreps random walks as above, and returns the percentage of them that were absorbed at *M* *)

BernoulliDistribution[p]   Statistics`DiscreteDistributions`
(* an object representing the Bernoulli distribution with success parameter *p* *)

Binomial[n, k]   kernel
(* returns the binomial coefficient *n* choose *k* *)

BinomialDistribution[n, p]   Statistics`DiscreteDistributions`
(* an object representing the binomial distribution with *n* trials and success parameter *p* *)

CDF[dist, x]   Statistics`DiscreteDistributions`
     Statistics`ContinuousDistributions`
(* returns the value of the cumulative distribution function of the given distribution at x *)

ChiSquareDistribution[r]   Statistics`ContinuousDistributions`
(* object representing the $\chi^2(r)$ distribution *)

Correlation[var1, var2]   Statistics`MultiDescriptiveStatistics`
(* returns the sample correlation between the two lists of data *)

CovarianceMatrix[datalist]   Statistics`MultiDescriptiveStatistics`
(* returns the sample covariance matrix of the given list of data, which are formatted as a list of *n*−tuples, one component for each variable *)

DensityHistogram[densityfn, n, a, b]   KnoxProb`Utilities`
(* displays a combined plot of a discrete density histogram and the given continuous density function on [*a*, *b*], with *n* + 1 points in the discrete approximation. It accepts options of GeneralizedBarChart in Graphics`Graphics`, and NumDigits→2 to control how many digits appear in the x-axis tick marks. *)

DepartureTimes[arrtimelist, servtimelist]   Section 6.2
(* given the lists of arrival and service times, returns list of customer departure
times *)

DiscreteUniformDistribution[n]   Statistics`DiscreteDistributions`
(* an object representing the uniform distribution on {1, 2, ..., n} *)

DotPlot[list]   KnoxProb`Utilities`
(* draws a dot plot of the given list of data. It takes some of the options of ListPlot,
although in the interest of getting a good dotplot it suppresses AspectRatio, AxesOrigin, Axes, PlotRange, and AxesLabel. It has four options of its own: Variable-
Name, set to be a string with which to label the horizontal axis, NumCategories,
initialized to 30, which gives the number of categories for stacking dots, DotSize,
initially .02, and DotColor, initially RGBColor[0,0,0], that is black. *)

DrawIntegerSample[popsize, n]   KnoxProb`Utilities`
(* selects a sample of size n from the set of integers 1,..., popsize with optional
Boolean arguments Ordered→True, and Replacement→False determining the four
possible sampling scenarios. *)

ExpectedHighTank[n, θ]   Section 2.1
(* computes the expected value of the highest tank number in a random sample of
size n from a uniformly distributed set of tank numbers in {1,2,...θ} *)

ExponentialDistribution[λ]   Statistics`ContinuousDistributions`
(* object representing the exp(λ) distribution *)

FRatioDistribution[m, n]   Statistics`ContinuousDistributions`
(* object representing the $F(m, n)$ distribution *)

Gamma[r]   kernel
(* returns the value of the gamma function at r *)

GammaDistribution[α, β]   Statistics`ContinuousDistributions`
(* object representing the gamma distribution with the given parameters *)

GeometricDistribution[p]   Statistics`DiscreteDistributions`
(* an object representing the geometric distribution with success parameter p *)

Histogram[datalist, numberofrectangles]   KnoxProb`Utilities`
(* plots a histogram of a list of data, with a desired number of rectangles. It inherits
some of the options of GeneralizedBarChart and has four of its own. The option
Type has any of the values Relative (default), Absolute, or Scaled depending on
whether you want bars to have heights which are relative frequencies, absolute
frequencies, or relative frequencies divided by interval length. The option Endpoints may be set to a list {a,b} of real numbers with a<b to force the histogram to
be plotted between these endpoints. Otherwise the command uses the min and max

of the datalist as endpoints. The option NumDigits (initialized to 2) can be used to set the number of decimal digits used in the tick marks on the x-axis. The option Distribution→Continuous may be reset to Discrete in order to force a histogram whose boxes are at the integer values between the lowest and highest integer data value. The user cannot override the PlotRange option, nor AxesOrigin, nor Ticks, nor BarOrientation, in the interest of having a well-formed graph.*)

HypergeometricDistribution[n, M, N]   Statistics`DiscreteDistributions`
(* an object representing the hypergeometric distribution, where a sample of size n is taken from a population of size N, in which M individuals are of a certain type *)

KPermutations[fromlist, k]   KnoxProb`Utilities`
(* returns all permutations of k objects from the given list *)

KSubsets[fromlist, k]   DiscreteMath`Combinatorica`
(* lists all subsets of size k from the given list *)

Mean[datalist]   Statistics`DiscreteDistributions`

Statistics`ContinuousDistributions`
(* returns the sample mean of the list of data *)

Mean[dist]   Statistics`ContinuousDistributions`
                            Statistics`DiscreteDistributions`
(* returns the mean of the given distribution *)

Multinomial[x1, x2, ... , xk]   kernel
(* returns the multinomial coefficient *)

MultinormalDistribution[meanvector,   Statistics`MultinormalDistribution`
covariancematrix]
(* an object representing the bivariate normal distribution with the given two-component vector of means, and the given covariance matrix. Generalizes to many dimensions *)

MyRandomArray[initseed, n]   Section 1.3
(* returns a list of uniform[0,1] random numbers of length n using the given initial seed value *)

NegativeBinomialDistribution[r, p]   Statistics`DiscreteDistributions`
(* an object representing the negative binomial distribution with r successes required and success parameter p *)

NormalDistribution[μ, σ]   Statistics`ContinuousDistributions`
(* an object representing the normal distribution with mean μ and variance σ^2 *)

NPlaces[number, places]   Section 2.5
(* rounds given real number to given number of places beyond the decimal point *)

OptionValue[t, x, T, EP, b, a, r]   Section 6.3
(* returns the present value at time t of a European call option on a stock whose current price is x. The exercise time is T, exercise price is EP, b and a are the up and down proportions for the binomial stock price model, and r is the riskless rate of interest *)

PDF[dist, x]   Statistics`DiscreteDistributions`
                                                          Statistics`ContinuousDistributions`
(* returns the value of the probability mass function or density function of the given distribution at state x. Also in Statistics`MultinormalDistribution` in the form PDF[multivariatedist,{x,y}] it gives the joint density function value at the point $\{x, y\}$ *)

PlotContsProb[density, domain, between]   KnoxProb`Utilities`
(* plots the area under the given function on the given domain between the points in the list between, which is assumed to consist of two points in increasing order. Options are the options that make sense for Show, and ShadingStyle→ RGBColor[1,0,0] which can be used to give a style to the shaded area region.*)

PlotStepFunction[F, endpoints, jumplist]   KnoxProb`Utilities`
(* plots a step function on domain specified, with jumps at the points in jumplist, which is a list of sorted numbers. The step function is assumed to be right continuous, as a c.d.f. is. It accepts option DotSize→.017 to change the size of the dots, StepStyle→RGBColor[0,0,0] to assign a style to the steps, and it inherits any options that make sense for Show.*)

PoissonDistribution[μ]   Statistics`DiscreteDistributions`
(* an object representing the Poisson distribution with parameter μ *)

ProbabilityHistogram[statelist, problist]   KnoxProb`Utilities`
(* takes a list of states, assumed to be in order, an associated list of probabilities for those states, assumed to have the same length, and graphic options for bars as in Histogram, and displays a bar chart for the probability distribution. The user cannot override the PlotRange option, nor AxesOrigin, nor Ticks, nor BarOrientation, in the interest of having a well-formed graph. *)

ProportionMInSample[n, k, m, numreps]   Section 1.4
(* returns the proportion of times among numreps replications that m was in a random sample without order or replacement of size k from $\{1, 2, ..., n\}$ *)

P4orfewer[λ]   Section 4.4
(* a function giving the probability of 4 or fewer arrivals of a Poisson process in the first unit of time, as a function of the arrival rate *)

Quantile[distribution, q]   Statistics`ContinuousDistributions`
(* returns the number x for which $P[X \leq x] = q$ for X having the given continuous distribution *)

Random[ ]   kernel
(* returns a random uniform[0,1] number *)

Random[distribution]   Statistics`ContinuousDistributions`
                                            Statistics`DiscreteDistributions`
(* returns a random number sampled from the given distribution *)

RandomArray[distribution, length]   Statistics`ContinuousDistributions`
                                             Statistics`DiscreteDistributions`
(* returns a list of the desired length of random numbers sampled from the given distribution. In Statistics`MultinormalDistribution`, returns a simulated list of the desired number of pairs from the given joint distribution *)

RandomKPermutation[list, size]   KnoxProb`Utilities`
(* returns a random permutation of the given size from the given list *)

RandomKSubset[list, k]   DiscreteMath`Combinatorica`
(* returns a randomly selected subset of size k of the given list *)

SeedRandom[seedvalue] or SeedRandom[ ]   kernel
(* if an integer seed value is supplied it resets the seed for the random number generator to that value, or if no argument is supplied it sets the seed randomly *)

SimArrivals[arrdist, numarrs]   Section 6.2
(* simulates desired number of interarrival times from the given distribution *)

SimContingencyTable[sampsize]   Section 1.6
(* for the particular 2 by 2 contingency table example 3, it simulates a random sample from the population, categorizes it and returns the table of category frequencies *)

SimMeanSequence[dist, nummeans, n]   Section 3.3
(* simulates a list of a desired number of running sample means from the given distribution, updating the list every n new observations *)

SimNormal[n, μ, σ]   Section 4.3
(* simulates a list of n pseudo-random observations from the $N(\mu, \sigma^2)$ distribution *)

SimPairs[dist, numpairs]   Section 3.2
(* two versions, the first simulates and plots the max and min of numpairs random pairs from the given distribution, and the second simulates and plots the random pairs themselves *)

SimSampleVariances[numvars, sampsize, μ, σ]   Section 4.5
(* simulates a desired number of sample variances, where samples of the given size are taken from the $N(\mu, \sigma^2)$ distribution *)

SimServiceTimes[servicedist, numservices]   Section 3.2
(* simulates desired number of service times from the given distribution *)

SimStockAndOption[initprice, T, EP, b, a, r]   Section 6.3
(*uses OptionValue to simulate and plot a stock price and its call option value for times from 0 to *T*. initprice is the initial stock price, and the other parameters are as in OptionValue *)

SimX[ ], SimY[ ]   Section 2.5
(*return random *X* and *Y* observations from the discrete distributions in Example 3 *)

SimXYPairs[numpairs]   Section 2.5
(* returns a list of a desired number of *X*, *Y* pairs from the discrete distributions in Example 3 *)

SimYGivenX[x]   Section 2.5
(* given a simulated *x* value in Example 3, returns a simulated *y* value *)

SimulateExp[n, λ]   Section 4.3
(* simulates a list of *n* pseudo−random observations from the exp(λ) distribution *)

SimulateGrades[n]   Section 2.1
(* simulates a random selection of *n* grades following the p.m.f. in formula (2) *)

SimulateMinima[m, popsize, sampsize]   Section 1.1
(* uses DrawIntegerSample to simulate *m* observations of the minimum value among the sample of the given size from the set {1, 2, ..., popsize} *)

SimulateMysteryX[m]   KnoxProb`Utilities`
(* see Exercise 1.1-11, simulates a list of *m* values of a mystery random variable *)

SimulateRandomWalk[p, n, M]   Section 1.3
(* simulates a list which is the sequence of states of a random walk with absorbing barriers at 0 and *M*, right move probability *p*, and initial position *n* *)

SimulateRange[numreps, popsize, sampsize]   Section 1.4
(* returns a list of a desired number of ranges, i.e., max−min, of samples of the given size taken in order and without replacement from {1, 2, ..., popsize} *)

SimulateRoulette[numreps]   Section 2.2
(*returns a list of numreps simulated roulette colors, red or other *)

SimulateSampleMeans[nummeans, distribution,   Section 2.6
          sampsize]
(* returns a list of a desired number of means of samples of the given size, taken from the given probability distribution *)

SimulateWaitingTimes[arrdist, servdist,   Section 6.2
          numcustomers]
(* simulates a list of waiting times for a single server queue with the given interarrival and service time distributions *)

StandardDeviation[datalist]   Statistics`DescriptiveStatistics`
(* returns the sample standard deviation of the given list of data *)

StepSize[p]   Section 1.3
(* returns 1 with probability p, else 0 *)

StudentTDistribution[r]   Statistics`ContinuousDistributions`
(* object representing the $t(r)$ distribution *)

Subsets[universe]   DiscreteMath`Combinatorica`
(* lists all subsets of any size from the given universe *)

UniformDistribution[a, b]   Statistics`ContinuousDistributions`
(* object representing the continuous uniform distribution on interval $[a, b]$ *)

Variance[datalist]   Statistics`DescriptiveStatistics`
(* returns the sample variance of the given list of data *)

Variance[dist]   Statistics`ContinuousDistributions`
          Statistics`DiscreteDistributions`
(* returns the variance of the given distribution *)

WeibullDistribution[α, β]   Statistics`ContinuousDistributions`
(* object representing the Weibull distribution with the given parameters *)

XYPairFrequencies[numpairs]   Section 2.5
(* in Example 3, returns a frequency table of X, Y pairs *)

References

Billingsley, Patrick, *Probability and Measure*. John Wiley & Sons, New York, 1979.

Burton, David M., *The History of Mathematics: An Introduction*. Allyn & Bacon Inc., Boston, 1985.

Chung, Kai Lai, *Elementary Probability Theory with Stochastic Processes*. Springer–Verlag,New York, 1979.

Çinlar, Erhan. *Introduction To Stochastic Processes*. Prentice-Hall, Inc., Englewood Cliffs, NJ, 1975.

Cooke, Roger, *The History of Mathematics: A Brief Course*. John Wiley & Sons, Inc., New York, 1997.

Feller, William, *An Introduction to Probability Theory and Its Applications, Vol. I and II, 3rd ed.*, John Wiley & Sons, Inc., New York, 1968.

Gross, Donald and Carl M. Harris, *Fundamentals of Queueing Theory, 2nd. ed.*, John Wiley & Sons, Inc., New York, 1985.

Hacking, Ian, *The Emergence of Probability*. Cambridge University Press, London, 1975.

Hoel, Paul G., Sidney C. Port, and Charles J. Stone, *Introduction to Probability Theory*. Houghton Mifflin, Boston, 1971.

Hogg, Robert V. and Allen T. Craig, *Introduction to Mathematical Statistics, 4th ed.*, Macmillan, New York, 1978.

Katz, Victor J., *A History of Mathematics: An Introduction*. HarperCollins, New York, 1993.

Lamberton, Damien and Bernard Lapeyre, *Introduction to Stochastic Calculus Applied to Finance*. Chapman & Hall, London, 1996.

Loè ve, Michel, *Probability Theory I, 4th ed.* Springer–Verlag,New York, 1977.

Parzen, Emanuel, *Modern Probability Theory and Its Applications*. John Wiley & Sons, New York, 1960.

Saaty, Thomas L., *Elements of Queueing Theory with Applications.* Dover Publications, Inc., New York, 1961.

Sharpe, William, *Portfolio Theory and Capital Markets.* McGraw-Hill, New York, 1970.

Snell, J. Laurie, *Introduction to Probability.* Random House, Inc., New York, 1988.

Index

(*Mathematica* commands in bold)